Orca

Orca

How We Came to Know and Love the Ocean's Greatest Predator

JASON M. COLBY

OXFORD
UNIVERSITY PRESS

Oxford University Press is a department of the University of Oxford. It furthers the University's objective of excellence in research, scholarship, and education by publishing worldwide. Oxford is a registered trade mark of Oxford University Press in the UK and certain other countries.

Published in the United States of America by Oxford University Press
198 Madison Avenue, New York, NY 10016, United States of America.

Library of Congress Cataloging-in-Publication Data
Names: Colby, Jason M. (Jason Michael), 1974– author.
Title: Orca : how we came to know and love the ocean's greatest predator / Jason M. Colby.
Description: New York : Oxford University Press, [2018] | Includes bibliographical references and index.
Identifiers: LCCN 2017042099 (print) | LCCN 2017053474 (ebook) |
ISBN 9780190673109 (Updf) | ISBN 9780190673116 (Epub) |
ISBN 9780190673093 (hardcover : alk. paper)
Subjects: LCSH: Killer whale. | Killer whale—Conservation. |
Whaling—History. | Whales—Anecdotes.
Classification: LCC QL737.C432 (ebook) | LCC QL737.C432 C558 2018 (print) |
DDC 599.53/6—dc23
LC record available at https://lccn.loc.gov/2017042099

9 8 7 6 5 4 3 2 1

Printed by Edwards Brothers Malloy, United States of America

For my father, John Colby—forever haunted by this story

We shall not cease from exploration, and the end of all our exploring will be to arrive where we started and know the place for the first time.

—*T. S. Eliot,* Four Quartets, *1943*

Contents

Orca

Introduction

AS A BOY, I saw my dad cry on only three occasions. One was his father's funeral. The other two involved dead orcas. In the 1970s, he worked as curator of Sealand of the Pacific, a small oceanarium near Victoria, British Columbia, and then for the Seattle Marine Aquarium and Sea World.[1] On both sides of the US-Canadian border, across the Salish Sea, he helped capture killer whales for sale and display—or, as he darkly joked, "for fun and profit."

Tell someone today that your father caught orcas for a living and you might as well declare him a slave trader. Killer whales are arguably the most recognized and beloved wild species on the planet. They are certainly the most profitable display animals in history, and with the 2013 release of *Blackfish*, their fate became an international cause célèbre. Broadcast and distributed by CNN, the film became one of the most influential documentaries of all time. Already years into my research for this book when the movie came out, I found little in it surprising. But *Blackfish* turned my father, long conflicted about his past, sharply against orca captivity. He wasn't alone. Almost overnight, viewers, politicians, and activists turned their sights on Sea World—a multibillion-dollar corporation famous for its killer whale shows. In this debate, it seemed there was no room for nuance or history. Millions around the world simply knew in their hearts that orcas had to be saved from captivity. What they didn't realize was that, decades earlier, captivity may have saved the world's orcas.

Orcinus orca is the apex predator of the ocean, but that ocean has changed rapidly in recent decades. Following World War II, rising populations and new technology drove humans to plunder the sea as never before, and many regarded killer whales as dangerous pests.[2] By the 1950s, whalers, scientists, and fishermen around the world were killing hundreds, perhaps thousands, per year. In a single expedition, celebrated by *Time* magazine, US soldiers slaughtered more than one hundred off Iceland.[3] But then a curious thing happened. In the mid-1960s, at the height of the violence, a few daring men caught and displayed live orcas for the first time. Captive killer whales in turn captivated the public, which would never view the species, or the ocean, in the same way again.

In hindsight, the reasons for the attraction seem obvious. In addition to their panda-like coloration, orcas boast the most stunning combination of size, power, and grace on the planet. Yet it was hardly inevitable that human beings would embrace the fearsome predators. Ancient mariners depicted orcas as sea monsters, and even in modern times the species' dark form and wolflike teeth seemed the stuff of nightmares. While covering Richard Byrd's first expedition to Antarctica in 1928, journalist Russell Owen described the "bloodthirsty" killer whales as "the meanest looking animals any of us have ever seen," and over the following decades other observers generally agreed.[4] But in the span of twenty years, roughly 1962 to 1982, the ominous "killer" became the lovable "orca."

The crucible of this transformation was the Pacific Northwest. These days, orcas are the sacred animal of this secular region. Their distinctive forms seem to appear everywhere—indigenous art, tourist posters, beer labels. Shops on the Seattle waterfront stock killer whale trading cards, and local ferries sell a candy called "Orca Poop." From Portland, Oregon, to Port Hardy, British Columbia, journalists breathlessly report the births and deaths of wild orcas, and networks of whale centers and museums post the lineages and comings and goings of various pods. Each year, five hundred thousand visitors from all over the world pay for whale-watching tours out of Seattle, Vancouver, Victoria, Friday Harbor, Telegraph Cove, and other ports—all to catch a glimpse of the region's killer whales.[5]

But there aren't many of them. While orcas range far and wide, nearly everywhere in the ocean, only a few hundred frequent the inner waters of the Pacific Northwest, and the seventy-six animals who presently make up the region's "southern resident killer whales" represent less than 1 percent of the world population. Put another way, the Salish Sea—a transborder ecosystem that includes the Strait of Georgia, Puget Sound, and the Strait of Juan de Fuca—boasts more than one hundred thousand people for every one resident killer whale. Yet the orca is the undisputed symbol of the region.

When pressed, Northwesterners tend to attribute this status to the influence of either indigenous culture or the whale-watching industry, but the chronology doesn't fit. Local Coast Salish tribes were focused on land and fishing rights in the 1960s and 1970s and played little role in promoting public affection for orcas, and commercial whale watching emerged only in the 1980s and 1990s, after the species had become a tourist draw.[6] The real reason killer whales are iconic is because Northwesterners were the first to catch, display, and sell them. Today, most observers associate orca captivity with Sea World, but it all started in the Salish Sea. From their first display in Seattle and Vancouver, killer whales were a hit. Just as elephants served as the signature animal of the modern zoo, orcas became the main attraction of the oceanarium industry, which was expanding across North America and Europe.[7] Between 1962 and 1976, the Pacific Northwest was the world's only source of captive orcas. In those years, some 270 were corralled behind nets (many of them multiple times) in the shared waters of Washington State and British Columbia. Of these, at least 12 died during capture, and more than 50 (mostly southern residents) were kept for display in the Northwest and around the world.[8]

These figures may seem ghastly, but they pale in comparison to the toll other industries took on cetaceans—the order of mammals that includes whales, dolphins, and porpoises. Between 1945 and 1982 alone, the world's whaling fleets harvested some two million great whales, while fishing vessels in the eastern Pacific knowingly killed more than six million dolphins in the process of catching yellowfin tuna—a method called "fishing on porpoise." During these same years, the whaling nations regularly targeted orcas, with Norway killing more than 2,000 and Japan another 1,500.[9]

Killer whale capture had little in common with commercial whaling. The Northwesterners who chased orcas didn't plan to render their bodies into oil and meat, and on only one occasion did they hunt to kill. Rather, it was curiosity and the display industry's hunger for a signature animal that drove the pursuit. Unlike industrial whaling, however, orca catching took place in public view, in a rapidly changing region. In the past, natives and newcomers alike had harvested otters, seals, and whales, and commercial fishing boats continued to ply the Salish Sea in the 1960s and 1970s. But the extractive economy of logging and fishing that shaped the values of the old Northwest was rapidly giving way to the middle-class culture of the new. More and more people worked in offices during the week and looked to forests and local waters for weekend recreation. These urban-based residents viewed marine life as a source of pleasure rather than livelihood, and it was no coincidence that many developed misgivings about the capture of a species that local aquariums had taught them to love.

Northwesterners were hardly the first to debate the wild animal trade, but they were the first to witness orca capture in their backyards. Zoo owners

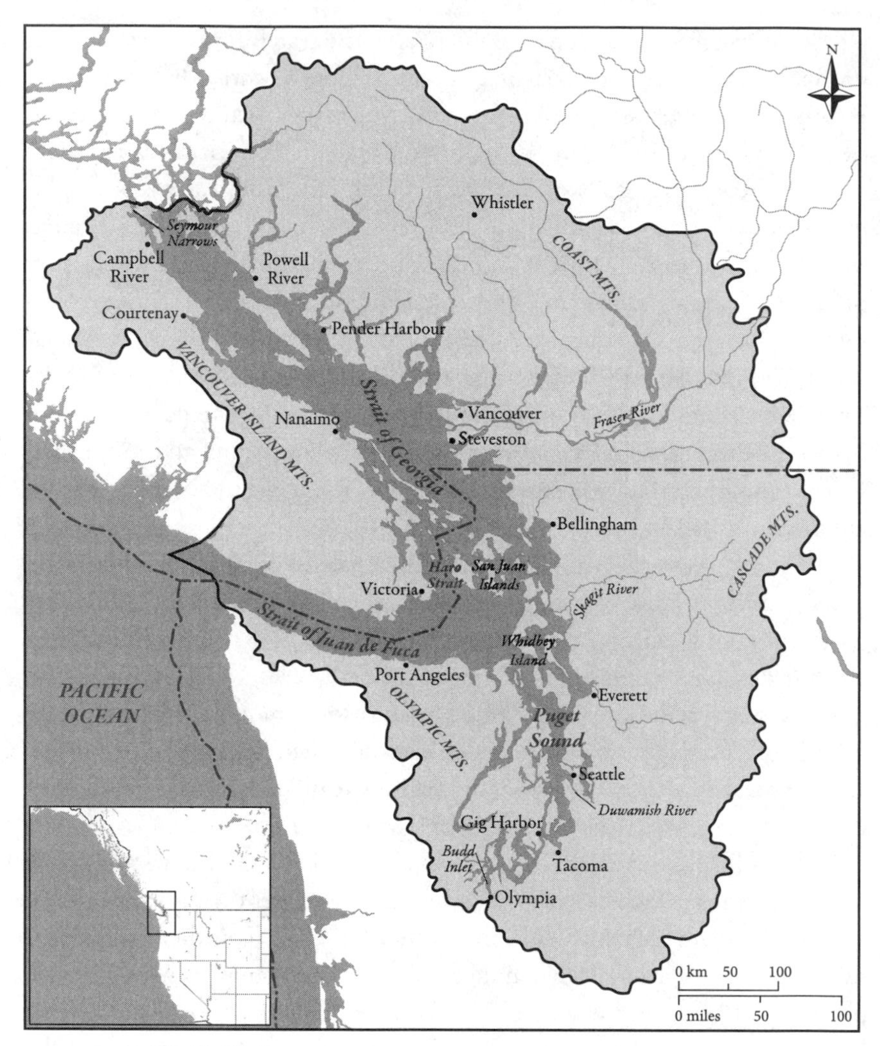

FIGURE I.1 The Salish Sea.

long knew that open collection of wildlife could spur public opposition, but geographical proximity meant that the killer whale captors of the Salish Sea could not avoid public scrutiny. Unlike their counterparts who collected wildlife in Africa and Asia, they operated where their facilities, and they themselves, resided. The personal costs could be high, as revealed by the life of Ted Griffin. After he dramatically delivered the orca Namu to his Seattle Marine Aquarium in 1965, Griffin became the first person to swim and perform with a killer whale. He was the talk and toast of the town—fêted by politicians, courted by Hollywood, and consulted by the Pentagon. Seven years later, his ongoing capture of orcas had made him the Northwest's pariah—villainized by journalists, harassed by activists, and alienated from the region he had helped change.

Yet controversial though their actions became, Griffin and others in the display industry enabled millions in the Northwest and around the world to experience orcas up close for the first time, and as often happens with unfamiliar species, close contact spurred curiosity and affection. It also made possible an intimate, physical knowledge. "Elephants . . . smell absolutely wonderful," observes writer Diana Starr Cooper, after researching the circus. "This is something that you are not apt to discover the details of from a book, a picture, a statue, or even the most artfully made movie."[10] The same dynamic was at work at oceanariums. Spectators who had never before glimpsed killer whales gazed at their striking coloration, heard their explosive breath, and got soaked by their powerful bodies. In the process, they came to see orcas as individuals and kindred beings rather than pests and perils. It was a transition that would help reframe human relations with whales around the world.

This history has hardly been ignored. Accounts of cetacean captivity, most of them critical, appeared long before the release of *Blackfish*. But few explore the broader cultural and political ramifications of live killer whale capture and display.[11] Historians who examine commercial whaling and international conservation efforts have given little attention to the influence of the oceanarium industry.[12] In his study of twentieth-century whale science, D. Graham Burnett marvels that, by the late 1960s, "cetaceans—an anomalous order of elusive, air-breathing marine mammals—had begun to serve as nothing less than 'a way of thinking about our planet.'" He attributes this development primarily to scientific debates within the International Whaling Commission (IWC) as well as to the influence of John C. Lilly, a controversial researcher who made sweeping claims about dolphin intelligence in the 1960s. For his part, historian of Greenpeace Frank Zelko credits the growing affection for whales to activists whose "environmentalism was infused with countercultural mysticism."[13] Yet a more likely factor in this change was the more than twenty million people who had seen captive killer whales by 1970. And that number continued to grow as more facilities in North America, Europe, and Asia acquired orcas. By the mid-1970s, Sea World's three franchises alone drew more than five million people per year .[14] Regardless of their feelings about captivity, few of these visitors would ever again regard the fate of the world's whales as an entirely abstract issue, and many of them became the activists whom Zelko celebrates.[15]

Yet live capture's greatest impact may have come at the intersection of science and government policy. For most of the twentieth century, cetacean research was conducted in collaboration with the whaling industry, and scientists' primary method was to kill and dissect. This was certainly true for killer whales. Between 1960 and 1967, scientists connected to the US government's Marine Mammal Biological Laboratory in Seattle intentionally killed at least ten orcas in order to examine their stomach contents.[16] Captivity changed all this, allowing researchers access to live orcas for the first time. Among them were Paul Spong, Michael Bigg, Graeme Ellis, and John Ford—figures who launched modern killer whale science and influenced the study and treatment of cetaceans around the world. It was their work, more than anything else, that spurred protection of orcas in US and Canadian waters, and when the Soviet Union reported killing more than nine hundred orcas in its 1979–1980 Antarctic hunt, it was these Pacific Northwest experts who convinced the International Whaling Commission (IWC) to halt the slaughter.

Like most historians, I rely heavily on the written record to tell this story, drawing upon published sources as well as public and private archives. But this has also been a journey through space and memory. My research took me around my native Northwest, to hidden nooks of the Salish Sea, and as far as Miami; San Diego; Washington, DC; and Sitka, Alaska, to meet and listen to the people who lived through this change. They are a varied lot: fishermen and entrepreneurs, scientists and trainers, activists and hippies—Canadians, Americans, and an outspoken New Zealander. Not all were eager to talk. Some had built careers as activists, but others, especially those who participated in captures, proved reluctant. This was certainly true of the book's central figure, Ted Griffin, who consented only after I told my father's story. In time, he and others shared documents, recordings, and photographs that no researcher had ever seen. As expected, the backgrounds and experiences of these people varied tremendously. Yet all found themselves transformed by their close encounters with killer whales.

Orcas, too, were actors in this drama. As complex social beings, they had cultures, challenges, and choices of their own and were not simply the victims of human action. Above all, the food preferences of separate killer whale populations shaped their interactions with people around the world. Whereas fish-eating orcas were largely seen as pests, those who preyed on other marine mammals raised fears that they might also attack people. In the case of the resident orcas of the Pacific Northwest, their search for salmon drew them across borders, to the mouths of rivers, and sometimes into conflict with people. Yet that same food culture made them reluctant to attack humans in the wild and tricky to feed in captivity. Their physical characteristics, too, shaped relations with people. Offering few clues to their sex prior to adolescence, orca bodies often caused confusion and embarrassment, while their sheer size dictated the means and viability of capture. Their distinctive markings provided a path for scientists to identify them, while their plaintive calls to one another touched human emotions and etched themselves into the memory of witnesses and captors alike. Ultimately the success of orcas as display animals drove the growth of the captivity industry and influenced the course of environmental politics.

Launched from docks on the Salish Sea, the quest to catch killer whales became big business with global implications. By the late 1970s, marine parks around the world demanded access to the charismatic predator, who became the centerpiece of Sea World's corporate identity and the key to its expansion. Eager to secure a reliable supply, the company looked to Puget Sound, then Iceland, then Alaska, before turning to captive breeding—literally making orcas of its own. By that time, many of the same governments that had targeted wild killer whales for elimination declared the need to protect them from harassment and capture. Nowhere was this reversal more striking than in the place where it all started. Once fearful of blackfish, Northwesterners turned orcas into icons and swept the story of their capture into a dark and shameful past. But only by delving into that past can we understand how we came to love these great creatures and how we might continue to live among them. As David Attenborough has taught us, "No one will protect what they don't care about, and no one will care about what they have never experienced."[17]

1

"The Most Terrible Jaws Afloat"

GAIUS PLINIUS SECUNDUS had witnessed a lot of violence in his life—war in Germania, Sicilian raids, Nero's reign of terror—but killer whales really seemed to scare him. Known to history as Pliny the Elder, he penned the first known description of *Orcinus orca* in his encyclopedic *Naturalis Historia*, completed shortly before his death in 79 CE. It painted a bloody picture. The orca "cannot be in any way adequately described," Pliny asserted, "but as an enormous mass of flesh armed with teeth." Whereas dolphins sometimes befriended people and even helped fishermen, the killer whale preyed on mother baleen whales and their vulnerable calves. "This animal attacks the balaena in its places of retirement," he wrote, "and with its teeth tears its young, or else attacks the females which have just brought forth, and, indeed, while they are still pregnant." Fleeing whales could expect no mercy from orcas, who "kill them either cooped up in a narrow passage, or else drive them on a shoal, or dash them to pieces against the rocks." So frightful were these battles to behold, Pliny noted, that it appeared "as though the sea were infuriate against itself."[1] In short, the destructive power of a killer whale had to be seen to be believed.

Pliny himself had seen one. Around 50 CE, an orca had wandered into the harbor of Ostia, Rome's port city. The animal had been drawn there, it seemed, by a ship from Gaul, which had run aground and spilled its cargo of hides. As the whale investigated, it became stuck in the shallows, unable to maneuver. Soon its back and dorsal fin were visible above the water, recounted Pliny, "very much resembling in appearance the keel of a vessel turned bottom upwards." Sensing an opportunity, the emperor Claudius arrived from Rome, ordering local fishermen to net off the harbor. After waiting for a crowd to gather, he led his praetorians into battle against the trapped whale. The result was "a spectacle to the Roman people," wrote Pliny. "Boats assailed the monster, while the soldiers on board showered lances upon it." But the orca fought back, sinking at least one vessel before it succumbed.[2]

At first glance, it seems an ancient and distant story. The Romans found great pleasure in watching animals die, most famously at the Coliseum, where specialized gladiators did battle with lions and other fearsome beasts. The emperor's foray offered spectators a chance to watch men face off against a rarely glimpsed sea monster, but the event appears to have little historical significance. The orca didn't eat Claudius, so Nero's rule would have to wait. Yet Pliny's description of killer whales echoed through the ages. Natural philosophers drew on it for centuries to describe the little-known species, and in the 1590s Dutch artist Jan van der Straet produced an engraving depicting the orca of Ostia as a hideous, violent monster, poised to devour anyone who ventures near it. That reputation would have remarkable staying power.

❧

The species has sported many labels—grampus, thrasher, blackfish, killer. Many today believe "orca" sounds friendlier, but *Orcinus orca* can be translated

FIGURE 1.1 Jan van der Straet, *An Orca at Ostia*, ca. 1595. Courtesy of Cooper Hewitt Smithsonian Design Museum.

roughly as "demon from hell." Until the early 1970s, most observers considered the term "orca" more ominous than "killer," but from the perspective of its marine prey, it hardly matters: no more terrifying creature stalks the sea.[3] Killer whales have been the ocean's apex predator since their appearance some ten million years ago, and in that time they have likely contributed to the extinction of several species and influenced the behavior and physiology of a great many more. Gray and humpback whales rear their calves in the tropics partly to avoid orcas, and they have developed evasive maneuvers to protect their young from attacks. Likewise, the great speed and agility of the Dall's porpoise may be an evolutionary adaptation to the threat of killer whales, which are, *Homo sapiens* aside, the most successful and widely distributed mammalian predators on earth.[4] Naturalist Carl Linnaeus made the first attempt to categorize the species in 1758, labeling it *Delphinus orca*— "demon dolphin."[5] Long before that, Basque fishermen had dubbed them *asesinas de ballena* (whale killers)—the likely origin of "killer whale." Because scientists classify orcas as the world's largest dolphin, perhaps "killer dolphin" would be more appropriate.[6]

Whatever we call them, killer whales are extraordinary animals. Like all cetaceans, they live and move in a three-dimensional aquatic world yet rise to the surface to extract oxygen. Of the more than seventy species of toothed whales, they are second in size only to the sperm whale. Full-grown males can reach nearly thirty feet in length and almost twenty thousand pounds—nearly three times the weight of the largest recorded great white shark. Despite their size, orcas are extraordinarily nimble, capable of acrobatic leaps and bursts of speed that can reach thirty miles per hour. Adults sport between forty-eight and fifty-two interlocking, conically shaped teeth, ideal for gripping prey. Before adolescence, males and females are difficult to distinguish. At around twelve years old, however, males begin to "sprout" larger dorsal fins, which can grow to six feet in height.[7] Females reach sexual maturity around the age of nine and reproduce well into their forties. In contrast to many mammalian species, female orcas can live long after menopause and often assume leadership roles. Scientists estimate that one recently deceased southern resident killer whale, nicknamed "Granny," was at least eighty years old, and some claim she was born before the *Titanic* sank.[8]

Despite possessing good eyesight above and below the water, orcas are acoustic creatures. By manipulating air in nasal passages beneath their blowholes, they generate a wide range of sound waves. As in the case of other toothed whales, this includes sophisticated biosonar that enables them to map their surroundings, locate prey, and even "see" inside the bodies of other animals—a cetacean version of ultrasound. In addition, they use a variety of whistles and pulsed calls unique to each community, and in many cases specific to each pod. These distinctive dialects frame orca identity, not unlike the role of language and accent in marking human ethnicity. As two whale researchers recently put it, "There is no room to doubt the cultural nature of killer whale communication."[9]

Although orca cultures vary greatly, most seem to be organized around interrelated matrilines in which older females hold authority. These matrilines, in turn, form the building blocks of larger pods—some of which are remarkably stable. With no fixed residence, the only home individual orcas know is their pod, which structures foraging, sleeping, breeding, and belonging, often for their entire lives. Like other cetacean young, killer whale calves are born into a borderless sea, with no refuge other than proximity to mother and family. With a gestation of about seventeen months followed by a similar period of nursing, breeding females give birth to a single calf only once every four or five years. As a result, pods invest greatly in the rearing of calves not only through protection but also by shared "babysitting," performed by both male and female relatives.

Feeding preferences shape the culture, behavior, and social structure of orca communities. Killer whales are specialists, with distinctive foraging communities that focus on particular prey in different locations throughout the year. In Punta Norte, Argentina, one killer whale pod arrives each summer to snatch sea lion and elephant seal pups off the beach, while in the waters of New Zealand, another orca community thrives by digging up and devouring stingrays. Off the coast of California, other groups focus largely on seals and sea lions, and at Alaska's Unimak Pass, a different population of some 160 orcas gathers each May to ambush young gray whales, whose carcasses they sometimes let sink to the seafloor for future feeding. For killer whales, as for people, observed the late researcher Eva Saulitis, "Food is place; food is culture."[10]

Whatever their prey preferences, killer whale communities display remarkable consistency, bordering on what scientist Lance Barrett-Lennard characterizes as "cultural conservatism."[11] This specialization makes killer whale pods vulnerable to prey shortages, but it is also the basis for the cooperative foraging that makes them such effective predators. In addition to sharing food with pod members, they refine and pass on hunting strategies from generation to generation. At times, their methods appear cruel to human eyes, such as when killer whales toy with a wounded seal pup or drown a gray whale calf within earshot of its mother. But such is the orca way. It is with good reason that many people—both admirers and detractors—have described them as "wolves of the sea."[12]

Like wolves, they have haunted the human psyche. For centuries, European whalers and fishermen brought home tales of orca violence, particularly toward great whales. "The grampus," declared sixteenth-century Swedish writer Olaus Magnus, "is armed with ferocious teeth, which it uses as brigantines do their prows, and rips at the whale's genitals or the body of its calf." In his 1874 account of marine mammals on the Pacific coast, former whaling skipper Charles Scammon—infamous for his own slaughter of gray whales in Mexican waters—described orcas as "marine beasts" who spread "terror and death" to every ocean.[13]

Rumor had it that people sometimes wound up on the menu. In the 1890s, sealers based in Victoria, British Columbia, had regular run-ins with orcas, including collisions that cost the lives of several men. Some claimed the killers had eaten crewmates.[14] It certainly seemed plausible. "The killer whale, or orca, is the demon of the seas," wrote Bronx Zoo director William T. Hornaday in 1910. "This creature has the appetite of a hog, the cruelty of a wolf, the courage of a bulldog and the most terrible jaws afloat."[15] Reports from Robert F. Scott's ill-fated expedition to the South Pole seemed to confirm these suspicions. In January 1911, soon after his ship reached Antarctica's Ross Island, Scott spotted a pod of killer whales approaching and urged journalist Herbert Ponting to take photos. But, the orcas broke apart the ice on which Ponting was standing, nearly causing him and two of the expedition's dogs to plunge into the sea. Scott suspected the orcas had mistaken the dogs for seals. Yet he also concluded that the predators were endowed with "singular intelligence" and "would undoubtedly snap up anyone who was unfortunate enough to fall into the water."[16]

Scott's journal became a bestseller, and readers everywhere seemed drawn to the frightening encounter with the orcas. A reporter in Clovis, New Mexico, declared that Scott's expedition revealed the killer whale as "a veritable demon," while a story by New York science writer Garrett Serviss entitled "The Dreadful Killer Whale" reached readers as far away as Omaha and El Paso.[17] For those accustomed to tales of men hunting whales on the high seas, the notion of whales stalking men seemed a terrifying inversion of the natural order.

FIGURE 1.2 Ernest Linzell, *Attacked by Killer Whales*, ca. 1920. Reproduced from Herbert G. Ponting, *The Great White South* (London: Duckworth, 1921).

Human-orca relations weren't always adversarial. In his time in the Russian Far East, German naturalist Georg Steller observed killer whales cooperating with indigenous Kamchadals in pursuit of baleen whales.[18] Much better documented was the interspecies partnership that developed in Australia's Twofold Bay. Beginning around 1840, an orca pod in the region adopted a unique hunting method. When a great whale passed by on its way to Antarctic feeding grounds, the pod would force the larger animal into the bay while several members swam near shore, breaching and slapping their tails on the water to alert the whalers in the town of Eden. The orcas would then weaken the larger whale, enabling the men to slay it with harpoon and lance. Following the kill, the whalers would anchor the carcass for a day, allowing their orca allies to feast on the tongue before towing the whale in for rendering. Locals called it "the law of the tongue." Over decades of working closely with the pod, the whalers of Eden came to recognize and name individual animals. One large male who lived until 1930, dubbed Old Tom, often took the harpoon line in his mouth to assist the men.[19]

It was an ideal arrangement for both parties. On the one hand, cooperation with the whalers gave the orcas access to large whales with less effort and danger; on the other hand, the work performed by the pod enabled the whalers, with only oar-powered skiffs, to harvest right whales, humpbacks, and even fast-swimming fin whales. At the time, the partnership received wide media coverage, with Australian newspapers likening the "killers" to hunting dogs. Yet it may have been the orcas who trained the men—to come when called, to follow them to the hunt, and to do the tough and bloody work of dispatching large whales.

❧

The interaction between people and orcas in the Pacific Northwest had its own unique contours. On the vast coastline that stretches from Northern California to Southeast Alaska, scientists have identified three distinct killer whale "ecotypes." The first, called "offshores," live in aggregations of between fifty and a hundred and hunt primarily for slow-moving sleeper sharks on the continental shelf.[20] They are the least observed, least understood, and least important to this story. The second have commonly been called "transients," though they are becoming known as "Bigg's killer whales"—after the late Canadian researcher Michael Bigg. Because they specialize in hunting other marine mammals, they often travel in silence, relying on passive listening to locate their prey. They have usually appeared in groups of six or fewer, and they seem to lack stable pods. Perhaps because of this social flux, Bigg's killer whales have one general set of calls, likely enabling them to strike up hunting cooperatives more easily.[21] Although Bigg's killer whales have become more numerous in the Salish Sea and around northern Vancouver Island in recent years, the orcas most commonly seen in the shared waters of Washington State and British Columbia belong to a third group, "residents."

The lives of resident killer whales hinge on the abundance of the region's second iconic creature: salmon. The Pacific Northwest boasts five species of anadromous salmon that hatch in rivers, spend their adult lives in the ocean, and return to their natal streams to spawn and die.[22] These include, from smallest to largest, pink, sockeye (or red), coho (or silver), and chum (or dog). The fifth, the iconic chinook (king, spring, or Tyee), is the biggest of all, weighing up to a hundred pounds, and it forms the foundation of these killer whales' diet. Indeed, some have argued that it would be more accurate to call resident orcas "chinook killer whales," as the abundance of this fatty, sweet-fleshed salmon dictates the itinerary, stability, and long-term survival of their pods. Although residents eat a broad range of fish, recent studies indicate that salmon make up 96 percent of their diet and that chinook account for at least 65 percent of those salmon.[23]

This reliance on salmon helps explain the social structure and acoustic culture of resident orcas. Likely due to the benefits of matriarchal knowledge and larger numbers for foraging, the pods of fish-eating killer whales are remarkably stable. Unlike Bigg's killer whales, who often split from their natal group when a younger sibling is born, residents remain with their mothers for their entire lives—a lack of offspring dispersal that is unique among mammals.[24] This intergenerational cohesion has in turn contributed to a more articulated culture. Each resident pod uses a distinctive set of calls, which have remained remarkably stable over time. The dialect used by J pod today, for example, is nearly identical to that recorded by the Royal Canadian Navy in 1958. Because many of their calls are outside the hearing range of salmon, residents are more vocal than their mammal-eating counterparts, with chatter flowing constantly within pods.[25]

Scientists divide the region's resident orcas into two communities, northern and southern. Northern resident killer whales range mostly from Southeast Alaska to Johnstone Strait, midway down Vancouver Island, and number about three hundred animals, grouped into three clans with a total of sixteen pods. In contrast, the southern residents roam from the west side of Vancouver Island to the coast of Northern California, but they spend much of their time in the Salish Sea. Southern resident killer whales form a small, tight-knit community consisting of one clan divided into three pods, which scientists have labeled J, K, and L. Although the three have distinct dialects, behaviors, and migration patterns, the similarities of their calls indicate that they originated from a single pod. As late as the 1870s, when industrial fishing first came to the region, southern residents may have totaled as many as 250 individuals.[26] Today they number fewer than 80.

By the mid-1970s, as killer whales became regional icons, some writers and activists urged people to look to them for lessons in social harmony, but in truth orcas are xenophobes.[27] While they display complex layers of connection within their pods, clans, and communities, they shun outsiders, particularly those from different foraging cultures. Indeed, the tendency toward specialization that makes killer whales such efficient predators hardens the divisions between them. In the Salish Sea, researchers have witnessed resident pods with young calves chasing away Bigg's killer whales—perhaps fearing cannibalism.[28] Some writers have even described these differences as "racial," though that term seems ill suited to describe the social organization of orcas. What is apparent is that resident and Bigg's killer whales never interbreed in the wild and have been genetically separate for some 250,000 years. It is a culturally driven segregation so pronounced that some scientists consider the two populations to be different species.[29]

It is likely that the Northwest's resident killer whale communities took shape at the end of the last ice age, about thirteen thousand years ago. In the wake of receding glaciers, salmon colonized the region's rivers. As chinook

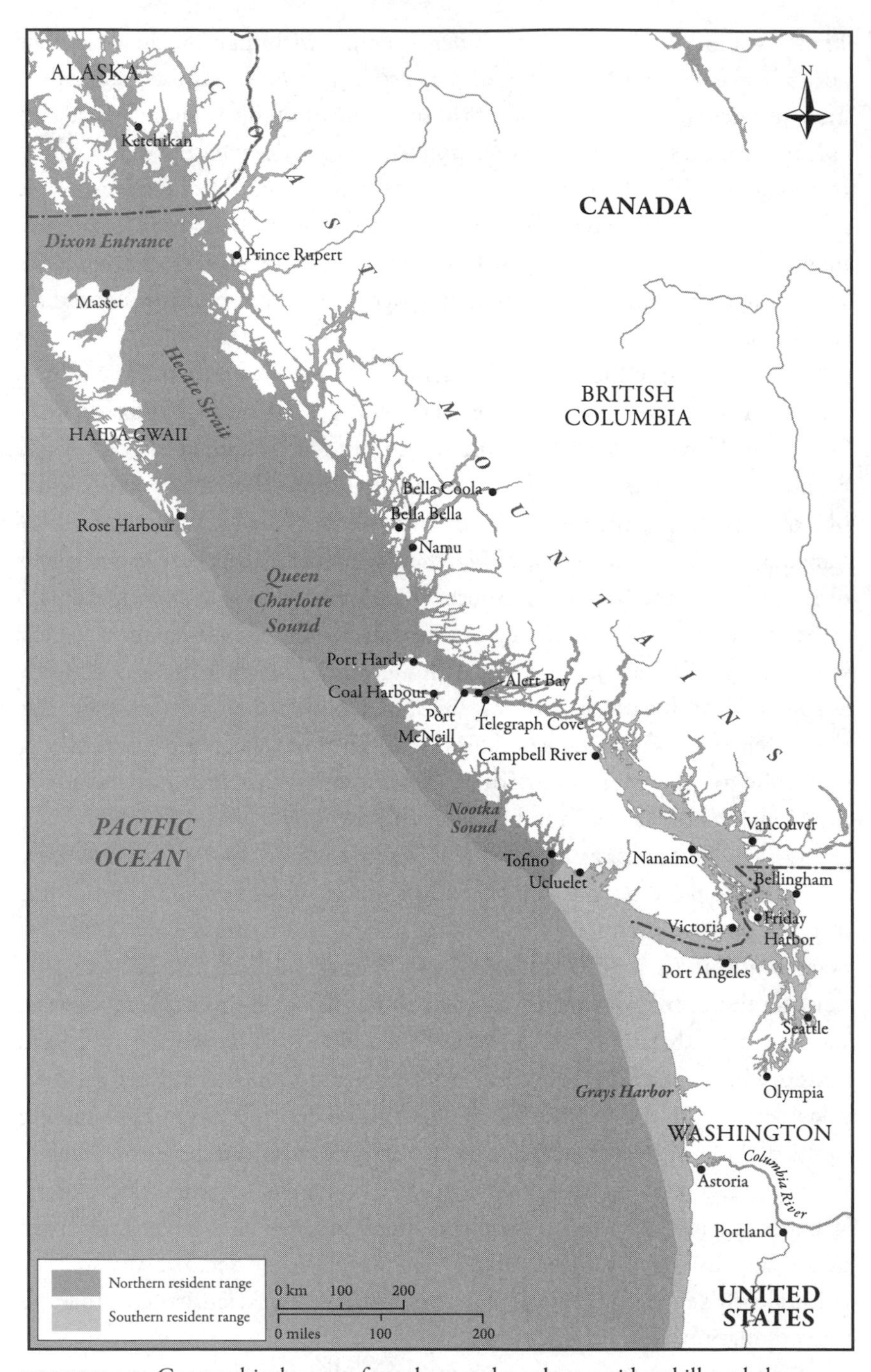

FIGURE 1.3 Geographical range of northern and southern resident killer whales.

runs flourished, orca pods learned the annual rhythms and routes of their preferred prey. In time, the northern and southern communities developed their own cultural practices. As northern residents linked their travels to the arrival of chinook to Johnstone Strait off northern Vancouver Island, for example, they grew fond of visiting Robson Bight to rub on the smooth pebbles found along the shoreline. Likewise, southern residents developed their own itineraries and rituals. In the early spring, they fed off the Sacramento River, where some one million chinook returned each year to spawn, and then moved north, gorging on Columbia River runs four times as large.[30] Summer brought them to Haro and Rosario Straits off the San Juan Islands, chasing salmon bound for the Fraser River. If hunger beckoned, they turned to Puget Sound, catching chinook off the Duwamish River and chum headed for the Skagit, before heading back to the Columbia for its autumn runs of chinook. Feasting on this bounty, the growing group split into three pods, which spent more and more time apart. To maintain their bonds and facilitate mating, they came to practice a unique "greeting ceremony" that has become famous among whale watchers. When members of separate pods encounter each other after a long separation, they form two lines facing one another on the surface, often less than 150 feet apart. After briefly holding this formation, they dive and frolic to celebrate their reunion.[31]

Also at the end of the last ice age, the first peoples came to the Salish Sea. Their lives, like those of the passing orca pods, soon followed the cycles of the ocean ecosystem. They learned to harvest camas bulbs in the spring, fish in the summer, and shellfish and berries in the fall. With salmon carrying protein from sea to land, human communities competed for fishing sites along rivers, which came to structure political and social power. Like all fishing peoples, they suffered lean times, sometimes caused by overharvesting. In the first millennium CE, however, their population stabilized, and by 1300 the Salish-speaking peoples controlled much of the region, fishing for many of the same species that sustained orcas. But human interaction with this apex marine predator depended on each society's economy. Because the Coast Salish were mainly freshwater fishers, their encounters with killer whales tended to occur where rivers met the sea.[32] The Cowichan people of Vancouver Island tell of a monstrous orca who blocked their namesake river, nearly driving the people to starvation. Yet in other tales, killer whales played the role of savior. The Salish tribes near the Snohomish River relate an episode in which orcas helped the people by killing sea lions that were devouring salmon at the river's mouth.[33] The pattern was different for those Coast Salish who fished in salt water. The Songhees, Saanich, and Lummi nations, for example, practiced reef-net fishing in the shallows of Haro and Rosario Straits, which brought frequent encounters with southern resident killer whales.[34]

Like other coastal peoples, the Salish hunted marine mammals, using sea otters for fur and porpoises and seals for meat and oil. But the region's most prolific whalers were the Nuu-chah-nulth of west Vancouver Island and the Makah of Cape Flattery in Washington State. The limitations of cedar canoes and human muscle prevented them from hunting deep-sea species such as sperm whales, but they regularly harpooned slow-moving grays and humpbacks. Only the most powerful chiefs harvested whales, whose meat and blubber they traded for salmon and slaves.[35] Although they mostly took baleen whales, these whaling peoples also killed a number of orcas. Young Nuu-chah-nulth and Makah pursued and sometimes harpooned the species as part of their training, and Charles Scammon reported that the Makah "consider their flesh and fat more luxurious food than the larger balaenas, or rorquals."[36] Most of the time, however, killer whales were simply too fast and elusive to hunt.

Perhaps for this reason, many of the region's peoples held orcas in awe. In one story, told from Oregon to Vancouver Island, Blackfish was eating the other whales and leaving none for the people until Thunderbird appeared and consumed the great predator. Renowned Kwakwaka'wakw artist Richard Hunt heard the tale often as a child. Born in Alert Bay in 1951, he grew up in Victoria amid the revival of indigenous carving led by his maternal grandfather, Mungo Martin (1879–1962). When I visited his Victoria studio in June 2017, Hunt was completing a totem pole featuring a killer whale eating a seal. Yet in his earlier work, he had often depicted Thunderbird and a similar creature, Kwa-Gulth Kulus, seizing orcas with their talons. "Killer Whale was Thunderbird's favorite food," Hunt told me. "Thunderbird was smaller, but he had magical power, so he could carry Killer Whale away."[37]

It was likely an indication of the power attributed to killer whales that only supernatural raptors could defeat them, and the possibility of orcas attacking humans often hovered in the background of such stories. In a well-known Tlingit tale, a future chief named Natsilane creates Blackfish to drown his treacherous brothers but afterward orders the creature never to harm another person. For their part, the Haida depicted killer whales as changelings, or people of the sea, living between two worlds, while other cultures claimed kinship to orcas. For the Nisga'a of the Nass River Valley, Blackfish was one of four clan heralds, along with Eagle, Raven, and Wolf.[38]

FIGURE 1.4 Richard Hunt, *Kwa-Gulth Kulus and Killer Whale*, 1979. Courtesy of University of Victoria Legacy Art Galleries, with permission from Richard Hunt.

Like their counterparts in eastern Russia, the indigenous peoples of the Northwest coast often associated killer whales with wolves. One Haida legend describes a wolf pack that learned to hunt whales but became trapped at sea, transforming into a pod of orcas. Another centers on Wasgo—a mythical creature, half-wolf and half-killer whale, who preyed on great whales. The Nuu-chah-nulth and Makah believed orcas could change into wolves at will and wander the land.[39] Wherever they were found, it seems, killer whales made a deep impression on the region's people.

Those same orca pods were tracing their time-worn paths when Europeans came to the Northwest. The first visits were sporadic. In June 1579, Sir Francis

Drake passed by in search of Spanish gold. While he may well have caught glimpses of grampuses in the fog, he missed the entrance to the Salish Sea, leaving it to other Europeans to discover what became known as the Strait of Juan de Fuca.[40] A century and a half later, Danish explorer Vitus Bering, sailing for Russia, followed the Aleutian Islands and reached the coast of Alaska, where his naturalist, Georg Stellar, marveled at the profusion of marine life, among them otters and whales. It was the urge to turn these creatures into commodities that drove colonial expansion in the Pacific Northwest. Nootka Sound was the early hub, with the Nuu-chah-nulth hosting the Spanish ship *Santiago* in 1774 and James Cook's expedition four years later. Although Cook failed to locate the fabled Northwest Passage, his crewmen traded for otter pelts, which they sold in China for a tidy profit. Thus was born the region's first marine mammal boom.

As natives killed otters to feed the demand for furs, European ships pushed into the Salish Sea, mapping its islands and inlets. In the summer of 1791, José María Narváez and his crew explored the Strait of Georgia, likely becoming the first Europeans to see orcas chasing Fraser River chinook. But it was the voyage of George Vancouver the following year that left a deeper legacy. In April 1792, when his ships reached the Northwest coast, Vancouver encountered the US merchant vessel *Columbia Rediviva*, commanded by Robert Gray. In time, Gray's ship would give its name to the great Columbia River, and its captain would lend his own to Grays Harbor—later Washington State's primary whaling port. As the two captains conversed, orcas may well have passed by, and soon after Vancouver followed the same route through the Strait of Juan de Fuca. He named the great estuary to the south Puget Sound, after his lieutenant Peter Puget, and then Vancouver visited the site of the future British Columbia city that would bear his name. After greeting a Spanish fleet, he sailed north, circumnavigating what would become known as Vancouver Island. It was August, and the waters were alive with killer whales, humpbacks, and other sea life. As he made his way north through what fishermen now call the Inside Passage, Vancouver marveled at the "numberless whales enjoying the season" and "playing about the ship in every direction."[41]

These whales inspired the Northwest's second marine mammal boom. By 1815, the otter harvest had given way to the land-based fur trade, and the first US settlers were arriving, many of them from New England.[42] As these "Boston Men" staked farms in Oregon, others plundered the sea. Using Hawaii as their base, whalers from Nantucket and New Bedford harvested thousands of sperm whales, whose oil made the best candles and industrial lubricant. They then turned to the North Pacific's population of right whales, whose baleen was prized for corsets. The whaling boom raised the hopes of settlers in the Salish Sea, who had long bought oil from the Makah and Nuu-chah-nulth. Eager to trade with US whaling ships as well as local Coast Salish peoples, the British Hudson's Bay Company founded Fort Victoria on southern Vancouver Island in 1843. In subsequent years, oil from visiting whalers and indigenous hunters greased the machinery and skids of the thriving timber industry. But the presence of foreign vessels also raised tensions. By 1844, some three hundred US whaling ships were hunting annually off the Northwest coast, even as the American presidential election spurred chants of "Fifty-four forty or fight!"—calling for US annexation of what would become British Columbia.[43]

All the while, the region's orcas went about their business. Their reputation remained fearsome, but they drew little interest except when they intersected with human activities. In June 1846, for example, the US whaling ship *William Hamilton* came upon a pod of killer whales attacking a blue whale. For the crew, this represented a rare opportunity. Blues were too fast for a ship under sail to catch, but on this occasion, the orcas had waylaid the great creature. Unlike his Australian counterparts in Twofold Bay, however, the captain refused to give these sea devils their due. Instead, he proudly recorded that his men seized the great whale "away from the rascals."[44] Yet at other times, orcas turned the tables. According to Charles Scammon, killer whales often stole carcasses from angry whalers. "Although they were frequently lanced and cut with boatspades," he reported, "they took the dead animals from their human captors, and hauled them under water, out of sight."[45] It was a revealing sign of things to come. The European and American newcomers who were taking control of the region viewed land and sea alike as a source of commodities. As this culture of extraction took hold in the Pacific Northwest, killer whales, like other predators, would figure only as threats to profit and progress.

2

The Old Northwest

ON THE MORNING of Monday, October 12, 1931, early risers in northern Portland noticed a strange creature with smooth black skin in Columbia Slough, right next to the Jantzen Beach Amusement Park. Locals debated its identity. Some argued it was a sturgeon, others a sei whale all the way from Japan. Finally, an old salt tagged it as a small "blackfish." News of the novelty spread like wildfire, drawing thousands of spectators and causing gridlock on the interstate bridge between Portland and Vancouver, Washington. A local newspaper warned that killer whales were one of the ocean's "most vicious" creatures, but this one promptly stole Portland's heart. "From the looks of things," declared Deputy Sheriff Martin T. Pratt, "nearly everyone in the city is determined to see the visitor," and when some locals began shooting at the animal, Pratt and his men arrested them.[1] The number of sightseers grew each day, and that weekend, tens of thousands crowded into Jantzen Beach to catch a glimpse of the whale, while enterprising fishermen charged twenty-five cents for whale-watching rides. By that time, someone had dubbed the orca Ethelbert, and the name stuck.[2]

Why the little whale had arrived there, a hundred miles up the Columbia, remains a mystery. It had probably become separated from its mother and lost its bearings, wandering up the great river that divides western Oregon from Washington State. But Columbia Slough was no place for an orca. In addition to lacking salt water, it was the main outlet for Portland's sewage. In summer, the waterway grew so foul that workers refused to handle timber passing through it. As the days passed, observers grew worried. The whale seemed sluggish, and its skin began to show unsightly blotches. The owner of Jantzen Beach proposed capturing the animal with a net and placing it in a saltwater tank. It would have been an extraordinary attraction for his amusement park—already known as the Coney Island of the West. But members of the Oregon Humane Society denounced the scheme as rank cruelty. Instead, they proposed blowing the young orca up with dynamite.[3]

An ex-whaler named Edward Lessard ended the debate. In the early morning of Saturday, October 24, Lessard and his son Joseph boarded a twenty-five-foot launch, approached Ethelbert, and lanced the little whale to death. They planned to display the dead orca to paying crowds outside the Pacific International Livestock Exhibition, which had just opened across from Jantzen Beach.[4] Portland's children were heartbroken and their parents outraged, but Lessard remained unrepentant. "It was the quickest killing I ever made," he boasted; whereas great whales sometimes fought for days, "this one was dead as a doornail in less than five minutes."[5] With no law against whaling within city limits, police arrested the two men on the charge of fishing with illegal gear as well as "outraging public decency and morals." Meanwhile, other entrepreneurs recovered the dead whale. It turned out that Ethelbert was a young female. Measured at thirteen and a half feet, she would have been about four years old, and she was coveted even in death.[6] After a bitter legal battle, the Lessards won acquittal and possession of the embalmed carcass. Yet they found themselves shunned for "murdering" Portland's pet whale.[7] So ended the tale of Ethelbert the friendly orca. Brief though it was, it hinted at the role killer whales would play in the passage from old Northwest to new.

The Lessards were men of the old Northwest. It was a rough-and-tumble place, shaped by dispossession and extraction, and its people knew wildlife through their work.[8] By 1860, officials in the new Washington Territory had forced many of the Coast Salish tribes onto reservations, and similar trends followed north of the border.[9] As settlers seized land, investors built sawmills in choice locations such as Port Gamble, Seattle, and Vancouver's Burrard Inlet. Even as timber and mining drove the economy, Northwesterners still looked to the sea for commodities. In the late 1860s, Scots-born James Dawson founded the region's first shore-based whaling operation in Victoria, hiring US harpooners from San Francisco.[10] But the new company quickly killed off the humpbacks of the Salish Sea, and its ships couldn't keep pace with faster fin and sei whales. Soon Dawson and other investors shifted to sealing.

Their prize was the northern fur seal—the best alternative to the now-scarce sea otter. Although most fur seals of the Pacific migrate annually to their breeding rookeries on Alaska's Pribilof Islands, they spend the majority of their lives in the open sea, and by the 1880s, Victoria was the staging ground for the pelagic harvest. Hiring Nuu-chah-nulth and Makah hunters,

FIGURE 2.1 Ethelbert on display in Portland, 1931. Courtesy of Oregon Historical Society Research Library, #bb016062.

the Victoria Sealing Company killed tens of thousands of fur seals annually, many of them breeding females. With the fur seal population plummeting, the United States convinced Japan, Russia, and Canada to sign the 1911 Northern Pacific Fur Seal Convention, under which the fur seal hunt was limited to a US government harvest on the Pribilofs.[11]

But diplomacy offered no solution to the orca problem. Each year, groups of Bigg's killer whales appeared off the Pribilof Islands to feed on young fur seals. As early as the 1880s, US officials expressed concern about the predators' impact on the seal population. A decade later, these fears seemed to be confirmed when the stomachs of two beached orcas yielded the remains of twenty-four and eighteen seal pups, respectively. Following the signing of the Fur Seal Convention, officials worried that killer whales would hamper the herd's recovery. In November 1913, researcher G. Dallas Hanna took matters into his own hands, firing at orcas off the Pribilofs and killing at least one of them.[12] Writing in *Science* magazine nine years later, Hanna estimated that killer whales had consumed at least three hundred thousand female fur seals in the past decade, amounting to an annual loss of $1 million for the US Treasury. "If the killer be found the great destroyer of fur seals," he concluded, "then methods for its destruction should be devised."[13] Observers in the Salish Sea took notice of the issue. "The 'orca' . . . is the most ferocious animal in the world, bar none," declared the *Seattle Times* in 1926, claiming that "something like 50 percent of the seal pups born on the Pribylofs [*sic*] are destroyed by killer whales." Fortunately, the newspaper added, the US Coast Guard had begun the practice of "shelling" the hated predators.[14]

By this time, however, Pacific Northwesterners were more concerned with salmon than seals. Having traded for or bought salmon from native peoples for decades, settlers began seizing control of the resource in the 1870s. Industrial fishing and canneries followed, first in the Columbia River, then spreading to the Salish Sea.[15] Meanwhile, the region's cities grew rapidly. In 1893, the Great Northern Railway reached the timber town of Everett, Washington, and then Seattle to the south. Soon after, the Klondike gold rush brought a new influx of migrants. From just 3,500 inhabitants in 1880, Seattle's population reached 237,000 by 1910.[16]

Nearby fishing villages grew with it. Among them was Gig Harbor, located near Tacoma in southern Puget Sound. Using gas engines and purse seine nets, the town's plucky Croatian and Norwegian immigrants built a thriving port known for its large salmon catches and custom-built boats. Similar trends came to southern British Columbia, where a rail connection completed in 1886 transformed Vancouver into a timber and fishing hub. By

1914, canneries dotted the shores of Vancouver Island as well as the mainland, and fishing communities took shape in Nanaimo, Prince Rupert, and Pender Harbour. On the Fraser River near Vancouver, the town of Steveston became an important fishing center, boasting hundreds of vessels and nearly twenty canneries.[17]

The industrial fishing boom reshaped the politics and ecology of the Salish Sea. With the number of commercial vessels growing each year, Canadian and American fishermen and customs officials clashed along their watery border. Massive US fish traps and aggressive American skippers infuriated Canadians by intercepting salmon headed for the Fraser River, while American canneries denounced the smuggling of US-caught fish to Canadian competitors.[18] Meanwhile, officials on both sides of the border waged war on predators viewed as threats to the fishing industry. Just as they paid bounties for the killing of cougars and bears on land, Canadian and US officials culled seals and sea lions to enhance salmon runs. In 1917, the Canadian government experimented with land mines at the mouth of the Fraser to eliminate seals.[19]

Although neither government targeted orcas specifically, the region's economic development had a profound impact on them. In 1913, railroad construction at Hell's Gate on the Fraser blocked passage to spawning beds upstream, sharply curtailing the runs of chinook and other salmon into the 1920s. Meanwhile, the number of motorized vessels fishing on the approaches to the river grew steadily.[20] On the US side of the border, industrial traps in Rosario and Haro Straits reduced the availability of chinook while presenting killer whales with new perils. In July 1908, for example, locals near a fish trap off San Juan Island pumped four hundred rifle shots into an orca and then took its body to Seattle's Luna Park for display. In the summer of 1929, a guard shot and killed a juvenile orca who had wandered into a fish trap off San Juan Island's False Bay, and the following year, locals killed another off the Skagit River.[21] Likewise, fishermen at the mouth of the Columbia River regularly shot passing orcas—a fate that may have befallen Ethelbert's mother.

At the same time, the Salish Sea was becoming a noisier place with the proliferation of motorized vessels. In Puget Sound, dozens of private steam-powered ferries—known as the Mosquito Fleet—carried passengers between nearby ports. Some of these vessels navigated much like orcas, using high-pitched horns and echoes to guide them through fog and darkness. It was a simple form of echolocation that nearby killer whales almost certainly heard along with the drone of engines.[22] Together, these environmental changes took a toll. Numbering perhaps 200 in 1900, southern resident orcas had likely fallen to around 150 by 1940.[23] And unlike local fishermen, who could search out new frontiers in Alaska and northern British Columbia when local salmon runs failed, resident killer whales remained confined to their home range.

These same years brought a resurgence of the whaling industry. With the advent of gas engines and exploding harpoons, whalers targeted fast-moving blue and fin whales just as the introduction of margarine and glycerin-based explosives raised the demand for their oil. While most of the era's whaling took place near Antarctica, the Pacific Northwest enjoyed its own modest boom. In 1904, the Canadian government issued its first whaling laws for the British Columbia coast, and three years later investors formed the Pacific Whaling Company. Soon shore stations appeared in the Salish Sea, including short-lived ventures in Nanaimo on Vancouver Island and Port Angeles, Washington—both of which sat on the migration routes of southern resident killer whales. Unlike their counterparts on the East Coast, who often targeted porpoises and pilot whales, Northwest whalers generally ignored orcas and other small cetaceans, focusing instead on baleen whales. In 1910, British Columbia whaling interests merged with the Canadian Northern Railroad. Soon after, the enterprise reached across the border, building a whaling station in Grays Harbor, Washington, and a US headquarters in Seattle. By the end of World War I, the Pacific coast of North America boasted ten whaling stations stretching from Alaska's Aleutian Islands to California's Bay Area.[24]

Like most whaling throughout the world, the actual killing took place offshore, far from the public eye. On rare occasions, however, the hunts clashed with rising middle-class values in the region's cities, particularly around the incipient tourist industry. The most striking example was the case of the Terminal Steamship Company, whose manager, J. A. Cates, launched the first whale-watching cruises in the Northwest (and possibly the world) in 1907. Departing from Vancouver, British Columbia, to nearby Howe Sound, these voyages catered to visitors and the urban well-to-do. Unlike present-day ventures, however, the selling point wasn't killer whales but rather humpbacks—the very species targeted by the Pacific Whaling Company's Nanaimo station. Worried the whalers would kill off his main attraction, Cates protested to federal officials. In the end, he failed to save the humpbacks of Howe Sound, but his whale-watching expeditions were an early glimmer of the new Northwest.[25]

Following World War I, US investor William Schupp merged the whaling interests of British Columbia and Washington State. He formed the Consolidated Whaling Company, based in Victoria, and then acquired the American Pacific Whaling Company, moving its headquarters from Seattle to Bellevue, Washington. For the next twenty years, these affiliated firms enjoyed a near monopoly on the Northwest harvest, funneling whale oil into the US market. By the late 1920s, however, global whale stocks were crashing.[26] In 1928, the Makah discontinued their small hunt off Cape Flattery, and three years later the United States and Canada signed a League of Nations convention protecting depleted species such as blues and humpbacks. Facing shrinking harvests and slumping prices, the Northwest whaling industry limped into the 1930s. A loan from the US government kept American Pacific afloat, but Consolidated Whaling closed its last West Coast station, Rose Harbor on Haida Gwaii, in 1947.[27]

By that time, depression and war had transformed the region. The economic crisis of the 1930s struck hard at the Pacific Northwest's extractive industries. On the US side of the border, Washington State enjoyed an infusion of federal funding, particularly through the Works Progress Administration and Civilian Conservation Corps, which built and upgraded many of the parks that would become tourist destinations. But the most profound transformation came from the completion of the Bonneville Dam (1938) and the Grand Coulee Dam (1942) on the Columbia River. These massive structures brought jobs, flood control, irrigation, and cheap electricity, but they also disrupted the great salmon runs that had long been the ecological and cultural foundation of the region.[28]

World War II only accelerated these trends, with western Washington receiving immense military spending, particularly through the Boeing Company. William Boeing had begun his career as a timber magnate, logging a fortune from lands surrounding Grays Harbor before shifting to aircraft. By the end of World War II, his company employed 50,000 workers in Puget Sound, half of them women. Boasting numerous military bases and a growing industrial capacity, the area proved attractive for migrants, with Seattle's population alone swelling from 368,000 to 530,000 between 1940 and 1944. The boom drew many Northwesterners into urban industrial and office jobs and away from the extractive industries, profoundly altering the region's culture. As historians William Robbins and Katrine Barber eloquently put it, the war "brought closure to the seasonal rhythms of summer and winter, the steady pace of life, and the intimate associations with known worlds."[29]

Although British Columbia lacked a manufacturing giant comparable to Boeing, it, too, experienced wartime changes. The Royal Canadian Air Force built a seaplane base in the northern Vancouver Island village of Coal Harbour, while the navy focused its Pacific resources at the Esquimalt

Naval Base in Victoria.[30] For its part, the Canadian army moved its Pacific Command headquarters to Vancouver's Jericho Beach in 1942. In the process, the province's maritime industries received a boost. At the height of the conflict, Vancouver's Burrard Dry Dock Company employed fourteen thousand laborers who worked around the clock to repair and refit ships for the war effort. At the same time, the demand for protein to feed Allied armies resulted in intensified fishing throughout British Columbia waters.[31]

Amid this boom, the shared world of fishermen and orcas was changing. The surge in fishing reduced not only salmon but also herring—the key prey species on which chinook and other salmon relied. Officials on both sides of the border sought to stem the decline. In Washington State, officials had outlawed industrial fish traps and expanded salmon hatchery programs. But decades of overfishing had taken their toll, and the combination of deforestation and dam construction reduced the capacity of once-productive streams. The decline was most evident in the Columbia River, where chinook numbers were crashing by the late 1940s. As a result, southern resident killer whales likely became more reliant on the Fraser and Puget Sound chinook runs. This meant spending more time in the Salish Sea, in close proximity to fishermen already nervous about falling salmon catches.

At the same time, whaling and government bounties on seals and sea lions left less food for Bigg's killer whales. Between 1914 and 1964, hunters in British Columbia alone annually killed an average of three thousand harbor seals—the preferred prey for these mammal-eating orcas.[32] Although little information exists on killer whale behavior in this period, anecdotal evidence suggests that Bigg's killer whales were becoming desperate. In 1934, for example, an indigenous man named Joe was towing a dead seal by canoe in Cowichan Bay when a large killer whale seized his prize, pulling the canoe and its occupant underwater. Although Joe survived, observers pondered the meaning of the frightening encounter.[33] And no one proved more fascinated than G. Clifford Carl.

Carl's life tracked the passage from old Northwest to new. Born in Vancouver in 1908, when the city was still a hard-bitten timber port, he earned his master's degree from the new University of British Columbia (UBC) before completing a doctorate in zoology at the University of Toronto. After serving in the Canadian Department of Fisheries for three years, he took a position at the Royal BC Museum in Victoria. As director from 1942 to 1970, Carl pursued broad interests in the natural and human history of the region. But he developed a particular fascination with killer whales.

The spark came near the end of World War II. In mid-June 1945, a pod of twenty orcas became stranded near Estevan Point, on the west side of Vancouver Island. When an indigenous man named Noah Paul discovered the animals, several were still alive, one female having apparently given premature birth while trying to free herself from the beach. Carl learned of the whales only two weeks later and came to examine their decaying bodies. In the process, he learned the stranding was not a unique occurrence. A local man informed him that two killer whales had recently run aground in Ucluelet. In that instance, reported Carl, "an Indian jumped on one's back while it was still alive and shot it with a rifle." Fascinated, the director spent two days measuring and studying the twenty carcasses. In addition to collecting a full skeleton for his museum, he took detailed measurements of the orcas and carefully sketched the pattern of each animal's eye patch—perhaps the first study of the markings that would one day be used to distinguish individual killer whales.[34]

Carl wanted to know about live orcas, too. Six months after examining the stranded pod, he sent the first of many requests to British Columbia lighthouses, asking keepers to report all sightings of killer whales. Where were they spotted? What direction were they headed? How many were there? What were they doing? He grew especially interested in a striking all-white orca often seen in local waters, whose distinctive body, he hoped, could help him track the animal's movements. When Carl had the opportunity, he also expressed concern about human violence toward the species. Noting that science had much to learn about killer whales, he argued that the animals "should not be molested."[35] Such calls had a limited impact at the time, but they represented an early step in the changing human view of the species.

Carl's interest in whales was hardly unique. By the 1930s, a growing number of scientists throughout the world had turned their attention to cetaceans. Although most were connected to the whaling industry, some viewed cetaceans as having intrinsic value, particularly for the study of evolution. In 1932, a subsidiary of the American Society of Mammalogists, known as the Council for the Conservation of Whales, called on the US State Department to protect "these wonderfully adapted creatures, the greatest mammals that have ever inhabited the globe." Eight years later, leading US cetologist A. Remington Kellogg published a celebratory article in *National Geographic* entitled "Whales: Giants of the Sea."[36] In the context of British Columbia's extractive economy, Carl's requests for information on killer whales must have seemed peculiar. Yet many lightkeepers participated, carefully logging orca sightings. Some limited themselves to raw numbers. In April 1946, for example, the Mayne Island keeper reported seeing eighty killer whales on a single day.[37] Others offered detailed, and perhaps embellished, accounts, such as one who described a pitched battle between the white orca and a young fin whale. Still others pondered the predators' impact on fishing. "After the killer whales have passed through Christie Pass," observed the Port Hardy light-keeper in July 1947, "the salmon seem to disappear entirely for a few days."[38]

Carl shared his findings with colleagues elsewhere, discussed killer whales with reporters, and mentioned them often on his regular radio program. In his mind, it was probably just a hobby, but in hindsight these efforts fostered a more complex view of the species among locals. Evidence for this change came in a May 1950 letter from Percy Pike, keeper of the Race Rocks lighthouse near Victoria. Pike had just read an article by Carl in the *Victoria Colonist* newspaper and wanted to contribute to the museum director's study, but he had concerns. What should he do if he found himself "surrounded by killer whales" while in his motorboat? Should he stop the engine or flee? Would the motor's noise frighten off the animals or spur them to attack? And, Pike wondered, what about skin divers? Killer whales were known to eat seals. Was there any danger of them mistaking divers for prey?[39]

Such questions hinted at how technology and recreational trends were reframing human interactions with orcas and other marine fauna. By the early 1950s, residents and visitors to the Northwest were using marine space in new ways, including scuba diving and sport fishing in small powerboats. Decades later, enthusiasts of these activities would view killer whales as a unique attraction, but at the time the predators seemed a potential threat. "The killers are savage things," wrote Muriel Wylie Blanchet in her classic memoir *The Curve of Time* (1961). Having cruised British Columbia waters with her children since the late 1920s, Blanchet had witnessed terrifying orca attacks on gray whales, and like many boaters, she gave killer whales a wide berth.[40]

In fact, although no attacks on people had been confirmed anywhere in the world, the species' fearsome reputation only grew in the postwar years. In December 1953, the cover of the popular men's magazine *Stag* featured a killer whale spilling two terrified men from a life raft. The following year, a group of US soldiers, some of whom had no doubt seen the *Stag* cover, participated in a violent campaign against orcas in the North Atlantic. The action came in response to a request from the Icelandic government, which believed killer whales threatened local fisheries. In September 1954, US soldiers armed with machine guns joined dozens of Icelandic fishing vessels in a coordinated assault near the Keflavik Naval Air Station, killing hundreds of orcas.[41] "The scene of destruction was terrible," observed one reporter from *Time* magazine. "I have never seen anything like it." Yet his account hardly raised sympathy for the species, claiming that the killer whales had feasted on injured companions. "As one was wounded, the others would set upon it and tear it to pieces with their jagged teeth," he claimed. In reality, the animals were likely trying to raise wounded companions to the surface, but the report reinforced the species' bloodthirsty image.[42] Over the following years, the US Navy launched regular missions to aid Icelandic fishermen, using orcas for target practice. Noting that the species was "feared as one of the deadliest of ocean creatures," the December 1956 issue of *Naval Aviation News* proudly reported that US forces annually destroyed "hundreds of killer whales with machine guns, rockets, and depth charges."[43]

The campaigns off Iceland were bloody, but they were hardly unique. Since the late 1930s, the government of Norway had encouraged whalers to kill dozens of orcas annually at the behest of fishermen, and after World War II Japan followed suit. From 1948 to 1957, Japanese vessels harvested 567 orcas, most of them in the Sea of Okhotsk. Although whalers processed the carcasses for food and oil, they also made them available to researchers. In September 1958, two Japanese scientists published the results of their study, noting the measurements, sex, and stomach contents of the slain orcas. They also offered observations on the social characteristics that made it possible to kill large numbers of the fast-swimming species. "In spite of their fierceness, killer whales [have] an ardent passion for their comrades," the authors noted. "When a member of their group is killed, it is possible to catch others without moving the catcher boat far away from the place [where] the first one was killed." Likewise, Soviet whalers harvested dozens of killer whales in the Antarctic each year, and in the 1958–1959 season they took 110.[44]

❧

There was no equivalent slaughter of orcas in the Pacific Northwest, but whales there were dying, too. It was a beautiful July day when I drove to Coal Harbour on northern Vancouver Island to talk to locals who remembered British Columbia's last whaling station. Seventy years earlier, in 1947, an investment group led by BC Packers had formed the Western Whaling Company, hailed as the first all-Canadian whaling firm on the Pacific coast. Acquiring the Coal Harbour base from the air force and hiring Norwegian skippers, the company harvested more than ten thousand whales over the next twenty years.[45] When I visited, the smell of boiling blubber no longer filled the air, but a little museum in the old air force hangar held a range of artifacts: flensing blades, whale meat crates, the enormous jawbones of a blue whale. The station mostly targeted sperm and fin whales, noted the museum's

FIGURE 2.2 Cover of *Stag* magazine, December 1953.

owner, Joel Eilertsen, as he pored over charts recording each season's harvest. The whalers didn't trouble themselves with orcas, he explained, "except for that one they killed for Disney."[46]

To hear the story, I sat down with Harry Hole, a longtime resident who had towed whale carcasses into Coal Harbour for the company. He confirmed that the whalers rarely targeted orcas. "They were too small, not worth what you would get out of them," he told me. But in early 1955, as it prepared to open Disneyland in Anaheim, the Walt Disney Company approached BC

Packers with a request for a killer whale, whose body it planned to display in California. After some effort, the whalers zeroed in on a twenty-one-foot male on May 19, 1955, just off the shore of Vancouver Island. Because Disney needed an intact specimen, the gunner struck the orca with a small, non-explosive harpoon. "They shot it through by the fin and then killed it with a rifle," recounted Hole, who later watched the flensers prepare the body for shipment. "They took the guts out, put a bunch of formaldehyde in the body, and filled the gut cavity with salt, and loaded it onto a boat," he explained.[47]

Disney's order seems an isolated episode, but it spoke to the larger perception of the species at the time. With little commercial value, orcas were regarded primarily as a pest in the Northwest, and indigenous and white residents alike thought little of slaying them. In August 1958, for example, locals at the Opitsat Indian Reserve immediately killed a pregnant female killer whale who ran aground nearby.[48] The hostility of fishermen posed the greatest threat. On both sides of the border, it had become common practice to shoot killer whales, and many of those captured in later years bore the scars of bullet wounds.

Such trends boded ill for the region's orcas. The postwar years had brought a concerted effort to eliminate predators and pests throughout North America. In 1963 alone, the US Fish and Wildlife Service killed nearly two hundred thousand mammals, among them three thousand gray wolves.[49] In the Pacific Northwest, much of this effort focused on species considered

FIGURE 2.3 Worker at the whaling station in Coal Harbour, British Columbia, prepares orca for shipment to California, May 1955. Courtesy of Maritime Museum of British Columbia and Coal Harbour History Museum.

threats to the fishing industry. Between 1945 and 1961, Washington State paid a five-dollar bounty for each harbor seal killed, reducing the Puget Sound population by perhaps 80 percent.[50] The Canadian government implemented similar policies in British Columbia, but it was the eradication of basking sharks that best exemplified the effort to eliminate marine pests. For years, British Columbia fishermen had complained that the large, slow-moving filter feeders tangled in nets, causing loss of valuable gear. The Department of Fisheries experimented with a range of methods to eliminate the species before settling on a bow blade, which could be mounted on a Fisheries patrol vessel and driven into unsuspecting sharks. In April 1956, *Vancouver Sun* columnist Jim Hazlewood joined the crew of the *Comox Post* for one of these expeditions. During its four-hour voyage in Barkley Sound, the vessel sliced apart more than thirty basking sharks in what Hazlewood called an "orgy of blood-letting."[51] The campaign was successful, and the great sharks remain a rare sight in the region.

Some locals called for similar action against killer whales.

~

The impetus for orca elimination originated with the complaints of tourists such as J. R. Hubbard. After World War II, Hubbard, owner of a lumber company in Spokane, Washington, had begun coming to Campbell River every summer to angle and unwind with friends and family. In August 1957, his daughter brought along a twelve-year-old friend, who was lucky enough to hook a large chinook. But as the girl battled to bring it in, Hubbard recounted, "a large school of Blackfish" appeared. Today, the sight of orcas near Campbell River attracts droves of whale watchers, but in the summer of 1957, the predators' appearance scattered salmon and tourists alike. Other vessels made quickly for shore, but the young girl was still in the rowboat with her guide when a killer whale tugged powerfully on the fish at the end of her line. In a letter sent to the Department of Fisheries months later, Hubbard admitted that the episode "still haunts me." After all, he noted, just one orca could "easily overturn a small boat with possible loss of life." In light of such perils, Hubbard and his friends had doubts about returning to Campbell River. If Canadian officials wanted to retain the "thousands of American dollars that are spent annually each year by the American salmon fisherman," he asserted, they needed to rid British Columbia of the blackfish scourge.[52]

It was hardly surprising that such complaints came from British Columbia's growing sport fishing economy. In contrast to the commercial fishing industry, which focused on the more abundant sockeye and pink salmon, sport fishermen sought the same prize as resident killer whales: the large chinook salmon often called Tyee. With tourism an increasingly important factor in the postwar economy, locals in Campbell River called for immediate action.

They had reason for hope. The Canadian government regularly culled seals and sea lions to protect salmon runs, and in April 1958 it destroyed the nearby marine hazard of Ripple Rock in the largest non-nuclear underwater explosion in human history. The blast surely had a devastating impact on nearby marine life, including a pod of orcas spotted shortly before detonation. If Ottawa had proved willing to remake the seafloor in the name of commerce, surely it could eliminate a few dozen killer whales. With pressure mounting, the Department of Fisheries held a public hearing in Campbell River in July 1960 to resolve the "blackfish problem." The eighteen attendees included department officials, commercial fishermen, members of the Campbell River Fish and Game Club, and the owners of local fishing lodges. All agreed that orcas were a menace, and many had solutions in mind—stationing riflemen along Seymour Narrows, perhaps, or dropping depth charges on passing whales. No one was more adamant than Alan MacLean, owner of Painter's Lodge, who argued that killer whales posed a threat to both the local economy and public safety. If the whales appeared in Tyee Pool, a small area where up to eighty boats sometimes fished at the same time, he had no doubt the animals could cause a "loss of life." MacLean urged officials to act fast, adding that "any assault on blackfish should be to kill and not to wound as otherwise results might be dangerous." Frank Wilson, a member of the Pacific Trollers Association, agreed that such an approach might work. Earlier in the year, he recalled, commercial fishermen had "declared war on the blackfish and shot at them whenever they appeared." The animals seemed to avoid the areas where they were attacked.[53]

Over the following days, Department of Fisheries marine mammalogist Gordon Pike held additional meetings and offered his recommendation. He agreed that killer whales posed a threat to commercial and sport fishing, but he also raised questions. First, was it even practical to eliminate them? Orcas seemed too smart to be driven onto beaches and slaughtered like pilot whales, and they were also difficult to hit with harpoons. Mortar shells lobbed from shore or patrol boats might work, but they threatened to cause collateral damage. Second, could the eradication of the species have unintended ecological consequences? Without killer whale predation, for example, would the number of seals and sea lions rise? Finally, might the killing of orcas actually end up hurting the economy? After all, Pike reflected, "a dead whale washed ashore in the hot summer weather would repel more tourists than would the appearance of a live whale." With such concerns in mind, he rejected general destruction of the species, instead opting for the elimination of "offending" animals and trusting that the survivors would "learn quickly to avoid areas where danger threatens them."[54]

Pike and his colleagues soon settled on their instrument of choice: a .50 caliber machine gun, to be mounted on a high bluff overlooking Seymour Narrows, just north of Campbell River. In hindsight, the decision might seem a product of the old Northwest. After all, many commercial fishermen supported elimination of orcas, and Pike himself worked closely with the Coal Harbour whaling station. But the main objective of this Killer Whale Control Program was to protect the growing tourist industry, and that priority remained foremost in the minds of Department of Fisheries officials as they prepared to use the weapon. Worried that tracer fire might ricochet off the water, they ordered ball ammunition from the Esquimalt Naval Base. They also poured a concrete pad and installed transverse stops to prevent errant fire, instructing gunners to use "short bursts." By July 1961, the weapon was mounted and ready, with a team of soldiers and Department of Fisheries personnel standing by. Campbell River local Stan Palmer assisted in the gun's installation and later had the opportunity to try out the weapon during its first test firing. For their target, the soldiers chose a group of cormorants sitting on a passing log, but neither the soldiers nor Palmer could land a shot within fifty feet of the confused birds. "This thing just sprayed bullets!" laughed Palmer. "I mean they didn't go *anywhere* they were supposed to."[55] Over the following weeks, the soldiers continued to practice, but they never fired on killer whales.

The reasons remain obscure. Perhaps, as some later claimed, officials worried the shells would start a fire if they struck brush on the opposite side of the narrows.[56] Perhaps Clifford Carl, in some letter now lost, raised concerns that the program threatened a species scientists had yet to study. Perhaps the whales simply didn't show up that year. Or, more likely, the Department of Fisheries heeded Pike's warning that machine-gun fire and rotting whale carcasses would repel rather than reassure tourists. Whatever the reason, the decision not to fire hinted at an incremental shift in regional values. Some locals and visitors were beginning to view the land and sea not only as sites of resource extraction but also as spaces for recreation and tourism. In the fall of 1961, as the soldiers at Seymour Narrows removed the machine gun, it was unclear what role killer whales would play in this new Northwest. But a young man in Puget Sound was determined to find out.

3

Griffin's Quest

BY THE EARLY spring of 1962, all of western Washington was abuzz with Seattle's upcoming world's fair—the first to be held in the United States since 1940. On April 21, with the push of a button in the Oval Office, President John F. Kennedy released a swarm of balloons 2,300 miles away in Seattle. Seconds later, warplanes from the naval air station on nearby Whidbey Island roared over the city, thrilling the throngs of eager fairgoers. Over the next six months, nearly ten million people passed through the turnstiles, among them Elvis Presley, to film his forgettable *It Happened at the World's Fair* (1963). Officially titled the Century 21 Exposition, the fair boasted exhibits from twenty-seven countries and a range of attractions. But with futuristic highlights such as the Monorail and Space Needle, it aimed above all to celebrate Seattle's new modern identity.[1] It seemed the perfect theme for the time and place. Just two months earlier, John Glenn had become the first American to orbit the earth, and the Cold War space race was in full swing. The Boeing Company, with its headquarters and three manufacturing plants in and around the city, was a leader in cutting-edge commercial and military aviation. If the 1962 world's fair didn't launch Seattle into the twenty-first century, it certainly signaled the city's move away from its nineteenth-century extractive economy.[2]

But these changes came at a cost. Seattle's maritime industries had been declining since World War II, even as Boeing jobs and freeway construction hastened flight to the suburbs. By the early 1960s, city leaders were pushing for urban renewal. The *Seattle Times* led the way, publishing a special feature in October 1961 that called for a "downtown for people."[3] To be sure, the Century 21 Exposition provided a short-term boost, drawing visitors to the fair site at the base of Queen Anne Hill and creating the new tourist hub of Seattle Center. But two months later, an event drew visitors to the waterfront itself. Though far less heralded than the fair, it would have greater ramifications for the region and the world. In June 1962, the Seattle Marine Aquarium opened its doors. Its owner was twenty-six-year-old Edward I. Griffin—Ted to his friends—and he was a young man in a hurry.

By May 2013, I had spent weeks searching for clues of Griffin's whereabouts when I came upon an old telephone number. The phone rang three times before a woman's voice answered. "I'm sorry to bother you," I explained. "I'm looking for Ted Griffin." Then came a long pause—it didn't sound promising.

"Who are you, and why do you want to talk to him?" she asked, polite but cautious. After I told her, there was another long pause. "OK, I'm going to discuss this with my husband. We'll look you up and decide whether he will speak to you. Call back in fifteen minutes."

Fifty years ago, it wasn't hard to find Ted Griffin. He was one of the Pacific Northwest's best-known figures—"the Tom Sawyer of Puget Sound,"

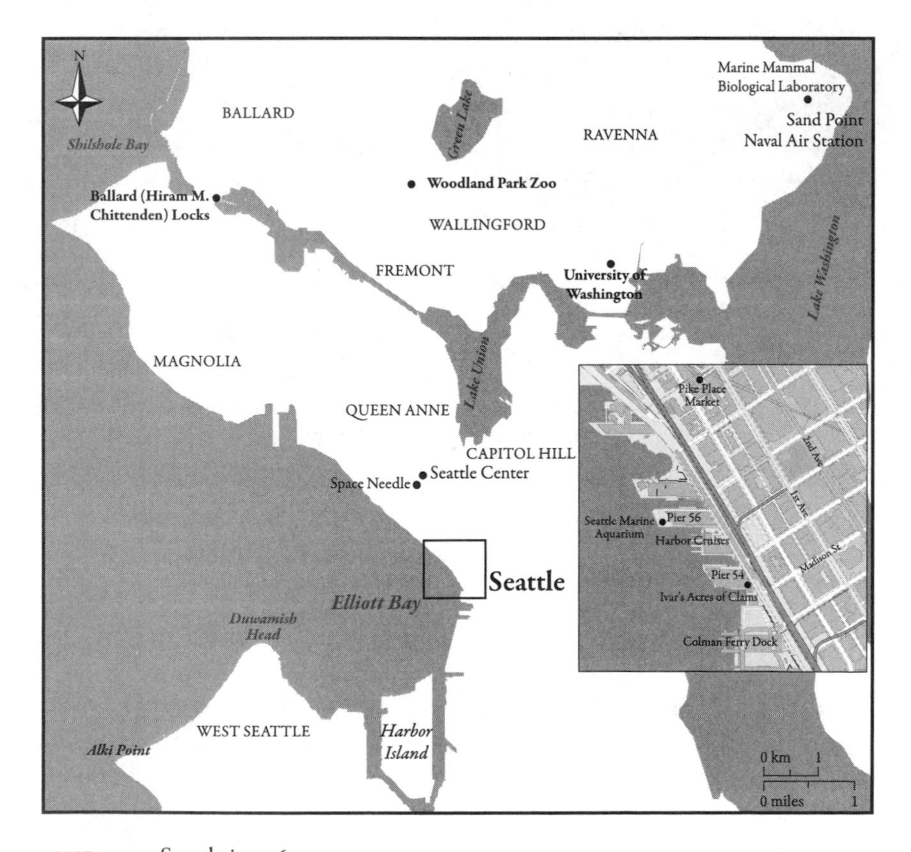

FIGURE 3.1 Seattle in 1963.

as Mark Keyes, a scientist at the National Oceanographic and Atmospheric Administration (NOAA), later called him.[4] From 1962 to 1972, his name was synonymous with Seattle, marine life, and especially killer whales. By 2013, however, he was a tough man to track down. My dad had joked that I would have better luck locating a Sasquatch, and I was starting to agree. Now I had found him, but he might refuse to talk to me. And if he did, I wasn't sure I could write this book. I phoned back sixteen minutes later, and the voice of an old man answered. "This is Ted," he said pleasantly. "What can I do for you?"

Ted Griffin has been called many things—dreamer, entrepreneur, carnival barker, orca killer—but he grew up as a little boy who loved animals. His family had deep roots in the old Northwest. One grandfather came to Tacoma in 1889—the year of Washington's statehood—and made his fortune transporting timber and coal; the other went into the Everett lumber business. Griffin's parents, Edwin and Nancy, met as students at the University of Washington and married after Ed completed a degree at Harvard Business School. They settled in Tacoma, where Ed shifted the family firm to home heating oil. It was there that Ted was born, in November 1935, followed fifteen months later by his brother, Jim. It was a volatile home. Though witty and charming, Ed was erratic, sometimes flying into frightening rages.[5] Even as a toddler, Ted found comfort in the company of animals, often sleeping in the doghouse with the family's English mastiffs. In time, he came to believe he could befriend any creature, no matter how large or fierce.

By 1940, Griffin's parents had divorced, and he and Jim split their time between Tacoma and Everett. Some of his fondest memories came from fishing trips with a stepfather, who gave him his first lessons in seamanship. One excursion to Whidbey Island in the summer of 1941 especially stood out. As Griffin fished from a dock in Glendale, local men warned the wide-eyed five-year-old that the dark shapes nearby belonged to blackfish, vicious predators who devoured salmon and little boys in equal measure. Minutes later, the lumber on which Griffin was sitting shifted, pitching him headlong into the cold water. Panicked, the boy thought of the "sea monsters" nearby. "Will they eat me?" he wondered. But as the men dragged him shivering to shore, he felt oddly peaceful. "It was as if the whale-talk and my plunge into the ocean had etched my subconscious," he later reflected, "forging a psychic link with killer whales."[6]

Few Northwesterners felt the same. Most seemed more eager to kill than to connect with cetaceans. In May 1937, a false killer whale wandered into southern Puget Sound near the capital of Olympia. Convinced it was a killer whale, state police fired at the animal from shore, while hunters chased it with rifle and harpoon. After two hours, the frightened creature beached itself on a mud bank, where pursuers shot it in the head. Six years later, in July 1943, a young female orca became entangled and drowned in a gill net off Hat Island—just five miles from where Griffin had fallen into the water. As the whale's mother lingered nearby looking for her calf, children posed for photos by its carcass. C. Howard Baltzo, a biologist for the Washington State Department of Fisheries, took the opportunity to carve off pieces of the youngster for friends to try. "They found the meat very palatable tasting when fresh killed, much like liver," reported the *Seattle Times*.[7] Yet locals sometimes worried that blackfish might take a bite out of them. In 1954, Mukilteo teenager Gary Spilman found himself surrounded by orcas in nearly the same spot where the calf had drowned a decade earlier. "They just circled around looking at me," Spilman recalled. "And I felt fear—nothing but fear."[8]

In these same years, Ted Griffin lived what seemed a privileged life in his father's mansion on Gravelly Lake near Tacoma. At home, Ed performed magic tricks for his sons and sometimes talked business and politics. He once took them to a friend's retreat in the San Juan Islands, where Ted befriended a wandering horse and marveled at passing orcas. Yet Ed struggled with alcoholism and depression, and young Ted found comfort tinkering on mechanical projects and exploring the natural world. A grandmother's gift of guppies sparked a lifelong fascination with fish, and after learning to weld, Griffin began building his own small aquarium tanks. At twelve, he used a makeshift diving helmet to explore Gravelly Lake, weighting his shoes and enlisting Jim to pump air to him by pedaling a mounted bicycle. Soon after, he ordered an Aqua Lung—Jacques Cousteau's new invention—and became, as a teenager, one of the first scuba divers in the Salish Sea.[9]

In the early 1950s, Griffin attended prep school in Vermont with his neighbor (and future Washington State governor) Booth Gardner, but his mind drifted back to the Salish Sea. In the summer, he haunted the Tacoma Municipal Aquarium, getting to know its director, Cecil Brosseau. The eager teenager followed Brosseau to the aquarium's back rooms, peppering him with questions about tank plumbing, marine life, and especially killer whales. What about all those stories of the species attacking boats and men? he once asked. "Probably just that, Ted," replied Brosseau, "stories."[10] In September 1953, Griffin enrolled in Colorado College, where he proved a gifted athlete but an unfocused student. Then, in March 1955, his father committed suicide. After attending to family matters, Griffin tried to return to college, but his heart wasn't in it. Instead, he took a job at a tropical fish hatchery and then at a local pet store. He learned the business quickly and in 1958 opened his own pet fish shop—Seattle Aquarium—near Green Lake, on Seventy-Sixth and Aurora.[11]

Yet the sea continued to call. In June 1961, Griffin married Joan Holloway, a Smith College graduate and daughter of a Seattle surgeon. The couple rented a cottage on Richmond Beach, just south of Edmonds, Washington, where the new husband spent much of his time underwater. With a whimsical sense of humor, Griffin enjoyed surprising boaters and frolicking with the cod and crabs below. But he kept his eye out for orcas. Like other divers, he was familiar with the species' reputation. The US Navy diving manual described killer whales as "ruthless and ferocious," instructing divers to "get out of the water immediately" if one was sighted. In December 1961, the "Ask Andy" column in the *Seattle Times* warned youngsters about the "terrifying" killer whale, which hunted in "bloodthisty" packs.[12] But something told Griffin the truth was different, if only he could get close enough to find out.

∞

Griffin's interest in the marine world was hardly unique. In the postwar years, the ocean captured the popular imagination. One of the signal events was the publication of US Fish and Wildlife biologist Rachel Carson's *The Sea around Us* (1951), which spent eighty-six weeks on the *New York Times* bestseller list.[13] In 1954, Disney's film adaptation of *20,000 Leagues under the Sea* proved a smash hit with its evocative visions of the undersea world. Two years later, director John Huston thrilled audiences with his rendition of *Moby Dick* (1956), starring Gregory Peck as the crazed Captain Ahab in doomed pursuit of a white sperm whale. Also in 1956, Jacques Cousteau released his influential documentary *The Silent World*, which gave viewers an intimate perspective on the sea and helped popularize scuba diving.[14] Meanwhile, governments throughout the world were investing in oceanographic science. Growing postwar populations raised the importance of fisheries, while advances in submarine technology made the sea a Cold War battleground. "Knowledge of the ocean is more than a matter of curiosity," declared President John F. Kennedy in 1961. "Our very survival may hinge upon it."[15]

It was in these same years that marine aquariums—often called oceanariums—rose to popularity. The first had opened in St. Augustine, Florida, in 1938 under the name Marine Studios. Initially intended to supply sets and sea life to the growing film industry, it soon became a tourist destination, eventually renamed Marineland. With trained bottlenose dolphins as its signature attraction, the park put on increasingly elaborate shows designed to draw crowds. From the beginning, it offered new possibilities not only for entertainment but also for research. As the only facility in North America with live cetaceans, Marineland attracted scientists from around the world. In 1953, it drew seven hundred thousand visitors and built a cutting-edge research facility.[16] The following year, Marineland's owners invested in a second park in Rancho Palos Verdes, California, near Los Angeles. Preceding Disneyland by a year and Sea World by a decade, Marineland of the Pacific became a major tourist attraction, particularly through its association with the popular television show *Sea Hunt* (1958–1961), starring Lloyd Bridges. For the visitors who crowded around tanks to watch sea lions and dolphins, Marineland offered a unique opportunity to see marine mammals close up, and in 1957, the park became the first to display a live pilot whale, Bubbles.[17]

Griffin was keenly aware of these trends. He had visited oceanariums in California and Florida, and he was already dreaming of building his own when he had a close encounter with a killer whale. On a clear day in September 1961, he spotted several orcas nosing around a kelp bed near his Richmond Beach home. As Griffin rowed out for a look, a large male surfaced near the boat, its explosive breath startling him. Like many before him, Griffin initially felt fear. Staring at the "saber-like" dorsal fin towering over him, Griffin conceded that the animal "looked like an executioner." But he felt a longing to connect, to "know this whale." The orca dove and rolled beneath the boat, taking a long look at the man above. It was high tide, and they were a hundred yards from shore. Mesmerized, Griffin imagined swimming alongside the graceful creature, breaking down the barriers between species, and he found himself questioning killer whales' fearsome reputation. "What if all those stories about being dangerous to man were false?" he wondered. "What if orcas were as friendly as dolphins?"[18]

Such questions were still on his mind in early 1962, when a friend suggested he visit a vacant waterfront warehouse as a possible location for his oceanarium. The location was superb—Pier 56, adjacent to the popular Harbor Tours and a short walk from the Colman ferry terminal. Realizing the upcoming world's fair would draw tourists to the city, Griffin leased the warehouse. With little capital of his own, he assumed he would have to do most of the work himself, but to his surprise, the project became a community effort. Friends and family pitched in, while electricians and carpenters he didn't even know showed up to help. With no state or federal regulations in place, the collection of sea life for display was a relatively simple process. Griffin thrilled at the capture of lingcod and octopuses, and he received help from members of the local Mud Sharks diving club, among them Gary Keffler.[19]

Born in north Seattle in 1934, Keffler had grown up swimming in Seattle's Green Lake and joined the Naval Air Reserve in 1953. Over the following years, he split time between selling gear at a dive shop and teaching diving classes to navy personnel at the Sand Point Naval Air Station on Lake Washington. A champion free diver and spear fisherman, Keffler sometimes worked as a stunt double for Lloyd Bridges on *Sea Hunt*, and he was well aware of human violence toward marine mammals. He had often seen indigenous and white fishermen firing at seals and sea lions. "We'd be out there diving in areas where commercial fishermen were shooting at them, and sometimes we worried that they'd shoot us!" he told me decades later. Keffler often noticed bullet holes in the skin of passing orcas, and he saw his first killer whale carcass on Lopez Island around 1961. "I was at the dock, and this guy pulls his boat up to get gas, and there's a baby whale in there, dead," he explained. "They were happy as hell—all the fishermen. They thought this was just great, because they thought killer whales ate all the salmon." Soon after, Keffler met Griffin.[20]

The aquarium owner had wandered into Keffler's Underwater Sports store hoping to complete his formal diving certification. Keffler was immediately struck by Griffin's singular personality. "Ted was a little different kind of a guy," he recalled. "He never went by the book on anything." While he taught Griffin much about diving, Keffler admitted that Griffin's judgment sometimes scared him. On one occasion, the two came upon a seven-foot wolf eel, which Griffin, on a whim, decided to capture. "With this net?" Keffler asked incredulously, but Griffin insisted. "By God, he grabbed the head, and I grabbed the body, and the thing just went nuts!" laughed Keffler. After wrestling the furious animal to the surface, they threw it into their small boat, where it snapped and writhed, nearly preventing them from climbing aboard.[21] On another expedition, Keffler and the Mud Shark divers were collecting animals off the San Juan Islands when a large number of orcas appeared around Griffin's boat—likely a summer gathering of southern resident orcas. Giving little thought to the safety of the divers below, Griffin stood transfixed by the whales. "So many," he thought. "Couldn't I get just one?"[22]

With the help of Keffler and his friends, Griffin had a range of sea life to display when the Seattle Marine Aquarium opened its doors on June 22, 1962. By today's standards, it was a small affair, but at the time there was nothing like it in Seattle. City leaders welcomed the new addition to the waterfront, and thousands of fairgoers visited the marine menagerie on Pier 56. When not tending to his tanks, Griffin met other business owners on the waterfront, among them Ivar Haglund. The colorful scion of a Swedish settler family, Haglund had made a name for himself as a local folk singer before opening a little aquarium of his own in 1938. His main attraction, Oscar the Octopus, had thrilled crowds, and Haglund sometimes generated publicity by pushing a baby seal through nearby Pike Place Market. By 1962, Haglund's aquarium had long since closed, but his restaurant, Ivar's Acres of Clams, had become a Seattle institution.[23] In time, he would prove a kindred spirit and supporter of Griffin.

After the world's fair closed, the aquarium's ticket sales plunged, and Griffin searched for a way to keep visitors coming through the gates. He acquired two sea lions, Gerti and Gus, and hired trainer Homer Snow, who taught the aquarium owner how to put on marine mammal shows. Handsome and charismatic, Griffin was a natural performer whose love of animals proved infectious. "I enjoyed helping people get connections to animals," he later reflected, "to share the experience I was having."[24] But he continued to ponder killer whales. He asked boaters and waterfront residents to call him with sightings, and he spent hours in his twenty-three-foot, high-speed runabout *Pegasus*, watching orcas and contemplating capture.[25] In the process, he spoke often with commercial and sport fishermen. Concern with declining salmon catches was running high, with the leaders of British Columbia, Washington, Oregon, and California organizing a conference on the future of Pacific salmon runs in January 1963. While a range of factors contributed to the decline, many fishermen blamed orcas and other predators. As one angler complained in a letter to the *Seattle Times*, "When the killer whales enter Deception Pass, the king salmon become frightened and head for the mouth of the Skagit River."[26]

Griffin himself heard stories of orcas stealing salmon and tearing through nets, and many fishermen admitted to shooting them. "They ought to bomb all them dumb whales out of here, use them for target practice," announced one man at the Thunderbird Boathouse in Port Angeles. "I cracked one good with my ought-six . . . heard the slug smack into him." Although such tales troubled Griffin, he recognized his own conflicted position. "I believe his behavior is wrong, but how can I pass judgment?" he reflected. "I hunt the whales. Some people have accused me of pursuing them for sport." Yet he worried that fishermen posed a growing threat to killer whales. Public display of an orca might reduce such violence, he reasoned, but he still had no idea how to catch one.[27]

He was not alone in contemplating the problem. In his celebratory account of the new oceanarium industry, *The Captive Sea* (1964), biologist Craig Phillips declared killer whales the "ultimate aquarium specimen," and he noted that collectors at Marineland of the Pacific were keen to capture one. In late 1957, Marineland curator Ken Norris traveled to Victoria to explore the possibility with Royal BC Museum director Clifford Carl. It was an ambitious project. Not only were orcas larger than Bubbles the pilot whale, but many believed them too vicious to capture or even to approach in a boat.[28] In the summer of 1958, two Long Island sport fishermen tried to harpoon a large killer whale and found themselves chased by the angry animal all the way back to Montauk. That September, the appearance of an orca off Malibu, California, set off a panic. Lifeguards at eleven stations relayed news of the prowling predator, spurring thousands of surfers and swimmers to flee the water. As one guard later explained, killer whales had a "reputation of attacking everything—even boats."[29] Nature writer Vinson Brown only added to this aura in an August 1961 column entitled "Lord of the Seas." Brown described orcas as gifted with a "savage intelligence," asserting that "the killers will attack in a frenzy of blood lust that is truly horrible to behold." While Brown acknowledged that the orca was "a beautiful animal," he noted that "the cruel death it brings to myriads of sea creatures hardly endears it to us."[30]

Yet Marineland remained determined to catch one, and three months later, its prize seemed to fall into its lap. On Friday, November 17, 1961, boaters spotted a lone killer whale swimming slowly in Newport Harbor, just south of Palos Verdes. At noon the following day, with thousands of spectators watching, Marineland collectors Frank Brocato and Frank Calandrino attempted to corral the animal. "Tragedy nearly struck once," reported the *Los Angeles Times*, when Calandrino fell overboard and "swam for his life" away from a predator that Marineland general manager William F. Monahan called "the most vicious animal on land or sea." After a nine-hour struggle, the team managed to net the agitated female orca and load her onto a truck. But as staffers lowered her into a pool, she slipped from the sling, smashing head first into the wall. Despite the accident, Monahan was ecstatic. "It is the first time in history a killer whale has been taken alive," he declared. "There is no way to put a monetary value on her."[31]

Throughout the following day, crowds poured in to see this "monster of the deep." As the first orca ever displayed publicly, she was a hit. But she wasn't well, and the next morning she showed signs of distress. "We'd suspected the animal was in trouble because of its erratic behavior in the harbor," recalled Brocato, "but the next day, she went crazy." Speeding in circles, the orca struck the tank repeatedly before convulsing and dying thirty-six hours after her capture.[32] In the ensuing necropsy, Norris and his staff concluded the seventeen-foot female had suffered from gastroenteritis and pneumonia, as well as brain trauma from the fall. They also noted, to their surprise, that her teeth were worn down to the gum line.[33] Posthumously named Wanda, she had shown what a unique attraction an orca could be, and Marineland was eager to acquire another.

Over the following months, Brocato and Calandrino made several more attempts to catch orcas off the California coast, on one occasion nearly netting a calf before its pod fled. In September 1962, they shifted their efforts to Washington State, and after a month of searching, they came upon a pair of killer whales in Haro Strait, off the west side of San Juan Island. Consisting of a female and a much larger male—likely her son—the two seemed to be chasing a harbor porpoise. Waiting until the frightened animal took refuge under the boat, Calandrino slipped a snare over the female's head, but then, as Brocato put it, "everything started to go wrong."[34] The whale dove and twisted, causing the nylon line to foul the boat's propeller. As the men struggled to untangle it, the tethered orca reached the 250-foot limit of the line and unleashed a series of high-pitch screeches, causing the large male to reappear. The two animals then charged the vessel at full speed, turning at the last moment to strike it with their tail flukes. Terrified the orcas would overturn the boat, Brocato grabbed his rifle and started shooting. One bullet struck the male, who retreated. Brocato then fired ten rounds into the female, killing her. That evening, the men towed her carcass to a rendering plant in Bellingham. Brocato kept her teeth for souvenirs; the rest became dog food.[35]

Brocato and Calandrino were undeterred, and the very next day they crossed paths with Griffin, who had driven *Pegasus* north to observe their operation. He was not impressed. "It was like talking to truck drivers," Griffin later told me. "They didn't know what they were doing, and that animal died as a result."[36] Accounts of the frightening encounter soon made the rounds of scientists and oceanarium staff throughout North America, with many concluding that killer whales were indeed too dangerous to capture. Months later, the publication of Joseph J. Cook and William J. Wisner's lowbrow *Killer*

Whale! (1963) reinforced the species' bloody reputation. "Most whales, dolphins, and porpoises are peaceful creatures," observed the authors. "How different the orca, which seems to be filled with a burning hatred! Nothing that lives or moves in or on the water is safe from its assaults."[37]

ꟹ

For his part, Griffin continued to believe it was possible to befriend a killer whale, but the puzzle of capture remained unsolved. Orcas were too large to wrangle onto boats by hand, and Marineland's experience showed that lassoing was perilous. Opportunity seemed to knock with the arrival of Bob O'Loughlin. A promoter based in Portland, Oregon, O'Loughlin hoped to display a live killer whale at one of his traveling sport shows, and he wanted Griffin's help. When a pod appeared at the mouth of the Duwamish River in south Seattle, Griffin joined the Portland team on its hunt. Riding aboard a Seattle Police Department helicopter, one of O'Loughlin's men fired a tranquilizer dart into a young orca, who seemed to relax on the water's surface. Just as Griffin moved in with a lasso, however, another killer whale appeared, knocking the dart out of the first orca's skin and pushing its podmate out of danger. After several more failed attempts, O'Loughlin left Seattle frustrated, but Griffin found himself moved by the whales' "compassion" for one another.[38]

By spring 1964, the Seattle Marine Aquarium was in trouble. Attendance had dropped steadily, while Griffin's attempts to catch an orca had left him deeply in debt. After hiring publicist Gary Boyker, he launched a series of high-profile animal captures in Puget Sound to draw public attention. Off Alki Point, his team netted an eighty-pound Pacific octopus, which became a popular attraction at the aquarium, and two weeks later, Griffin and a fellow diver encountered a large sixgill shark, which they rode to the surface. Griffin's capture of three more of these "giant sharks of Puget Sound" generated still more publicity, and a feature story on an "orphaned" harbor seal named Snorki brought additional crowds to the aquarium.[39] But Griffin remained focused on catching a killer whale, and he sought help at the federal government's Marine Mammal Biological Laboratory.

ꟹ

Located on the Sand Point Naval Air Station near the University of Washington, the lab was the only facility in North America devoted exclusively to the study of marine mammals. But its research focused on their commercial use. The lab's primary mandate was managing the northern fur seal

FIGURE 3.2 Ted Griffin cradles Snorki, a young harbor seal, at Seattle Marine Aquarium, July 1964. In author's possession.

population, which migrated annually to the rookeries on Alaska's Pribilof Islands. In addition to overseeing the commercial harvest, this meant conducting research on the seals' anatomies, reproductive cycle, and predators. Funded largely by the seal hunt, the lab's researchers supervised the annual slaughter of nearly eighty thousand animals, whose hides and meat came to Seattle for processing.[40]

The facility's most distinguished scientist was Victor B. Scheffer. A veteran of the Fish and Wildlife Service raised in nearby Puyallup, Washington, Scheffer was nearing sixty when he met Griffin, and he had spent most of his career studying marine mammals, primarily in connection to the fur seal harvest. He had conducted research on the Pribilof Islands since 1940, and between 1956 and 1963, he participated in the targeted killing of 230,000 female fur seals in a failed attempt to boost the herd's fertility. By 1964, he had shifted his focus to improving the curing procedure for hides, in close cooperation with the Fouke Fur Company.[41] At the time, Griffin didn't fit into Scheffer's utilitarian worldview. With his waterfront aquarium, headlines of orphaned seal pups, and quest to obtain a live killer whale, the young entrepreneur likely struck the old-school scientist as quixotic. Scheffer himself considered orcas dangerous, having seen them "crazed by blood" off the California coast.[42] Yet he was happy to discuss the species, which the lab often killed for research, and he introduced Griffin to his colleague Dale Rice.

Hired at the laboratory in 1958, Rice specialized in cetaceans. Each whaling season, he traveled to Richmond, California, where he examined the carcasses towed in by the last two US whaling firms: the Del Monte Fishing and Golden Gate Fishing Companies. During the off-season, he and his researchers chartered whaling vessels to collect fur seals, but they also targeted orcas. As Rice later explained, in 1960 the lab began instructing its researchers to harpoon killer whales "whenever the opportunity arose during other marine mammal investigations."[43] In April 1961, the *Seattle Times* celebrated the scientists' killing of a large male orca off the coast of San Francisco, reminding readers that killer whales "prey on fur seals, a valuable resource." Over the following years, the lab continued this initiative, and in January 1964, just months before meeting Griffin, Rice himself directed the killing of an eighteen-foot female orca off San Miguel Island.[44]

Griffin insisted that he wanted to catch a killer whale alive, and Rice believed he had just the tool, offering the aquarium owner a Greener harpoon rifle. The aquarium owner was gracious but skeptical. Wouldn't it injure the animal? Accustomed to seeing whales slaughtered, Rice likely found such concerns puzzling. But he reassured Griffin that the foot-long harpoons couldn't possibly hurt killer whales. Their blubber was too thick.[45]

Weeks later, Griffin had his first chance to try out the rifle. With reports of orcas headed north through Puget Sound, he leapt into a chartered G-2 Bell helicopter and headed off in pursuit. Midway down Colvos Passage, on the west side of Vashon Island, he spotted the pod. The plan was to harpoon a large male and track it using two attached buoys. Once the whale tired, it would be herded into a shallow bay for capture. The operation started well, with Griffin firing into a surfacing whale's left flank. "I felt a momentary remorse at shooting him," he later recalled, "yet it was the only way I thought possible to capture a killer whale." Seconds later, as the animal submerged, the helicopter suddenly pitched to starboard. Looking down, Griffin realized to his horror that a buoy had caught near his feet, leaving the aircraft attached to the harpoon line. In that instant, the raw physical power of a killer whale became all too real. As the bull orca dove deeper, he dragged the helicopter and its helpless occupants toward the water. Only after Griffin managed to kick the buoy out of the cabin did he pause to consider "how close to death we had come."[46]

The pod swam south, nearing a group of sport fishing vessels off Point Defiance near Tacoma. Prior to Griffin's pursuit, the whales had likely come to the area for the same reason as the fishermen—to catch the young "blackmouth" chinook that often lingered in Dalco Passage. Now pursued by the noisy helicopter and Griffin's assistant aboard *Pegasus*, the orcas stopped just off Gig Harbor and in unison sunk out of sight. Fifteen minutes passed with no sign of them. Griffin was flummoxed. Where did they go? Had they somehow escaped?[47]

Suddenly, the whales burst from the water in a perfect phalanx. Racing past the startled fishermen, they evaded *Pegasus* and fled north. Watching from the helicopter, Griffin could only marvel at their coordinated breakout and envy their freedom and power. Although he remained determined to possess an orca, he admitted privately that "my dream of holding one captive, for interaction, was a paradox."[48]

4

Murray Newman and Moby Doll

FOR TWO MONTHS, Sam Burich had sat at East Point, on the jagged edge of Saturna Island, staring across the Strait of Georgia and waiting for killer whales. The thirty-eight-year-old Croatian immigrant was not a whaler but a sculptor by trade, though he dabbled in fishing to make ends meet. He had come to East Point at the behest of the Vancouver Aquarium to harvest a whale, not for its blubber or meat but for its body—its form. To pass the time, he played his harmonica, carved hieroglyphs in the sandstone, and chatted with his assistant, twenty-six-year-old fisherman Josef Bauer. On this Thursday, July 16, 1964, just after 11:00 a.m., orcas came into sight. It was J pod, making its way north in pursuit of Fraser River chinook. Bauer saw them first and raced over to load their mounted harpoon gun. Then Burich took over. He knew he would have only one shot as the orcas passed. Most of them were out of range, and Burich needed to hit a whale small enough to kill. Soon he spotted his target, a youngster bobbing along playfully just ninety feet away. Burich took aim and pulled the trigger, detonating the one-and-a-half-ounce powder charge. The steel harpoon, four feet long and two inches thick, hit home.[1]

At first, Burich couldn't believe it, but Bauer's shouts convinced him. Struck just behind the head, the animal appeared stunned. Burich and Bauer rushed to their boat, the forty-foot *Corsair II*, moored in a nearby cove. As the men approached the whale, they found a surprising scene. Rather than fleeing from danger, the other orcas had surrounded their injured podmate, raising it to the surface to breathe. It was a touching sight, but Burich knew the reputation of these beasts. Worried the whales would retaliate, like those shot by Marineland's collectors two years earlier, he and Bauer hung back. Within an hour, the pod had left, and the two men moved in for the kill. But the little orca didn't want to die.

As Burich recovered the harpoon line, the whale tried desperately to expel the cold metal from its body. It leapt, twisted, and shrieked, repeatedly slapping its tail on the water. Despite the resistance of three buoys and hundreds of feet of heavy nylon line, the animal tried to pull away. Eager to end the struggle, Burich fired his .30-.30 rifle at the whale, to little effect. Meanwhile, the keeper of the nearby lighthouse spread news of the struggle to other residents of Saturna, who came by land and sea to watch. By 2:30 p.m., the whale's strength was flagging, but so was its tormentors' resolve.[2] When armed men arrived to finish off the orca, Bauer decided he couldn't allow it. He knew how most fishermen felt about blackfish, but the little whale's cries, and the empathy showed by the others, tugged at him. Against Burich's warnings, Bauer rowed a small skiff out to the wounded creature, shielding it from the nearby rifles.

The two men now faced an unexpected dilemma: What should they do with a wounded, but live, killer whale? The answer arrived by floatplane in the form of Murray Newman, curator of the Vancouver Aquarium.[3]

Vancouver's new aquarium was a symbol of the city's up-and-coming status. Since taking office in 1952, Liberal premier W. A. C. Bennett had generously supported British Columbia's timber and mining industries as well as the construction of hydroelectric dams. Although many of these projects threatened the province's salmon runs, business in Vancouver was booming, and the barons of extraction wanted to polish their rough-hewn city.[4] In addition to world-class theaters and museums, they envisioned a public aquarium. Vancouver already had one of sorts. Ivar Haglund, the Seattle businessman famous for his seafood, had turned the basement of the bathhouse at English Bay Beach into a small, soggy aquarium. But city leaders had grander visions. Led by insurance executive Carl Lietze, they formed the Vancouver Public Aquarium Association in 1950 and enlisted the support of timber magnate H. R. MacMillan and BC Packers president John Buchanan—two of the most powerful men in the province. In addition to making their own donations, these prominent members helped secure public support. Government grants helped fund the construction, and the Vancouver Parks Board allocated land in Stanley Park, the city's most treasured green space.[5] In addition to good access to seawater, the site placed the aquarium near the Stanley Park Zoo.

For the position of curator, the association in 1955 hired thirty-one-year-old Murray Newman. Slight, bookish, and balding, Newman didn't impress at first glance, but he was well-suited to the position. Born and raised in Chicago, he had been an avid aquarist from an early age. After serving in the South Pacific during World War II, he earned degrees at the University of Chicago and University of California, Berkeley, before entering the PhD program in zoology at UBC on a fellowship funded by MacMillan himself. Although his scientific training helped him win the job, Newman found his calling as a fundraiser and publicist, working tirelessly to bring money and attention to the institution that defined his career. The Vancouver Aquarium opened its doors on June 15, 1956, charging an admission fee of twenty-five cents. It drew ten thousand people in its first two days.[6]

In its early years, the aquarium was a small operation, collecting most of its specimens from nearby waters. In addition to fishermen who donated eels, skates, and sculpins, the federal Department of Fisheries helped stock the facility's tanks. To catch more reclusive creatures, Newman hired divers such as Terry McLeod—a Vancouver teenager who started as a floor sweeper in 1960 and moved up to animal collection. Among McLeod's most thrilling memories was spraying the irritant copper sulfate into the dens of octopuses. "They would come shooting out," he recalled, "and we'd grab them." McLeod was a natural with the aquarium's animals, and Newman began grooming him for bigger things, including training the facility's first marine mammals.[7]

By the early 1960s, the aquarium was in the midst of a publicly subsidized $850,000 expansion. Eager for highlights for its new British Columbia Hall, Newman commissioned life-size models of the region's two most striking megafauna: the basking shark and the killer whale. To craft them, however, the sculptors needed specimens, and those proved hard to obtain. By now, the Department of Fisheries had eradicated basking sharks from local waters. After months of searching, the aquarium staff finally managed to harpoon one in April 1962 aboard the *Comox Post*—the same vessel that had earlier carried out the slaughter.[8] Newman and his staff then turned to the task of orca hunting. Marineland's frightening fiasco had convinced them that capture was impractical. By early 1964, Newman was aware of Ted Griffin's quest to catch an orca alive in Puget Sound but thought it a fool's errand. Decades later, he would explain that he "envisioned the killer whale sculpture as a symbol of the marine life in the waters of British Columbia."[9] But the only way to obtain that symbol of life, he believed, was to kill one. It was this thinking that landed Sam Burich at East Point with his harmonica and harpoon gun.

❧

At the time, there was nothing odd about a man hunting whales. Japan and the Soviet Union were leading an explosive growth in commercial whaling, which the International Whaling Commission (IWC), founded in 1946,

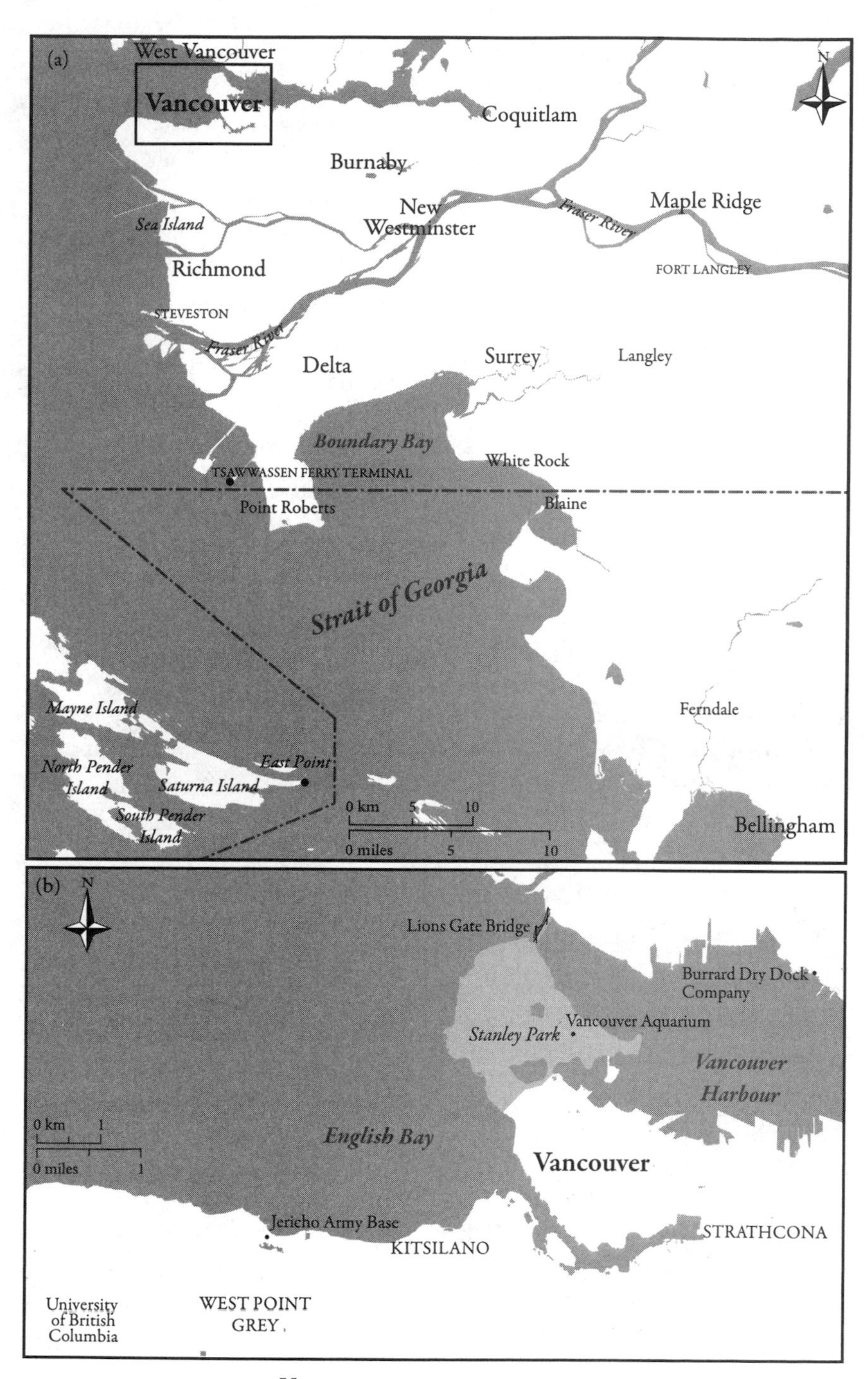

FIGURE 4.1A AND 4.1B Vancouver in 1964.

failed to curb. In the 1963–1964 season, the world's whaling fleets reported killing 13,583 fin whales, and tensions were rising over declining whale stocks. British Columbia played its own small role in the debate. Vancouver hosted the IWC's annual meeting in 1961, and the Coal Harbour station on Vancouver Island continued to operate until 1967. Canadian whalers didn't target orcas, but their competitors did, and the IWC made no attempt to regulate the hunt with Japan alone harvesting ninety-nine killer whales in 1964.[10] In this context, the Vancouver Aquarium's attempt to harpoon a single orca could hardly be expected to raise controversy. And it likely wouldn't have if Joe Bauer had allowed the gunmen at East Point to kill the whale.

Bauer was a latecomer to the expedition. When Burich set up camp at East Point in May 1964, his primary assistant was Ronald Sparrow. A member of the indigenous Musqueam band, Sparrow had experience with harpoons, having installed an old Norwegian harpoon gun on his fishing boat to shoot killer whales who threatened his catch.[11] Assuming the expedition would be short, journalists gathered to cover the hunt, while the Coast Guard vessel *Chilco Post* stood by to finish off any orca Burich managed to wound. A pod passed by that first morning, before Sparrow's harpoon gun could be mounted, but over the following weeks, orcas proved scarce. Eventually, reporters, scientists, and the Coast Guard drifted away, and Sparrow left to go halibut fishing. With his own boat in dry dock for repairs, Bauer agreed to help Burich at East Point. By midsummer, however, it seemed no more killer whales would come, and Newman instructed them to end the effort. On the morning of July 16, the two men were nursing hangovers and packing up to leave when Bauer noticed shapes in the distance. "Whales!" he shouted from an outcropping above their tents, to which Burich grumbled, "Bullshit!" Seconds later came the fateful shot, and not long after that Bauer made his decision to protect the injured orca. As I would learn, the choice had come naturally to the young fisherman.[12]

It was a sunny day in May 2016 when I drove to Steveston to meet Bauer. Perched at the mouth of the Fraser River's South Arm, fifteen miles south of Vancouver, Steveston revealed just how much the region had changed. Once a fishing hub with nearly twenty canneries, it was now a bedroom community that catered to summer tourists. Visitors came to see the old Gulf of Georgia Cannery, stroll Fisherman's Wharf, and perhaps take a whale-watching cruise. These days Steveston billed itself as the "Gateway to the Orca," but fifty years ago its fishermen cursed and shot at the species. When I visited, few of the old-timers were still around to reflect on this transformation—few except Joe Bauer. He lived just a few blocks from the cannery historical site, on a lot that seemed a tribute to earlier times. In contrast to the tidy yards of his neighbors, a riot of evergreens and honeysuckle enclosed Bauer's home, and an old fishing net lay in a heap below his address marker, which featured a killer whale. At seventy-eight, Bauer was still a bear of a man, with a light German accent and the thick hands of a lifelong fisherman.

He had a rough start in life. Born to German immigrants in Steveston in 1938, he was visiting family in Bavaria with his parents when World War II broke out. Soon after, the Nazis imprisoned his mother and sent his father to the concentration camp at Dachau. British officials reunited the boy with his parents in 1945 and moved them to a hostel in Scotland, where seven-year-old Joe befriended a Gaelic fisherman. Although he understood little of the man's speech, the boy joined him in his dory, learning to seine herring. When the Bauers returned to Steveston, they discovered that the government had sold their home, as it had those of their Japanese Canadian neighbors. Joe's father went to work for BC Packers, sometimes bringing whale meat home for supper, while his son struggled to learn English and defend himself from classmates. "The popular thing was to beat up a 'kraut' because their dads had fought the Germans," Bauer told me.

Young Joe took to fishing, which fed both his family and his interest in marine life. He donated live specimens to Haglund's collection at the bathhouse and later to the Vancouver Aquarium. In the process, he had frequent encounters with marine mammals, but unlike many locals, he refrained from violence. On one occasion, he rescued a baby seal who had become tangled in his net. "By that time, we still had that five-dollar per nose bounty on them, and the mother may have been shot," he noted. "I brought the baby home and put her in the bathtub, and eventually I took her to the aquarium, and they named her Josephine." Killer whales, too, sometimes collided with his net. Bauer knew that fellow fishermen shot orcas, but he felt differently. "I never had any bad feelings about them," he told me. "I always admired them." When Murray Newman asked him to help harpoon a killer whale, Bauer had misgivings. "I had things that bothered me [about the expedition], but I also had the feeling of the importance," he reflected. "Newman wanted to put this animal's image in front of the public so they could see." Yet when the moment came, Bauer stopped Burich and the other men at East Point from killing the whale. Instead, he rowed out to the wounded creature against Burich's warnings. "He was petrified," recalled Bauer. "He figured I would get attacked."

Yet the orca remained calm, allowing Bauer to come within a few feet. By that time, residents of Saturna Island were ready with their guns. But Bauer stayed by the whale, urging them not to shoot.[13]

Initial news coverage depicted the event as a cinematic clash between man and leviathan. The *Vancouver Sun* breathlessly reported that two local fishermen were locked in a "life-and-death struggle on the misty Strait of Georgia." Newman heightened the drama, warning that the deadly orca might surface under Burich and Bauer's boat at any moment, hurling them into the sea. He assured reporters that an "extremely lifelike" sculpture of the beast would soon hang in the aquarium foyer, but the whale's desire to go on living forced him to delay those plans.[14] Landing by floatplane near East Point, Newman quickly assessed the situation. The harpoon wound was ugly, but the whale seemed to be in good health, and he estimated it was no longer than seventeen feet. Burich had targeted a small orca for an easier kill, but his choice now made captivity seem feasible. After weighing his options, Newman phoned the Burrard Dry Dock Company—still the city's largest shipyard—and convinced manager David Wallace to accommodate a live killer whale. He then instructed Burich to bring the wounded animal to Vancouver. It was 7:30 p.m.[15]

There seemed no other way but to tow the creature, and the seas were choppy. After rigging a tire to act as a shock absorber, Burich and Bauer steered north for Vancouver. To their relief, towing proved unnecessary, as the little orca followed dutifully behind the boat. Convinced he had seen the animal's penis during the struggle, Bauer concluded it was male, and the two men affectionately dubbed him Hound Dog.[16] They puttered slowly through the night, sometimes catching glimpses of another orca who seemed to be following. At 11:30 a.m., the *Chilco Post* escorted the *Corsair II* and its trailing whale under the Lions Gate Bridge, where hundreds of spectators had gathered to watch. The flotilla passed the aquarium in Stanley Park, and then, more than twenty-four hours after being harpooned, the exhausted orca swam into a submerged berth at Burrard Dry Dock.

As Newman and his staff worked to keep the animal alive, news of Vancouver's captive whale made headlines around the world. Exercising great caution, assistant curator Vince Penfold and neuroscientist Patrick McGeer removed the harpoon and used a pole to inject large doses of penicillin. Throughout the process, observers marveled that the killer whale, presumed to be vicious, made no move to attack its captors. Perhaps, these men speculated, the lack of aggression meant the animal was female. With first aid

FIGURE 4.2 Harpooned orca arrives at Burrard Dry Dock in Vancouver, British Columbia, July 1964. Courtesy of Terry McLeod.

administered, Newman faced the pressing task of finding a permanent home for the whale. From the beginning, he admitted he was "weighing scientific reasons with patriotic ones."[17] The aquarium lacked facilities to accommodate its prize, and while Newman hoped taxpayers would fund construction of a new pool, he conceded the animal might better serve science elsewhere. Yet he could hardly ignore the whale's potential as an attraction for his aquarium. That appeal was obvious on Saturday, July 18, when Burrard Dry Dock opened its doors to visitors. In the span of eight hours, some twenty thousand people came to see the city's famous killer whale—the second in history to be displayed publicly.[18]

The Vancouver Aquarium's windfall caught the attention of oceanariums elsewhere. Almost immediately, Marineland of the Pacific offered to purchase the whale. Newman responded by suggesting a price of US$20,000 ($150,000 in 2018 dollars). At the time, he meant the figure to sound ludicrous. "I thought if I asked a lot, they would go away," he later explained, "but they were willing to pay it!"[19] Regardless of his intentions, the figure circulated publicly, playing a key role in transforming killer whales from perceived pests into potential commodities. Soon after, Newman faced a more aggressive suitor in Charlie White. The owner of the new Pacific Undersea Gardens attraction in nearby Victoria, White lobbied publicly to bring the animal to the provincial capital. "It would be a tremendous attraction for Victoria," asserted White. "Think of it—the only killer whale in captivity, and performing at that."[20] Newman deflected such efforts, declaring it his duty to keep the animal in Vancouver, and he also took pains to distinguish his "public" aquarium from a "commercial outfit" like the Undersea Gardens, arguing that the whale was a scientific treasure, not a tourist attraction. "We know Mr. White would get rich if he had it," Newman observed, "but we're not interested in that."[21]

In subsequent decades, Newman and his admirers would argue that the little orca's harpooning, however violent it later seemed, initiated a shift in public views toward killer whales.[22] Yet the truth was more complicated. Over the previous fifteen years, many in southern British Columbia had heard or read Clifford Carl's thoughts on the species, and some had misgivings about the aquarium's expedition at East Point. No environmental movement yet existed in the Pacific Northwest, but both the Humane Society and the Society for the Prevention of Cruelty to Animals (SPCA) voiced their concerns with the capture. Although neither organization had protested the British Columbia whaling industry, they criticized the aquarium's methods as "too cruel," with some members calling for the animal to be put out of its misery. Victoria resident and former SPCA officer Florence G. Barr launched a petition to have the orca released. "A whale is a very intelligent mammal," she asserted. "It's no different than keeping a man with a spear in his back in captivity." And she considered the public enthusiasm for the capture even more appalling, declaring Vancouverites "a bunch of savages."[23] Others found more ominous meaning in the event. Claiming to draw on indigenous lore, social activist Maisie Hurley warned that disaster would befall the city if the whale died in captivity. Well-known Haida artist Bill Reid proved less alarmist. "We're probably having the disaster right now," he quipped, "all this rain."[24]

As people debated its fate, the little orca sought to make sense of its new surroundings. Accustomed to traveling with its pod in the open sea, it now found itself in a confined space in a busy port. Hydrophones installed by a Canadian navy acoustician recorded the whale's responses. During the first hours of captivity, its vocalizations revealed "signs of panic" as commercial vessels passed through Burrard Inlet. The following day, with thousands of sightseers present, the orca began sending pulsed long-distance calls.[25] Observers tried to make sense of them. Perhaps they were the cries of a female trying to "entice a brave or amorous young bull" or "the plaintive voice of an expectant mummy whale." Or maybe this was simply a juvenile animal "calling for Poppa." Even more intriguing was the fact that another orca answered. Late Saturday afternoon, as the crowds dispersed, UBC researcher H. Dean Fisher picked up vocalizations from a killer whale two miles away, near Lions Gate Bridge. When the captive animal heard the response, Fisher reported, it sent out a "louder, rapid chatter sound."[26] Had a passing orca simply stopped to investigate? Or had a member of J pod—perhaps the whale's mother—followed the *Corsair II* and entered the harbor to check on the little whale?

Whatever the explanation, the calls intrigued Newman. In response to the dry dock company's insistence that he move his whale elsewhere, the curator arranged to transfer the animal to the pier at the Jericho Army Base. Located in English Bay, six miles to the west, the new pen would be within earshot of passing killer whales, which Newman hoped to "lure" into the harbor using the captive's calls. Although local newspapers cheered this "love trap," not everyone embraced the scheme. Members of the nearby Royal Vancouver Yacht Club protested vehemently. If the captive's calls succeeded in attracting other killer whales to English Bay, they warned, the predators could threaten the lives and property of boat owners.[27]

Scientists saw other opportunities in the whale's capture. Within days, researchers from the University of Washington arrived to examine the animal's circulatory system. From his post at the Royal BC Museum, Clifford Carl urged British Columbians to embrace their new prize. After spending two decades collecting information on *Orcinus orca*, he viewed the captive animal as a "living laboratory" that could unlock a range of engineering and medical mysteries.[28] Its flexible skin could help improve vessel design, while its biosonar and respiratory systems might spur breakthroughs in submarine and diving technology. For his part, Patrick McGeer, a neuroscientist at UBC, spoke confidently of translating the whale's calls.[29]

Privately, however, Newman and his staff admitted they knew little about their new prize. They had not even been able to determine something as basic as the whale's sex. Because male orcas sprout their larger dorsal fins in their teens, the sex of a young whale is difficult to determine.[30] Joe Bauer remained adamant that he had seen the orca's penis, and some still called the animal Hound Dog, but Penfold and McGeer announced publicly that the whale was female, and pressure mounted to name it. Dismissing suggestions such as Calamity Jane and Jericho Jenny, Newman dubbed the whale Moby Doll. Privately, however, he admitted his staff remained unsure of its sex. "It's what you might call a delicate situation," quipped one reporter. "Moby Doll just may turn out to be Moby Dick."[31]

On July 24, the aquarium moved the newly named orca to the Jericho Army Base, with the help of government agencies and local businesses. Navy divers from Esquimalt prepared the pen, Burrard Dry Dock allowed its berth to be detached, a towing company provided six tugboats, and Vancouver police vessels escorted the strange flotilla.[32] Although it was only a six-mile journey to Jericho Beach, it took nearly ten hours to complete, due largely to Moby Doll's resistance. Throughout the voyage, she vocalized loudly, repeatedly probing for an escape route. Finally, the tugs positioned the dry dock for the transfer, but still the whale refused to move, and when staff in a rowboat tried to coax her in, observed journalist Fred Allgood, the whale "attacked the boat and fought for her freedom" in what he described as a fit of "feminine temper." A police boat broke the stalemate by sounding its siren, jolting the nervous whale. At that moment, Penfold grabbed the animal's dorsal fin and coaxed her into the enclosure, to the applause of spectators. Three days later, he undertook a close examination to settle lingering doubts about the whale's sex. The result was conclusive. "Moby," wrote one reporter, "is definitely a doll."[33]

Newman was pleased with the move to the Jericho base. In later years, he would celebrate Moby Doll as "the first orca to be exhibited live by an aquarium," but it was a dubious claim.[34] Wanda at Marineland had been the first, and apart from the single day at Burrard Dry Dock, Newman seemed determined to avoid displaying Moby Doll. Despite winning Vancouver's Man of the Year Award for the orca's capture, Newman was never a man of the people. The chaotic atmosphere at the dry dock and during the transfer had troubled him, and he was determined to keep the public away. Not only did Moby Doll need to recover from her wound, but Newman and his staff worried that too much human contact with a "sulking" killer whale would ruin the animal's potential as a star attraction. "Later, as a friendly, clever whale," explained the *Vancouver Sun*, "she could be a sensation, attracting tourists from all over the world."[35] To Newman's dismay, however, Vancouverites proved unwilling to wait. In the days following the move, would-be whale watchers laid siege to the Jericho base. The guardhouse turned away most sightseers, but others came by water—on boats, rafts, even air mattresses. One man buzzed the dock in a small airplane hoping to get a look.[36]

Among the few who succeeded in slipping through was seventeen-year-old Mark Perry. Like many in the Pacific Northwest, Perry was familiar with killer whales' unsavory reputation, and his stepfather, a longtime employee of BC Packers, had little use for Vancouver's new pet. "Killer whales and fishermen were like oil and water," Perry told me. "The fishermen thought the killer whales ate all the salmon, and every time they had a chance to take a shot at one, I think they did." But Perry himself was intrigued by Moby Doll and made plans with a friend to sneak onto the base to see her. When his friend chickened out, Perry decided to go alone, driving to adjoining Locarno Beach and squeezing through an opening in the base's wire fence. "It was low tide," he recounted, "so I stayed down by the water, out of sight of the MPs [military police]." After finding the pier, he made his way toward Moby Doll's pen, picking his way between pilings thick with barnacles and mussels. Less than a hundred feet away, he heard an explosive breath. "Scared the heck out of me," Perry chuckled. But rather than flee, he lingered, mesmerized by the sight of the whale's circling dorsal fin. After all the stories about killer whales, he noted, "I couldn't believe how placid it was." Arriving home, Perry tried to relate his experience to his stepfather, but he wasn't interested. "I couldn't convince him," reflected Perry, "but it sure changed my attitude."[37] Indeed, that encounter shaped much of his life's path.

Moby Doll had a similar impact on Sam Burich. In early reports, journalists had hailed Burich as "Vancouver's Ahab," but his accidental capture of the little orca affected him deeply. With Newman's permission, he took up the task of befriending the animal following her move to Jericho. Like many after him, he sought acoustic connection, first blowing a police whistle and then playing his harmonica to the little whale. "[Sam] changed quite a bit," observed Joe Bauer. "That's why he started to go down there on that float playing music."[38] After several such sessions, Moby Doll stunned everyone by whistling back, perhaps attempting to mimic the instrument. "It may not have been a polite thing for a lady to do," joked one reporter, "but Burich was delighted." So was Newman, who hoped a bond with the sculptor might convince Moby Doll to eat. Unaware that the whale hailed from a strictly fish-eating culture, scientists and aquarium staffers offered her a range of fare, including seal meat and even whale tongue shipped in from Coal Harbour—all to her evident disgust.[39]

In the following weeks, Moby Doll fell out of the news, and aquarium staffers grew more concerned with her health. By late August, the little orca had developed an unsightly skin condition, which began to spread. Some believed it was an infection caused by pollution in the harbor. Others argued it was the result of insufficient salt water. Washed by the runoff of the Fraser River, English Bay has extremely low salinity much of the year, particularly near Jericho Beach. Whatever the cause, the animal was weakening. On September 8, Burich and trainer Terry McLeod believed they saw Moby Doll take a fish from a line, but no one could be sure.

The next day, Moby Doll had unexpected visitors. Intrigued by reports of the captive orca, Ted Griffin and his wife, Joan, drove *Pegasus* to Vancouver on September 9. Thinking nothing of sneaking onto a Canadian army base, Griffin tied up at the Jericho pier. At first, he was disappointed by the orca's gaunt and lethargic appearance. "In light of the whale's traumatic capture," he recalled, "it was pure folly to have expected anything else." Griffin climbed down to the small float in the pen, where he noticed several live lingcod that aquarium staffers had tied to strings. He cut one fish loose and held it by the head, slapping its body on the water. "Soon Moby Doll was excitedly zooming by looking at the fish which I held in my hand," explained Griffin. "Finally, the whale came up to the float, stopped, rolled partly on its side, opened its mouth, and yanked the fish away." Ten minutes later, Moby Doll took a second lingcod from Griffin. For the young entrepreneur, hand-feeding a killer whale was a dream come true, and he may have briefly connected to orca culture. Years later, researchers would discover the central role of food sharing for the species. In the wild, killer whales regularly share prey with podmates, including the young and ill, and the gesture of offering food may have sparked recognition in the little whale. In Griffin's eyes, Moby Doll seemed like "a lost child." She needed attention and love.[40]

Newman's sudden arrival broke the idyll. "You're doing a very naughty thing," he scolded, as Griffin scrambled up the ladder.

"The whale is eating, Murray, right out of my hand!" exclaimed Griffin. Newman was annoyed, but he allowed his Seattle-based rival to stay. Soon after, the whale took another cod on a line held by Allan Williams, chairman of the West Vancouver Parks Commission. Hoping to get Williams's support for the location of a sea pen off West Vancouver, Newman publicly claimed that the chairman had been the first to feed Moby Doll. But in fact it was Griffin, who achieved the feat before Newman and his party arrived.[41] It wasn't the last time he would upstage Newman or display unique courage around the ocean's greatest predator. Indeed, the encounter made Griffin even more determined to have his own killer whale. "I thought of taking her home," he admitted. "In fact, I wanted Moby Doll so much I considered stealing her."[42]

In the following days, Burich, McLeod, and even Newman himself hand-fed the animal. Moby Doll displayed a ravenous appetite, gaining weight steadily. But although her handlers grew more comfortable, none mustered the courage to swim with the still-feared creature. When Penfold sought a closer look at the orca's skin condition, he did so in a wire cage to prevent attack.[43] For his part, Newman certainly didn't envision putting humans in the same pool as the feared predator. He did, however, look forward to displaying Moby Doll to the paying public once she fully recovered.

She didn't. On October 8, Moby seemed vigorous, eating a hundred pounds of fish, but the next day something was wrong. At her 2:00 p.m. feeding, the whale took only a single herring, and an hour later she sank to the bottom of the pen. Among those who rushed to the scene was Bauer, the

FIGURE 4.3 Terry McLeod feeds Moby Doll at Jericho Army Base, September 1964. Courtesy of Terry McLeod.

man most responsible for saving the whale's life during her capture. There he found army divers reluctant to enter the pen. "I told them, 'It has been down there for two hours now, and it breathes air like we do. It has got to be dead,'" explained Bauer, "but those guys were petrified. They didn't really want to go into the pond with the whale."[44] When they finally raised the lifeless body at 5:15 p.m., the aquarium staff was devastated. "I loved that whale," Newman told reporters. "Capturing it was the best thing I ever did." He barred the press from the ensuing necropsy. "I don't want any pictures taken of the corpse," he explained. "It would be like photographing the body of your poor old maiden aunt."[45]

At first, many linked Moby Doll's death to her skin condition. Yet the necropsy revealed that the harpoon had caused more injury than previously thought, clipping the base of the skull and likely causing brain damage. The wound had not healed well, and the area behind the head was teeming with worms. Yet the ultimate cause of death was probably the low salinity, and hence reduced buoyancy, of the water in her pen. "It was like throwing two to three hundred pounds on the whale's back and expecting [it] to swim 24/7," reflected McLeod.[46] Immersed in the freshwater runoff of the Fraser, Moby Doll had slowly been smothered by the same river that sustained her salmon-eating pod.

Death held one more surprise for Moby's keepers, courtesy of Joe Bauer. As reporters and aquarium staff milled around the carcass suspended from a crane, the young fisherman marched past Penfold, reached into the whale's genital slit, and pulled out its penis. "What do you think now, Vince?" he teased. Moby Doll was a male after all. Measuring fifteen feet, four inches, he was only five years old—six at the most.[47] Now that he was dead, scientists coveted him even more than before. Requests poured in for various parts of the body. In the end, the aquarium retained the skull, and Pat McGeer kept the brain for his lab at UBC. The rest of Moby Doll was, as the *Vancouver Sun* put it, "distributed" for science.[48]

Vancouver mourned its loss, but Moby Doll faded quickly from the news. Just days after the whale's death, the retirement of Soviet premier Nikita Khrushchev captured headlines. The following year, Burich completed his model of the whale for the aquarium's foyer, and he made a special casting of the animal's penis for Bauer. Moby Doll's time in Vancouver was brief, but he had made his mark. Although the whale generated little revenue for the aquarium, his unplanned capture proved the viability of holding a killer whale in captivity, and it hinted at the potential of live orcas as tourist attractions. It also revealed the emotional attachment the species could generate. When the aquarium announced plans to capture a replacement whale, some residents expressed concern that another animal might suffer the same fate as Moby Doll. Yet the whale's impact on general perceptions of his species was limited. Only a small group of researchers and handlers had spent time with him, and even many of them continued to regard orcas as dangerous. This included Newman himself. Although the aquarium director had likened Moby Doll to a "poor old maiden aunt," the public mourning for the whale puzzled him. "I worry about this sentimentalizing," he admitted. "It was a nice whale, but it was still a predatory, carnivorous creature. It would swallow you alive."[49]

5

Namu's Journey

THE CALL CAME by ship-to-shore radio from a Washington State ferry. The skipper on the Seattle-Bremerton route had just spotted killer whales headed south, and he thought Ted Griffin should know. Shouting his thanks, the aquarium owner raced down the dock, leapt into *Pegasus*, and tore off in the direction of the sighting. Clocked at sixty miles per hour, the shallow-draft runabout may have been the fastest boat on Puget Sound, and it overtook the orcas near Vashon Island. But as Griffin throttled down, he realized to his disbelief that someone else was already chasing them. There, clear as day, was a blue helicopter hovering over the whales. Incensed, Griffin steered *Pegasus* closer, until he could almost touch the helicopter's pontoons. Looking up, he spotted a burly man leaning out the cabin door and eying the pod.

"Get away from my whales!" Griffin shouted.

"Your whales?" the man laughed. "You'll have to catch them first." It was the first time Griffin had met Don Goldsberry, ex-fisherman and animal collector for the Point Defiance Aquarium (formerly the Tacoma Aquarium). The two men's shared pursuit of orcas would soon bind them together. On this day, however, Griffin left feeling a bit embarrassed, having behaved, as he put it, "like a rancher possessive of his herd."[1]

Some part of him knew his quest to capture and befriend a killer whale was becoming unhealthy. He had a struggling aquarium in Seattle and a growing family on Bainbridge Island. Orcas were his obsession, but they weren't paying the bills. At home, he still talked and laughed with Joan and played with his little sons, Jay and John. But he had whales on the brain. He dreamed of them when asleep and sometimes mumbled about them when awake. With each reported sighting, he dropped everything—to Joan's increased annoyance.[2] In time, Griffin had come to see patterns in the animals' migrations and behavior. He noted that they appeared when chinook salmon were running and that they seemed to cling to the west side of Puget Sound when headed south and to the east side when swimming north. When he got them in his sights, he followed like a bloodhound, often for several days in a row. Like scientists and fishermen at the time, Griffin assumed the orca population of the North Pacific was fairly large—five thousand by his estimate. Yet he also came to recognize some individuals by the nicks on their fins and white patches on their backs, "as unique as nose and eye shapes to humans." At night, he often drifted near them, listening to their calls. "I could make out a mother's whistle, a calf's answer with quick squeals," he later wrote. "The chattering sounds were melodious."[3] But the whales came to know him, too.

Having evolved with no natural predators, orcas rarely felt hunted—except by trigger-happy fishermen. Yet here was a man who appeared whenever they entered Puget Sound, who pursued them day and night. They came to recognize the sound of *Pegasus* and found ways to avoid the bothersome vessel. Griffin noted that some mature males learned to tilt sideways as they surfaced to prevent their large dorsal fins from giving away their pod's position. At the time, he assumed the species' social organization was patriarchal, with a breeding male accompanied by female mates and offspring, and he often viewed the pursuit as a contest with a dominant bull. He named one large and especially cagey male "Hook-fin." But he also noticed that mature females would often approach and then race away at high speed, attempting to draw the boat away from the pod.[4] Griffin was impressed with the animals' intuition. He was learning quickly, but so were the whales, and they were becoming more difficult to track with each passing month. If the orcas continued to find ways of evading him, how would he ever catch one?

As it turned out, he didn't have to.

~

On the afternoon of Tuesday, June 22, 1965, two fishermen from Steveston, British Columbia—Willie Lechkobit and Robert "Lonnie" McGarvey—were working their small gill net boats near the cannery town of Namu. After pulling into Warrior Cove, Lechkobit climbed into his bunk for a rest, but just after midnight a gale kicked up, pushing his net onto a nearby reef. Fearing he would lose the entire vessel, Lechkobit cut the net loose and tied up in Namu. But McGarvey stayed out, and in the morning, when he checked on his friend's net, he discovered two killer whales behind it. One was a very young calf, ten feet long at most. The other was a male at least twice as large. McGarvey radioed Lechkobit, who came for a look. Sure enough, the detached net had trapped two orcas between the reef and a rocky outcropping. Lechkobit's first impulse was to release them, but McGarvey stopped him, reminding his friend of the recent stir caused by Moby Doll. Surely these two blackfish were worth more than a load of salmon. As the men edged their boats closer, however, they noticed a large gap between the net and the rocks. The animals weren't trapped at all, and the big one knew it.[5]

The two whales were northern residents, members of what scientists later labeled C1 pod, and they were likely siblings, though born some fifteen years apart. Like Lechkobit and McGarvey, they had come to the area to intercept salmon headed to the Bella Coola River. Among resident orcas, it is common for older siblings to mind younger ones while their mothers forage. In this case, the two animals had likely been swimming together when they encountered the snagged net. Regardless of whether the big male found himself trapped initially or later moved through the gap to join the calf, one thing was certain: he realized that he didn't have to stay. Indeed, Lechkobit and McGarvey watched in wonder as the large whale tried to nudge the calf through the opening. Failing that, he swam through the gap himself, demonstrating the escape route. But the nervous youngster refused to follow. Faced with the choice of freeing himself or returning to his frightened sibling, the big brother opted for the latter.[6]

After placing two more nets to reinforce the enclosure, Lechkobit and McGarvey started making calls. As expected, they had several eager customers. Both Marineland of the Pacific in California and the Pacific Undersea Gardens in Victoria expressed interest, as did the organizers of the New York World's Fair, which had just opened for its second season. Initially, however, it seemed Murray Newman had the inside track. As residents of the Vancouver area, Lechkobit and McGarvey had followed the story of Moby Doll, and they knew the Vancouver Aquarium wanted another whale. They also recalled Newman's suggested price when Marineland had offered to buy the harpooned orca. As the only people in the world with killer whales for sale, the fishermen hoped to sell the animals for at least $20,000 each. Now on the buyer's end of a market he had helped create, Newman balked at the high figure, and he demanded more information on the whales. How big were they? What were their sexes?[7]

"You're the expert," retorted McGarvey, "but then you didn't do too well on the last one." Soon after, Newman flew to Namu to see the whales. At first glance, the scene at the reef didn't make sense. Why didn't the powerful animals escape? Surely they could tear through or swim over the nets. Convinced the whales wouldn't be contained for long, he was especially eager to acquire the calf. But the Vancouver Aquarium Association had limited funds, and he could offer only $2,000 for the two animals together, along with a bottle of whiskey. The fishermen scoffed. "He tried to steal them from us," McGarvey told a reporter. "The nets alone are worth $3,600."[8]

Soon after, two US purse seine vessels were passing north through Queen Charlotte Sound when they picked up radio chatter about the captive whales. Skippers Adam Ross of the *Chinook* and Peter Babich of the *Pacific Maid* were based in Gig Harbor, Washington, but like many Puget Sound fishermen, they traversed British Columbia's Inside Passage each summer to fish in Alaskan waters. They had good reasons to ignore the conversation. Not only were they eager to reach Ketchikan, but tension often crackled between US and Canadian fishermen. Yet Ross and Babich knew of Ted Griffin's quest for an orca, and they decided to hail Lechkobit and McGarvey.[9]

Soon after, Griffin received a call from Steveston on behalf of the Canadian fishermen. He couldn't believe it. Two killer whales? In nets? Surely they must be drowned. When the caller assured him the animals were fine, held near the town of Namu, Griffin turned to logistics. Even if the whales were alive, how could he pay for them? His failed capture attempts had left him near bankruptcy. Surely Marineland and the Vancouver Aquarium could make higher offers, and even if he managed to win a bidding war, how would he transport the whales from a distant cove in Canada? Still, Joan urged him to try. Although six months pregnant, she assured him she could manage the aquarium and the children.

After refueling in Alert Bay, British Columbia, Griffin's floatplane splashed down in Namu. Minutes later, Lechkobit and McGarvey picked him up and began the short trip to Warrior Cove. But after learning that Griffin had brought no money, the fishermen turned the boat around and deposited him back on the dock. They weren't in the business of selling whales on credit, they explained. Marineland of the Pacific had promised to pay cash.[10]

By this time, most assumed Marineland would win this cetacean lottery. The California oceanarium was the most profitable in the world, and despite Brocato and Caladrino's previous failure, Marineland remained determined to acquire a killer whale. In fact, just a few months earlier, the collectors had tried to use tranquilizers to capture orcas in British Columbia's Johnstone Strait.[11] Like Newman, Marineland officials were especially keen to buy the calf. But unlike the Vancouver Aquarium, they had the capital to pay for both animals and had made arrangements to fly the calf out of Port Hardy.[12] But then the whales changed the equation.

Since the capture, Lechkobit and McGarvey had noticed that other orcas made daily appearances at Warrior Cove. The most frequent visitors were an adult female and a juvenile who often lingered and even touched noses with the animals behind the nets. Finally, on the night of Friday, June 25, after three days trapped against the reef, the little calf slipped away and with it the sellers' market the two men had enjoyed. Believing the big male likely to escape and too large to transport, Marineland withdrew its offer. The despondent fishermen then phoned Griffin. They had lost too many fishing days, they explained—they needed to recover their gear and start working again. When Griffin offered $5,000 for the remaining whale and $3,000 to replace their nets, they agreed but set the deadline for midnight the following day. After that, the fishermen warned, they would cut the whale loose.[13]

Griffin hung up the phone in a daze of adrenaline. It was a sizable sum (about $70,000 in 2018 dollars), and he didn't have anything close to it. Even if he had, it was Saturday afternoon, and the banks were closed. After gathering what cash he could and borrowing from friends and family, he spent Sunday morning visiting businesses along the Seattle waterfront. If the owners were willing to lend him cash, Griffin promised, he would not only repay them but bring back an attraction unlike any in the world. Trident Imports and Harbor Tours opened their registers, but the largest sum came from Ivar Haglund. With the money stuffed into a backpack, Griffin boarded a floatplane for Namu. The plane stopped in Vancouver to pick up a former Royal Canadian Mounted Police (RCMP) officer for protection. That evening, Griffin climbed aboard McGarvey's boat bearing $8,000 in small bills. He departed with the world's first bill of sale for "one live and healthy killer whale."[14]

Griffin didn't anticipate that his purchase of an orca would cause diplomatic problems. After all, as he reminded reporters, *Orcinus orca* "comes under no international treaties or conventions and is not protected in any way."[15] But the transaction didn't sit well with some Canadians. Homer Stevens, head of the United Fishermen and Allied Workers Union (UFAWU), lobbied Ottawa to block the sale. Although many of his members had shot orcas over the years, Stevens demanded that the potential tourist attraction be kept in Canada. For their part, the Vancouver Aquarium and Victoria's Pacific Undersea Gardens joined together to make one last bid, but it came too late. "I think it should stay in Canada," lamented the manager of the Undersea Gardens. "It seems that every time we get something good, something really worthwhile, it gets funneled off to the US."[16] With such resentment simmering, Griffin decided he needed someone to watch over the whale as he arranged for transport to Seattle. In addition to flying trainer Homer Snow to Namu, he asked Cecil Brosseau at the Point Defiance Aquarium to lend him Don Goldsberry, the very man Griffin had confronted aboard a helicopter months earlier. "You won't find a harder worker with commercial fishing experience," Brosseau assured him.[17]

Although also born in 1935, within months and miles of Griffin, Goldsberry had grown up in very different circumstances on the industrial side of Tacoma. As a teenager, he found work on local purse seine boats, coming to know the fishing families in nearby Gig Harbor. Old Spiro Babich, a Croatian immigrant, took a liking to him, and by the early 1950s Goldsberry was a regular on his crew, along with Spiro's son Peter—the man who had called the Canadian fishermen to inform them of Griffin. On several occasions, Goldsberry fished with the Babiches off Alaska's False Pass, where he saw many killer whales. After years of fishing, Goldsberry took a job as animal collector at the Point Defiance Aquarium, where he first contemplated orca capture. Eager to join Griffin's adventure, he took up an armed vigil at the reef.

Meanwhile, Griffin struggled with the question of how to get his newly purchased whale to Pier 56. No one had ever transported such a large cetacean alive before, and scientists at the Marine Mammal Biological Laboratory in Seattle warned him that lifting the animal by sling could damage its lungs. As a result, he decided the best option was to tow the whale in a floating sea pen. Since no such thing was available, he would have to build it himself. After having two tons of material delivered to Seattle's Boeing Field, Griffin flew the load on a chartered DC-3 plane to Port Hardy on northern Vancouver Island, where he transferred it to a freighter bound for Namu. He then visited his new whale at Warrior Cove, where Snow and Goldsberry had surprising news. First, they reported offering some two hundred pounds of chinook salmon per day to the whale. Some of it came free from visiting fishing vessels, which stopped to allow their crews to see the already famous orca. Considering the region's fishermen had long viewed killer whales as pests, this seemed a striking shift. But Snow had bought most of the salmon, and at fifty cents a pound, that meant $100 per day just to feed the whale.[18] Second, the two men reported that the animal had nearly escaped a few nights earlier, when the net's lead line had snagged on the reef. As the tide rose, this left a ten-foot gap between the cork line and the water's surface. The big orca could have slipped out at any time. "This may sound crazy," Snow told the whale's owner, "but a couple times, when it seemed he was thinking about swimming away, we talked him out of it."[19]

Griffin quickly turned to construction of the pen. To his relief, BC Packers offered the use of the equipment and pier at the Namu cannery. Hiring every welder he could find, he built the pen in four days. Fashioned out of structural steel and equipped with forty-one empty oil drums for flotation, it measured sixty feet long, forty feet wide, and twenty feet deep. The project caught the attention of locals, who debated whether it would float, or even hold together. "You cannot keep a whale's spirit confined," warned one indigenous worker. "Just you try to put him in your cage, you'll see."[20] But when the time came, nearly a hundred people helped Griffin lower the pen into the water. Meanwhile, outsiders hatched plans to catch the orcas visiting the netted whale. Merrill Spencer, director of the Virginia Mason Research Center in Seattle, had launched an expedition with researchers from the Marine Mammal Biological Laboratory. Along with the Vancouver Aquarium, they urged Griffin to keep his whale in place until other orcas could be captured.[21]

With costs and risks mounting, however, Griffin decided he couldn't wait. After towing his steel creation into place, he donned a wetsuit and began stitching the pen to the net for the animal's transfer. As Griffin worked his way down, the orca watched him, chirping inquisitively. In a rush, Griffin accidentally jabbed himself with his knife, exhaling a muzzled "ouch" into his mouthpiece. The whale immediately mimicked a response. Did he know the man in front of him had been hurt? Was he searching for an assurance that the diver meant him no harm? Whatever his intent, the orca's effort to connect touched Griffin, who felt tears flowing behind his face mask.[22] For his part, Newman watched the operation with interest. He later admitted feeling "very jealous" of Griffin's prize, but he also marveled at the courage of this young entrepreneur, who seemed to be "entirely without fear of these potentially dangerous beasts."[23]

Soon after, Goldsberry directed the transfer, pulling up the nets and coaxing the nervous animal into the pen. The team had planned to tow it using the small tugboat *Robert E. Lee*, owned by Seattle radio personality Bob Hardwick. When Hardwick's vessel struggled to pull the pen the two miles to Namu, however, Griffin hired a Canadian purse seiner for the first leg to Port Hardy. As the flotilla prepared to depart, local workers lined the pier to catch a glimpse of the now-famous whale. Indigenous workers shook Griffin's hand, and Stephen Hunt, a native artist from Bella Bella, presented the aquarium owner with a painting of a killer whale. Appreciative, Griffin decided to name the whale Namu, after the cannery town that had made his dream possible.[24]

~

The 450-mile journey to Seattle held many obstacles. The first was Queen Charlotte Sound, a thirty-mile stretch of open sea whose swells had sent many a vessel to the bottom. For twenty-four hours, the seiner struggled southward while the crew on the *Robert E. Lee* tried to keep the pen from breaking up. No sooner had they passed this gauntlet than they heard warnings that unknown persons had hired Seattle private investigator Jack Hazzard to free Namu. "We have the whale towing operation under surveillance by our own people," Hazzard told reporters, and he declared that his divers would cut the animal loose "at the opportune time."[25] As the director of the improbably named International Spies Ltd., Hazzard may simply have been carrying out a publicity stunt. But Griffin took the threat seriously, arming his crew with small depth charges to ward off divers and alerting Canadian and US officials. The RCMP ordered Hazzard to stay away, and Washington State's US senator Warren G. Magnuson warned that the US Navy would protect Namu and his handlers from sabotage or piracy.[26]

When the flotilla reached Port Hardy, all seemed well, as about half of the town's one thousand residents turned out to see the famous orca. Griffin immediately set about mending the pen, which was coming apart in several places. "Namu watched us work," he noted. "He reminded me of my two-year-old son, Jay, following me around the house, curious about everything I did." In town, Griffin received the welcome news that his brother had contracted the Seattle tug *Iver Foss* to complete the voyage and convinced Lloyds of London to approve an insurance policy on the whale. "All risks are covered," Jim reported, "except escape."[27]

Meanwhile, *Seattle Times* reporter Stanton Patty, who was riding aboard the *Robert E. Lee*, took the opportunity to visit the Coal Harbour whaling station. Located just ten miles west of Port Hardy, the operation was a stark reminder of the region's historical relationship with whales. As he choked on the stench of boiling blubber, Patty reflected on the two contrasting spectacles. "As Namu was being admired at Port Hardy," he noted, "the seventy-man crew at Coal Harbour was turning eight whales into shipments of whale meat for Japanese families, meal for livestock feed and oils for industrial lubricant."[28]

For Patty, as for the rest of the *Robert E. Lee*'s crew, the quest to deliver Namu had become the adventure of a lifetime. The men swapped stories and bonded in their cramped quarters. While subsisting on pasta and chocolate bars, they fed Namu fresh chinook salmon, bought from passing fishing boats. In previous years, some of these same fishermen had likely taken shots at blackfish; now they took detours to visit one. The crew experienced changes as well. At a time when it was uncommon to sport facial hair, the men vowed not to shave until they completed their journey, giving them the appearance of protohippies on a psychedelic voyage. Already a popular Seattle personality, thirty-four-year-old Bob Hardwick broadcast his radio show from the pilothouse, peppering it with anecdotes about Namu and his handlers. In addition to Snow and Goldsberry, care of the whale fell to Gil Hewlett, a young biologist from the Vancouver Aquarium. Hewlett occupied himself by observing and recording Namu, as well as contemplating the animal's future in captivity. The orca "may have the instincts of a wolf," he told reporters, "but I'm convinced he can be trained."[29] In fact, Namu's gentle demeanor contributed to the crew's sense of shared mission. "The whale's magnetism cast a spell over the diverse individuals," Griffin later reflected. "Somehow it all worked."[30]

❧

But not all went smoothly. Unbeknownst to Griffin, his publicist, Gary Boyker, had signed an exclusive deal with the *Seattle Times* to cover the story. With Patty and his photographer riding aboard the *Robert E. Lee*, the newspaper gave the adventure daily front-page placement, and the *Seattle Post-Intelligencer* (*P-I*) scrambled to keep up. Forced to observe mostly from a floatplane, *P-I* reporter Emmett Watson offered caustic commentary. One

FIGURE 5.1 Namu in travel pen, July 1965. Courtesy of Ted Griffin.

article began by conceding that several members of the expedition were formidable. The barrel-chested Goldsberry "looks like he could have caught Namu bare-handed," wrote Watson, and the daring Griffin, already known for wrangling sharks, might well be serious about swimming with his killer whale. But characters such as Hardwick and Patty, he needled, "couldn't track an elephant with a nosebleed through the snow."[31]

The trip also brought tensions for Namu himself. Confined to a floating cage, the four-ton animal held up well at first. In addition to nimbly avoiding the edges of the pen amid shifting currents and rough seas, he remained remarkably calm. One reason was that the female and juvenile whales who had visited him at the reef, along with the calf who escaped, had followed the flotilla, calling to him constantly and often coming within feet of the pen. Most observers perceived this grouping as a nuclear family: Namu the father was being following by his mate and two children. "It's fantastic—the warmest, friendliest scene I've ever seen," Hardwick told his listeners. "Mama and the two babies are out there, not eight or 10 feet from the boat, cheeping at Namu, having a wonderful time."[32]

But the apparent harmony ended abruptly the evening of July 13 as the flotilla passed through Johnstone Strait. As Hewlett recorded in his diary, when a group of orcas approached the pen, Namu let out "a terrifying squeal, almost like a throttled cat." For the first time on the voyage, the captive whale seemed to panic. He leapt from the water and crashed into the side of the pen, tangling his pectoral fins in the net. As Hewlett wrote in his journal, "The family unit circles around towards the end of the pen. Those of us on the pen are yelling and screaming at the top of our lungs. This is an incredible experience. The excitement is almost overwhelming."[33]

At the time, the crew believed the ruckus stemmed from a large male approaching too close to the pen. Assuming Namu was reacting to a rival attempting to steal his mate, the men used explosive charges to drive off the animal. Griffin dubbed the intruder "Oil Can Harry," after the nemesis of the cartoon character Might Mouse, and Patty reported that "Mrs. Namu" had spurned the "Home Wrecker."[34] Yet the actual cause of the commotion was likely geography. Although none of the crew realized it, they were approaching the dividing line between the northern and southern resident killer whales. In its pursuit of chinook, Namu's C1 pod ranged far, from Southeast Alaska to Johnstone Strait, but it rarely ventured south of Seymour Narrows. Up to this point, the captive orca may have allowed himself to believe that, wherever he was going, his mother and siblings were going, too. At this moment, he may have realized they would follow no farther.[35]

Meanwhile, plans were afoot to capture Namu's family. Homer Stevens and the UFAWU publicly called on Canadian officials to halt the flotilla for this purpose. "We strongly urge the federal government to declare an embargo," he proclaimed, "until the Vancouver Aquarium, together with the federal and provincial governments, can impound the other three whales for study and exhibition in Canada." For his part, Murray Newman seemed to encourage the scheme. Although he referred to the three orcas as "strays," questioning whether they were the same whales from Warrior Cove, he argued that the animals could be caught if the pen were pulled into a shallow bay.[36]

Griffin rejected such plans. Not only did his insurance policy require Namu's immediate transport to Seattle, but he feared that another capture would further harm the family. "It is a distinct possibility that one or both of the calves following Namu are still nursing," he warned. "In the event of the cow's capture, or even the capture of all three, it is almost certain that the two young ones would die."[37] Griffin's stance annoyed Newman. "Whales are international in character," he declared, warning that "we may come down and capture our killer whale in Puget Sound."[38] Only the sudden disappearance of the three whales ended the debate.

Meanwhile, the crew prepared to contend with Seymour Narrows. Although just two miles long, the passage boasted wicked crosscurrents and eddies, as well as tidal flows reaching eighteen knots. After waiting out the tide, the flotilla made a safe run in the early morning of July 17, 1965—passing the site where four years earlier the Canadian Department of Fisheries had mounted its machine gun to kill orcas just like Namu. By the time they tied up near the Royal Canadian Air Force base at Comox, the crews of both vessels were exhausted. "We're all beginning to think there must be something to these Indian legends about the elements taking revenge if you bother killer whales," commented tug skipper George Losey.[39]

By this time, the journey had taken a visible toll on the pen. Eager to help, diver Gary Keffler flew up from Seattle and spent hours mending the structure. In the process, he found himself surprised by Namu's docile behavior. "I was underwater all the time wiring those things back together right next to the whale," Keffler recalled. "He would just turn his head a little and look at me to see what I was doing. Never made any kind of move."[40] But Namu didn't look well. In addition to losing weight, he was sunburned from spending too much time on the surface. As Keffler worked, Gil Hewlett scoured local pharmacies, buying all the tubes of zinc oxide he could find and applying it to the whale's back and dorsal fin.[41] Although the balm helped, some indigenous leaders expressed concern about the whale's fate. Near Victoria, Chief

Edwin Underwood of the Tsawout band deplored Namu's separation from his family. "I do not think it is right that these people should be bothering the black fish," he declared. "They are meant to live, to have the freedom of the oceans—not to be kept in a cage."[42]

As the procession neared the border, Griffin had more pressing concerns. Although Canadian officials had waived all fees, their US counterparts were still debating how to treat Namu's entry. "We've been told by our customs here the duty on a wild animal is seven percent," Gary Boyker explained to the press, "but the import book says a fish for exhibition purposes can get in free." Of course, a whale isn't a fish, he conceded, "but the question to consider is: is the whale a wild animal?" Senator Warren Magnuson again came to Griffin's aid, ensuring that the US customs station in Friday Harbor charged no fees.[43] Washington State wasn't so kind. In addition to imposing sales tax on the whale, officials informed Griffin that state law prohibited feeding Namu any fish fit for human consumption.[44] In other words, the captive orca was banned from eating the chinook salmon he had hunted in the wild.

Meanwhile, media attention dwarfed that given to Moby Doll a year earlier. Newspapers around the world followed Namu's journey. *Sports Illustrated* ran a seven-page spread on Griffin and his whale, while *Life* magazine and *National Geographic* jockeyed for feature stories. Even before the flotilla entered US waters, Seattleites were buying up Namu buttons and T-shirts. Local teenagers learned a new dance called "The Namu," with wiggly moves such as "sounding," "blowing," and "the fin." As one reporter marveled, locals seemed determined "to turn Namu into a Seattle fetish bigger than the Space Needle."[45] Even as the Voting Rights Act made its way through the US Congress and thousands of US troops deployed weekly for South Vietnam, Griffin and his orca captured headlines in Seattle. Perhaps it was the spectacle of a great ocean predator tamed by man. Perhaps the colorful characters in this drama, both human and cetacean, offered readers a feel-good story in ominous times. Whatever the reason, the Namu craze swept the Pacific Northwest.

Public excitement was on full display when word spread that "Namu's Navy" would enter Puget Sound through Deception Pass. Spanned by a quarter-mile bridge connecting Whidbey and Fidalgo Islands, the pass is one of the area's most spectacular locations, and a crowd began forming in the early afternoon of July 25. Soon thousands of spectators had gathered on the bridge, creating an eight-hour traffic jam. Leaning over the unnervingly low railings, they scanned the water 180 feet below until Namu passed underneath just before dusk.

Among the curious was journalist Wallie Funk. Raised in nearby Anacortes, the forty-three-year-old Funk had recently moved his family to Whidbey Island after buying its two local papers, and he followed Namu's story closely. When the flotilla appeared outside Deception Pass, he took his young sons, Mark and Carl, aboard a boat to see the famous whale. As they launched from Alexander Beach, ten-year-old Mark was excited. He had previously given little thought to orcas or other marine life, but the front-page coverage sparked his interest. "The budding consciousness was when the first story came out about Namu's capture and Ted Griffin's movement in that direction," he later told me. "You got caught up in it. It was exciting."[46]

Another eager spectator on the bridge was twenty-two-year-old Ralph Munro. Like Funk, Munro had deep roots in the region. His grandfather, a master stonecutter, had helped build the parliament building in Victoria, and his father worked at a Puget Sound shipyard. As a boy growing up on Bainbridge Island in the 1950s, Munro had often shot at orcas as they passed his beachfront home. By the 1960s, he was an undergraduate student at Western Washington University, but he was captivated by news stories about Namu and wanted to catch a glimpse. "There were ten thousand people on the bridge," he recalled. "The cars were backed up for miles." Like others there

FIGURE 5.2 Crowd on Deception Pass Bridge in Washington State, July 25, 1965. Photo by Wallie V. Funk. Courtesy of Center for Pacific Northwest Studies, Western Washington University.

that day, he felt only admiration for Griffin. "He was a hero," Munro reflected. "Here was a guy who got this *killer* whale."[47]

As the procession headed south, excitement mounted. The mayor of Langley, on the southeastern end of Whidbey Island, motored out to present Namu with a key to the town, and the mayor of Everett offered Griffin $10,000 just to have the whale visit his city for a few hours. Local ferries and pleasure boats stopped so passengers could gawk at the famous orca while the *Iver Foss*, escorted by Coast Guard and Seattle police vessels, approached Elliott Bay. As *Seattle P-I* reporter Don Page put it, the city was "daft over the whale's arrival."[48]

The tug delivered Namu to Pier 56 on the morning of July 28—one month to the day after his sale to Griffin. The public outpouring was extraordinary. With Ivar Haglund acting as master of ceremonies, thousands gathered on the waterfront, entertained by a Dixieland band and performers from the city's upcoming Seafair celebration. Among the crowd were Lieutenant Governor John Cherberg and Acting Mayor Clarence Massart, who proclaimed it "Ted Griffin and Namu Day." "Today we mark an important breakthrough in our knowledge of one of the least-known and most-feared of the sea animals," declared Massart:

> In the course of the last month, Namu has become a part of the life of millions of people around the world. As the best-known resident of Seattle's waterfront, Namu will be a source of wonder and enjoyment for everyone. As a major tourist attraction, Namu could well become the spark which Seattle needs to begin work on our Waterfront Development Plan.

With the television cameras rolling, he proudly presented Griffin with a key to the city, inscribed to "a gallant man and a magnificent feat."[49]

In the eyes of many, Griffin seemed a cross between rock star and conquering hero. As visitors lined up to see his famous whale, reporters jostled for interviews, young women asked for autographs, and salesmen pushed ideas for Namu toys and memorabilia. For his part, the aquarium owner could take a moment to savor his accomplishment. After years of pursuit, Griffin finally had his prize. But as he was discovering, everyone seemed to want a piece of Namu.

6

A Boy and His Whale

IT WAS FEBRUARY 1966, and Richard Stroud had seal sex on his mind. A recent graduate of Oregon State University, the Portland-born Stroud had taken a job at the Marine Mammal Biological Laboratory in Seattle. For one of his first field assignments, he had come to Morro Bay to study the reproduction of northern fur seals in their wintering area off the California coast. The primary focus of the lab, which was still administered by the US Fish and Wildlife Service, remained the fur seal harvest on the Pribilof Islands, and its scientists retained close ties to US whaling firms, often chartering their vessels for research. For this seventy-day expedition, Stroud and his colleagues hired the 136-foot whaler *Lynnann* for the purpose of shooting and dissecting five hundred fur seals under the terms of the 1957 treaty. Stroud also had instructions to kill and examine killer whales when possible. So when the *Lynnann* passed six orcas off Morro Bay just before noon on February 12, he asked Captain Roy J. "Bud" Newton to follow them.[1]

Ordinarily, Newton wouldn't have bothered with killers. His employer, the Del Monte Fishing Company, focused on fin, sperm, and humpback whales. Located in Richmond, a short drive from Berkeley, the station processed nearly two hundred whales per year. But the whaling season was months away, and the US government was paying for this voyage. Newton wheeled the *Lynnann* around, and after an hour-long chase, his crew harpooned and killed a large male killer whale. Measuring just under twenty-one feet, it was a healthy specimen, though its teeth seemed unusually worn.[2]

Stroud planned to examine the orca's stomach contents and send its skull and organs to the Seattle lab. Yet he chose not to dissect the carcass in port. Instead, as one reporter explained, Stroud and his fellow researchers "planned to butcher their killer whale Sunday while far out at sea." The reasons for this decision are unclear. Perhaps they hoped to spare Morro Bay residents the stench of orca innards. Perhaps they believed the mess would be easier to clean up at sea. Most likely, they sought to keep the operation out of the public eye. Stroud himself warned that killer whales could be dangerous to humans.[3] But he knew that popular views of the species were changing, particularly with the display of Namu at the Seattle Marine Aquarium. He and his supervisors still viewed orcas as predatory pests, but live display was spurring a rapid shift in public opinion.

The Seattle waterfront was the epicenter of that change. "The town has gone nutty for Namu," observed the *Victoria Times* in late July 1965. "The health and diet of Namu, the world's only captive killer whale, are debated more fervently than Viet Nam."[4] At first this was welcome news for Ted Griffin. The purchase and transport of the animal had put him $60,000 in debt, and he needed to find ways to make his new attraction pay. In addition to forming the company Namu Inc. to handle T-shirts, toys, and other merchandising, he copyrighted the name "Namu the Whale." Yet admissions proved the biggest source of revenue. Coinciding with Seattle's annual Seafair celebration, Namu's arrival brought a tenfold increase in visitors to the aquarium. On the first Sunday five thousand people came to see the whale, and more than one

FIGURE 6.1 Richard Stroud with slain orca in Morro Bay, California, February 1966. Courtesy of *San Luis Obispo Tribune*.

hundred thousand paid for admission in the five weeks following his arrival.[5] For Northwesterners, most of whom had seen orcas only from a distance, if at all, it was an entirely new way to observe the species.

Scientists seemed equally excited, among them Dixy Lee Ray. A zoologist at the University of Washington, Ray was the well-known host of the local television show *Animals of the Seashore*, and since 1963 she had served as director of the Pacific Science Center, located on the former grounds of the world's fair. Ray considered Griffin's acquisition of Namu critical to public education. "We are a timid animal," she observed. "Ignorance breeds fear, and what we fear, we destroy." As such, she had little patience for those who called for the whale's release. "By understanding Namu, learning more about him, there would be less wanton killing of his kind," she asserted. "So his capture is, of course, an advance in conservation." Like Clifford Carl in Victoria, Ray also predicted that the study of orca bodies could lead to breakthroughs in sonar technology as well as submarine and aircraft design.[6]

In fact, such research had already begun. Technicians from Boeing initiated studies of Namu's biosonar and physical characteristics while he was en route to Seattle, and after the whale settled in at Pier 56, researcher Merrill Spencer launched his own studies of the whale with funds donated by Griffin. The scientist was especially interested in the orca's circulatory system during dives, believing such research could lead to advances in human cardiovascular medicine.[7] Also active was physicist Thomas C. Poulter. The head of the Biological Sonar Laboratory at Stanford Research Institute, Poulter had served as second in command on Admiral Richard Byrd's second Antarctic expedition in the 1930s. In addition to famously saving Byrd from carbon monoxide poisoning on the trip, Poulter had become fascinated with killer whales.[8] Thirty years later, he jumped at the opportunity to study one in captivity. In addition to the animal's echolocation clicks, he recorded Griffin's attempts to "talk" to Namu by imitating the whale's calls. To his astonishment, Poulter discovered that, instead of the aquarium owner improving his vocalizations to match Namu, the whale was adjusting his calls to mimic Griffin.[9]

By 1965, dolphin communication and intelligence had become a hot topic, spurred by the work of neuroscientist John C. Lilly. A decade earlier, Lilly had conducted experiments on bottlenose dolphins at Marineland Research Laboratory in St. Augustine. Despite inadvertently killing five of the animals under anesthesia, he became fascinated and won funding from the US Navy and NASA to build a research facility in St. Thomas in the US Virgin Islands.

In 1961, Lilly published *Man and Dolphin*, which explored the possibilities of dolphin language and learning and suggested the potential military uses for marine mammals.[10] Lilly's experiments soon took a bizarre turn. In addition to administering LSD to captive dolphins, he encouraged a female assistant to gratify the animals sexually. By late 1962, the navy had rescinded Lilly's funding and launched its own Marine Mammal Program at Point Mugu, California. There, staff trained captive animals to perform a range of tasks. In its early years, the program focused on bottlenose dolphins, taking its cue from the display industry, and its applications were fairly simple. In late August 1965, for example, trainers used a dolphin named Tuffy to carry supplies to aquanauts in Sealab II, an underwater habitat off La Jolla. But they contemplated more complex tasks and the training of other marine mammal species.[11]

Hollywood, too, had become enamored with dolphins. In 1963, MGM released the film *Flipper*, about a friendship between an injured bottlenose dolphin and a young Florida boy, whose fisherman father—played by Chuck Connors of *The Rifleman*—learns to accept a species previously considered a pest. The following year, the *Flipper* television series (1964–1967) began its successful run. Filmed in close cooperation with the Miami Seaquarium, which provided the animals and trainers for the production, the show proved popular among young people and made bottlenose dolphins the charismatic face of marine life. Always on the lookout for a new star, Hollywood took notice of "Namu the Whale" in August 1965.

By this time, Griffin had grown frustrated. Despite paying off most of his debt, he felt he had too little time to spend with Namu. In addition to managing the aquarium, he struggled with his new celebrity status, which included endless demands for interviews and media appearances. "Public life had consumed me," he later wrote. "My original purpose in obtaining a whale was getting pushed aside."[12] When time permitted, he worked hard to bond with the orca. In addition to his attempts to communicate, he regularly rowed a small boat into the pen to scratch Namu with a pole brush. Because the skin of orcas sloughs off continuously, they frequently rub against each other and other objects. Likely missing the physical contact of his podmates, Namu seemed to enjoy these sessions and became comfortable with the boat's presence. Although Griffin had dreamed of swimming with a killer whale for years, he couldn't yet bring himself to jump in.

Just by being in close proximity, Griffin was breaking new ground, and many believed he was courting danger. As late as December 1965, writer Gil

FIGURE 6.2 Griffin scratches Namu at Seattle Marine Aquarium, August 1965. Courtesy of Ted Griffin.

Paust, of the New York magazine *Argosy*, mocked the "sentimental public" of Seattle for fawning over a species that had devoured "an untold number of shipwrecked sailors and even swimmers." The total body count remained unknown, Paust asserted, only because "few witnesses survive to tell the story."[13] In this light, the prospect of swimming with Seattle's captive orca was no joking matter. There was no denying the immense predatory power of Namu's twenty-three-foot, four-ton body. Anyone who saw him knew he *could* kill Griffin. The question was: *Would* he?

The final nudge into the water came from Hollywood. In late August, shortly after filming the second season of *Flipper*, producer Ivan Tors arrived in Seattle with an offer to make Namu the star of a feature-length movie. Before United Artists could extend a contract, however, it needed proof that the whale wouldn't attack people. Although Namu had accepted hand feeding, Griffin had yet to join the big whale in his pen. Goaded by his brother, Jim, who jumped into the water first, Griffin finally braved a swim with Namu, brushing his skin and even daring to grab hold of the orca as he passed by. Fresh off filming scenes with sharks and divers for the James Bond movie *Thunderball* (1965), cinematographer Lamar Boren joined them in the water, shooting the first footage of a person in the water with a killer whale. Watched by nervous spectators, the interaction sealed the deal with United Artists, and in the following days newspapers reported that the aquarium owner was actually swimming with his orca.[14]

It was a transformational moment. As NOAA scientist Mark Keyes would declare years later, "By the single act of going into the water with Namu, Ted Griffin contributed more to the conservation and appreciation of killer whales by societies of the world than all the biologists and conservationists put together, from the dawn of time to that moment."[15] The statement might seem an exaggeration, until one realizes that neither conservationists nor the International Whaling Commission had shown interest in orcas and that scientific research on the species was virtually nonexistent.

Within a month, the tourist season had ended, and Griffin was scouting for a healthy wintering spot for Namu that was suitable for filming. He settled on a tiny bay near Port Orchard, which he dubbed "Rich Cove." Days later, a Harbor Tours vessel delivered Namu to his new home in his travel pen. As the orca happily explored the cove, Griffin found himself conflicted. "I wanted Namu to be free, yet couldn't part with him," he reflected. "I had to fulfill my dream of interacting with a killer whale."[16] And with the animal sequestered, Griffin now had the perfect opportunity. Hiring Don Goldsberry to manage the aquarium's day-to-day operations, he devoted as much time as possible to Namu, but he also invited locals to meet the whale. "We used to come out almost every day and rub his head," one resident later recounted. "My daughter was in high school, and she was just thrilled to see him."[17]

Initially, Griffin focused his efforts on feeding and training. To his relief, state officials had exempted Namu from the ban on feeding animals fish fit for human consumption. But his orca had expensive taste. Like other resident killer whales, Namu preferred fresh chinook salmon, and he tended to shun other offerings. Even with the film studio paying for some of the orca's upkeep, his $100 daily tab was too steep. As Griffin later put it, "We could hire ten people per day for the cost of the salmon." He finally hit on a solution at the nearby fish hatcheries, where state officials auctioned off cheaper "spawned out" salmon.[18] As Griffin shifted Namu to these lower-cost fish, he began training him for performance. In addition to coaching the animal to perform leaps and other stunts on demand, he spent more and more time in the water swimming with, and eventually riding, the powerful predator. News of this breakthrough traveled quickly. At college in Bellingham, Washington, Ralph Munro received word from his father that a man was swimming with a killer whale near their Bainbridge

Island home. "We could not believe that," Munro laughed decades later. "That was unbelievable."[19]

❧

Like people, killer whales have unique personalities, and by all accounts Namu was a sweet soul. He quickly learned to balance Griffin on his back, even stopping to scoop him up when he fell off. Namu clearly came to relish close contact with his owner, and Griffin found their sessions so entrancing that he used any excuse to linger in Rich Cove. In the process, Griffin grew so trusting that he often skirted the boundaries of common sense. On one occasion, he found his fingers stuck in the whale's blowhole, and at other times he refused to let go of fish already in Namu's mouth—in effect, daring the animal to bite down on him. Yet none of these actions provoked the gentle giant, and Griffin found himself wondering if this "delicate sensitivity" was common to all killer whales.[20] Decades later, when reflecting on the 2010 death of Sea World trainer Dawn Brancheau, Griffin admitted that he sometimes acted recklessly with Namu, but he had his reasons. "I took chances with Namu, believing he and I had developed a connection, a relationship beyond just acceptance," he observed. "I began to regard him with a trust which I can now recognize and admit exceeded the level of trust I was willing to offer [to] or accept from any human."[21]

❧

In mid-October 1965, as the film crew shot footage of Griffin swimming with Namu, Tors approached the aquarium owner with another offer: he wanted to film a killer whale capture, and he was willing to finance it. Keen to acquire a mate for Namu, Griffin accepted, but the means of capture remained in question.[22] The only two orcas kept for any time in captivity—Moby Doll and Namu—had been caught by blind luck. Yet the events in Warrior Cove proved that even flimsy gill nets could contain a killer whale. Perhaps in cooperation with local fishermen, Griffin reasoned, he could drive a pod into a shallow bay and encircle it. Of course, tracking the animals remained a problem. Experience had shown that orcas were adept at evading pursuit, often changing directions mid-dive. But Griffin still had the Greener harpoon rifle that the Marine Mammal Biological Laboratory had lent him. Assuming pods had a harem-like structure, he planned to fire the small projectile into the lead bull and use attached buoys to track the pod.

The capture team and camera crew based their operations out of Gig Harbor, where Griffin and Goldsberry hired purse seine skippers Adam Ross and Vince Naterlin to help. After several days of searching, word came on

Friday, October 29, of whales headed west toward nearby Carr Inlet. With Ross's *Chinook* and Naterlin's *Golden Gate* following behind, Griffin pursued the whales by helicopter with his harpoon rifle at the ready. Failing to find a suitable male to target, he focused on a group of three orcas, aiming for the shoulder of a large female as she rounded Fox Island. The instant he fired, the animal rolled sideways—perhaps to get a look at the helicopter—and the harpoon struck her underside, leaving the buoys trailing behind her. Soon after, the *Chinook* launched its skiff, which pulled the net into position. At the last moment, the three whales dove under the sinking lead line, but the harpooned animal couldn't complete her escape. The attached buoys snagged on the net, causing her to tug sharply against the harpoon lodged in her abdomen. Griffin watched helplessly from the helicopter, feeling "emotionally tangled in her traumatic struggle, inexplicably wanting her to get away." He breathed a sigh of relief when she tore free.[23]

But he wasn't the only one on the hunt. Now funded by a federal grant, Merrill Spencer was attempting his own capture just to the north in Henderson Bay. He had announced his plans months before, during the hullabaloo surrounding Namu's arrival. "It shouldn't be too difficult," he predicted. "We've developed effective techniques for capturing killer whales, including the tranquilizer dart."[24] At the time, tranquilizers were an exciting new technology biologists were using to capture and tag dangerous predators such as wolves and bears, but they are ill-suited to orcas. Like all cetaceans, killer whales are voluntary respirators, actively deciding when to take each breath. If they lose consciousness, they will asphyxiate, and for this reason they never fully sleep. Unaware of this, Spencer fired a powerful dose into a large female orca from a helicopter, but before his crew could secure her, the animal started to lose consciousness. Realizing something was wrong, Spencer radioed Griffin. The aquarium owner had little trouble locating the animal. Spencer's helicopter was still hovering near her, just a mile south of Purdy. "When we reached the whale," Griffin recalled, "it was floundering at the surface, unable to breathe." With the help of a nearby fishing vessel, he and Goldsberry secured the large female with a line, holding her blowhole above water. "We pounded on the animal to keep her awake the way you would wake a person overdosed on sleeping pills," he later told me. "Nothing worked. Within a few minutes she stopped swimming and just sank away." That night, they raised the carcass and turned it over to the film crew, which shot footage of the body and then transported it to the Seattle Rendering Works.[25]

The following day, Griffin returned to his own operation, but he soon received an urgent call that Joan was going into labor. He jumped into a floatplane and stepped onto his Bainbridge Island dock minutes later. The next morning, he and Joan caught the 5:00 a.m. ferry to Seattle, where, shortly after arriving at Swedish Hospital, Joan gave birth to the couple's first daughter, Gaye. Assured by doctors that mother and child were well, Griffin hurried back to Henderson Bay, where his team was in hot pursuit of two females from J pod—the same family that had lost Moby Doll one year earlier. They were very likely a mother and her young daughter.[26]

On Monday, November 1, after twenty-four hours of jockeying, the team finally made its catch. Working in tandem with the *Golden Gate*, Ross maneuvered the *Chinook* into position and cordoned off the two whales. Soon after, Gary Keffler and his divers arrived to transfer the animals to Namu's travel pen. Meanwhile, reporters had gotten wind of the operation, printing stories that Griffin had captured a "Bride for Namu."[27] But as one of the seine boats began towing the pen toward Rich Cove, the large female's swimming grew erratic, and Griffin noticed bubbles escaping from her side. At that moment, he realized that she was the same whale he had struck with the harpoon days earlier. "The cow is dying!" he cried out in panic, watching the orca sink to the bottom of the pen. He and Goldsberry leapt into water and tried to lift her to the surface. In hindsight, it was an act of courage. People had long known to avoid large, wounded predators, and the usual practice in exotic animal collection was to kill the mothers of young mammals, not rescue them. But the scene was heart-wrenching. With the two men struggling to get a line around her, the wounded whale convulsed and died within feet of a juvenile orca who was probably her daughter.[28]

❧

It was a bittersweet moment for Griffin. His newborn baby girl would soon arrive home from the hospital, and he had just accomplished the first intentional live capture of a healthy killer whale in history. But he had inadvertently orphaned the young animal, and he had difficulty facing that reality. Over the years, he had told himself that his obsession with orcas would reduce violence toward the species, and in the last few months he had enjoyed his celebrity status as "the killer's greatest friend." Unwilling to tarnish that image or face reporters' questions, he decided to conceal the death. After weighting the carcass, his divers sunk it off Fox Island. When the flotilla arrived in Rich Cove with only one whale, Griffin told reporters that the larger animal had escaped.[29] The lie didn't sit well with him, and then he received a call that made matters worse.

It was Merrill Spencer. The *Seattle P-I* had just published a photo of workers at the rendering plant processing the body of the whale who had suffocated the previous day.[30] Worried he would lose his funding, as well as a pending contract with *National Geographic*, Spencer begged Griffin to take the blame. The aquarium owner agreed to help his friend, claiming publicly that he, not Spencer, was responsible for the dead whale at the plant. Soon after, he admitted that the large female he had caught with the juvenile orca had died rather than escaped. Although Griffin had in fact caused only one of the two deaths, the news tarnished his public image.[31]

~

By that time, Griffin had introduced the young, fourteen-foot female to Namu in Rich Cove. From two different populations, northern and southern resident, the animals seemed shy at first. But Namu quickly taught the newcomer to take dead salmon by breaking the fish apart for her to eat. At first, Griffin hoped the youngster might make a suitable mate for the big male, but she soon became aggressive toward Namu and especially toward Griffin. Traumatized by her capture, which had likely involved watching her mother die, the little whale seemed unwilling to accept Griffin's bond with Namu. After several dangerous interactions in which the young female attempted to knock him off Namu, Griffin decided to sell her to a new marine park in California called Sea World, which was eager for a killer whale of its own.[32]

The ensuing transaction proved momentous for the history of the oceanarium industry. Unsure how long the whale would live, Sea World arranged to lease the animal for five years, and it requested permission to name her Namu, but Griffin refused. As a result, after confirming the orca was female, Sea World opted for Shamu ("She-Namu"). As had been the case in Seattle months earlier, the arrival of a captive orca stirred a media frenzy in San Diego. Sea World posted advertisements of the coming attraction across town, and the mayor declared Monday, December 20, 1965, "Shamu Day." Although Griffin couldn't have known it, he had just provided the signature animal to a marine park that would become a corporate giant.[33]

~

Griffin had ambitions of his own. He had long admired Marineland of the Pacific, and during his trip to deliver Shamu he found Sea World's facilities equally impressive. By the time he returned to Seattle, he was determined to build his own modern oceanarium. Lending urgency to the plan was his concern about Namu's future health at the Seattle waterfront. At the time, pollution was a growing concern in Puget Sound. In addition to decades of pulp waste and farm runoff, local waters continued to receive raw sewage from

FIGURE 6.3 Divers prepare Shamu for the journey to Sea World, December 1965. Courtesy of Ted Griffin.

the region's growing cities. By 1965, officials had declared many local beaches unsafe for swimming, and over the next three years Seattleites debated a massive bond issue to build a sewage treatment plant. During Namu's initial stint at Pier 56, Griffin had used nylon lining to limit the toxins entering the whale's pen, but it was only a temporary fix. If Namu was going to live long in captivity, he needed to be removed from Elliott Bay, and Griffin proposed construction of a new oceanarium, complete with a one-million-gallon whale pool.[34]

He filed his petition to the city on January 10, 1966. To be called Northwest Marineland, the marine park would occupy part of the northwestern corner of Seattle Center and would rely exclusively on private financing.[35] Using the same engineering firm that had designed Seattle's Space Needle and Oahu's Sea Life Park, the project received support from the Seattle Center Advisory Commission as well as Mayor Dorm Braman, who declared it "a very desirable addition to the features of Seattle Center for the entertainment and enjoyment of our own people and our visitors."[36] Nonetheless, Griffin's proposal met fierce opposition. *Seattle P-I* columnist Lou Guzzo lambasted it as a threat to the civic values of Seattle Center. Although Guzzo had recently called on city leaders to "give us a Marineland!," he opposed Griffin's proposal to do just that. Other opponents argued that it would be unnatural to display Namu away from the waterfront, even as many of them warned that the water there threatened the whale's health. To block Griffin's plan, critics cobbled together a proposal to build a teen center on the same spot.[37]

Mayor Braman was flabbergasted. As he noted in a letter to Clarence Massart, president of the city council, "For the three years that this corner has been available, no other suggestion has come forward," until the sudden appearance of the "undeveloped" plan for the teen center. He also enclosed photos of the large pool construction then underway at the Vancouver Aquarium. "It would appear that the City of Vancouver and its citizens think that the installation of a porpoise pool—remote from salt water, and in their beautiful Stanley Park—in which they hope to acquire and display a killer whale, is quite in order," Braman observed. "Strange, how what is acceptable and aesthetic in one area is totally unacceptable in another." And in contrast to the Vancouver Aquarium, he noted, Griffin's project would cost taxpayers nothing.[38]

By mid-February, opposition was growing. The local KING television and radio station ran a biting editorial. Dismissing Griffin's plan as "a whale bathtub and a feeding pond for seals," the station warned that "Namu and his friends will—to put it bluntly—smell bad."[39] Others argued that the private facility should not be built because Seattle needed a publicly funded oceanographic research center. Yet they admitted that such an institution couldn't be located in Seattle's polluted waters. Among them was Dixy Lee Ray. Just a few months earlier, Ray had warned that the whale wouldn't last long in Elliott Bay, but she viewed Griffin's marine park as a threat to her Pacific Science Center. She was also determined to promote a research-oriented public aquarium, and she worried a private marine park would muddy the issue. As the *Wall Street Journal* put it, Ray and her allies envisioned "a big, federally financed facility devoted more to science than to show."[40]

Yet few of Griffin's critics acknowledged that his aquarium was the only facility in the world where research on a live killer whale was actually being conducted. In addition to Spencer and Poulter, Theodore D. Walker of the Scripps Institute had come to study the whale, and Namu also drew the interest of the US Navy. After swimming with the imposing animal in Rich Cove, Admiral Noel Gaylor convinced the Office of Naval Research to fly Griffin to Washington, DC. Up to that time, researchers in the navy's Marine Mammal Program had considered orcas too dangerous to work with humans. Now program leaders picked Griffin's brain about the species' temperament and capabilities.[41]

Indeed, Namu's bond with his owner had upended scientific as well as popular understandings of orcas. In March 1966, *National Geographic* published a feature article by Griffin entitled "Making Friends with a Killer Whale." In addition to detailing the orca's capture and training, Griffin explained the trust he had placed in a predator reputed to be the "most bloodthirsty" in the ocean. True, Namu was confined by nets, noted Griffin, "but it was always up to the whale, in the last analysis, to decide if I would emerge from the pool alive and unmarked." The fact that the orca didn't harm him shocked many experts. "I'm afraid we must toss away some of our earlier preconceptions about these animals," admitted A. Remington Kellogg, the nation's preeminent whale scientist. "This behavior of Namu is entirely contrary to what anyone could have expected."[42]

Actor Robert Lansing was still skeptical. By late March, filming on San Juan Island had wrapped up, and the time had come for the star of the movie to shoot his scenes swimming with Namu. Yet despite Griffin's reassurance, Lansing had the jitters. Born in San Diego, he was familiar with the violent reputation of killer whales. "According to the reference books," Lansing told reporters, "the killer whale is the most vicious animal on our planet."[43] Recognizing the actor's reluctance, director László Benedek decided to splice footage of Namu and Griffin into the movie, effectively using the whale's owner as Lansing's stunt double. Finally, near the end of filming, Griffin coaxed the actor to take a ride on Namu. "It was really something!" remarked Lansing. "You don't think about it when you're doing it, but afterward, the thrill begins to soak in."[44]

Yet Griffin's relationship with the captive orca remained unique. Some evenings, he fell asleep on Namu's back as the big animal rested. At other times, he lay still on the seafloor until Namu raised him to the surface for a breath and even embraced him with his pectoral fins. Over the previous months, Griffin had come to know Namu's moods, contours, and body language as one might know a lover's. By his own admission, his time with the whale had "become an addiction," causing him to drift away from family and friends. And Namu seemed to long for his owner's companionship, at times refusing to allow him to leave the cove. "Namu holds me hostage for his pleasure," Griffin reflected, "as I have held him captive for mine."[45]

The bond between the two could charm even hardened skeptics. Perhaps the best example was Jim Halpin of *Seattle Magazine*. "Frankly, I had had Namu up to *here*," Halpin wrote. "I could get along very well, thank you, if I never heard Namu's name again." But after spending an afternoon with the captive orca and his "oddly tormented master," Halpin declared himself a convert. The pivotal moment came when Griffin challenged the young reporter to swim with the whale. "The first moment was the worst," Halpin admitted. "As soon as I went under, Namu whirled and rushed at me head-on," appearing like a "live, malevolent locomotive." Yet at the last second, marveled the writer, "that impossibly great body swept under mine, and the next minute I was hoisted above water." By the end of the session, Halpin found himself entranced by the "almost eerie rapport" between Griffin and Namu. "Whatever happens to this amazing man and his mammal," he predicted, "we have not heard the last of either."[46]

By the time Halpin's article appeared, Namu was back at the Seattle waterfront, and Griffin had introduced the first choreographed killer whale show in history. That spring, the aquarium enjoyed a surge of visitors, many of them children. Among them was eleven-year-old Mark Funk, who had seen Namu's arrival at Deception Pass and now thrilled at the sight of Griffin performing with his orca. "It was almost like he was our Jacques Cousteau," recalled Funk. "He became the *whale* rider."[47] On Monday, May 2, 1966, Griffin hosted four hundred pre-K children from a local Head Start program free of charge. "Come here Namu! Please come here!" cried one little girl, as she ran around the whale's pen.[48] These shows gave spectators, young and old, their first close

FIGURE 6.4 Griffin swims with Namu in Rich Cove, March 1966. Courtesy of Ted Griffin.

glimpses of an orca in all its grace and glory, and many marveled at the rapport between the whale and his owner. In this sense, the display had much in common with that of circus elephants, featuring an enormous animal clearly able to smash a human body, yet choosing not to do so. At the same time, visitors could feel the vicarious power of a man asserting control over a dangerous creature and even take part in feeding and petting Namu.[49] Like the *National Geographic* article, the performances ultimately presented the killer whale not as a fearsome predator but as a domesticated companion.

But the show wouldn't last. In early July 1966, the usually playful Namu grew lethargic and refused food. On the evening of Saturday, July 9, Griffin received the call he had been dreading, and he raced to Pier 56 to watch his friend die. Falling into delirium, Namu tore around the pen, smashing into a steel walkway so violently that he shook the pier. Finally, when the whale sank out of view, the helpless Griffin donned his diving gear and entered the pen. As he came upon the beautiful black-and-white form tangled in wires below, he caught sight of "a distant ray of light reflected in Namu's half-open eye." It seemed a flicker of life, and for a brief moment Griffin allowed himself to think "perhaps . . ." Then he burst into tears.[50]

At a press conference the next day, Griffin mused that Namu had been lonely for a mate and may have been attempting to escape. "There is no question it was a break for freedom," he told reporters. "He almost made it. I wish he had." As Griffin walked back through the aquarium lobby, he passed newspaper clippings and letters from the late whale's adoring young fans. "Thank you for letting us see Namu," read one note in crayon. "I like you, Ted Griffin."[51]

A few weeks after the famous whale's death, United Artists released *Namu, The Killer Whale* (1966). It received mixed reviews in Seattle and elsewhere. The film was "not likely to win any prizes," observed *Seattle Times* arts editor Wayne Johnson, but it revealed "what a prize Namu himself actually was." Margaret Harford of the *Los Angeles Times* agreed. "The story is unfortunately a bit slow, made stodgy by the interference of human actors," she wrote, but the whale himself came off as a "jolly good fellow," helping to make *Namu* a "fine picture for the family."[52]

Five decades later, it was difficult to assess the movie's impact on audiences. I had heard of it as a boy, but I didn't see *Namu* until 2013, when I sat down to watch it with my older son, then seven. On the surface, it seemed a simple plot. When the fishermen of a small Pacific Northwest village fatally shoot a female orca, her "mate"—the eponymous Namu—is netted off in a small cove, to be studied by a marine biologist played by Robert Lansing. Surprisingly friendly and gentle, the whale charms most of the town's inhabitants, especially the children. Yet local fishermen remain determined to kill the creature. In truth, I was less interested in the script than in the underwater footage of Griffin and Namu. But the story riveted my son, who grew ever more agitated near the end, when Namu's admirers release him from the cove. When the fishermen give chase with their guns, the escaped whale is forced to spill them from their boat to protect himself, and as one of the men flails in the water, Namu races toward him. At this moment in the film, my son startled me by bellowing "Eat him!" So fond of the whale had he become, and so enraged by the fishermen, that he seemed disappointed when Namu rescues them.

Hokey as it now seems, *Namu* the movie mirrored aspects of the actual whale's life. The real Namu repeatedly surprised spectators by caring for, rather than eating, people in the water with him. Griffin himself grew so enamored that he sometimes considered dropping the net at Rich Cove and letting his investment swim away, and he never forgave himself for the orca's death. Yet the subsequent necropsy revealed it was not loneliness but rather pollution that had killed Namu. The whale had been suffering from an acute bacterial infection, likely contracted from sewage runoff in Elliott Bay. And the examination revealed something else: a slug from an Enfield rifle embedded in Namu's body.[53] Like his movie version, the real whale had suffered from human violence. But Namu's year with Griffin proved a watershed, and his time in Seattle, in the news, and on the big screen had offered a new vision of the ocean's greatest predator.

7

Fishing for Orcas

GRIFFIN HAD BEEN receiving letters for weeks, and they painted a vivid picture of Namu's impact on those who had seen the famous orca. "We are sorry that Namu is dead," wrote seven-year-old Christopher. "I wish that you will get another whale." A little girl declared, "I will mention in my prayer tonight for God to send Namu safely to Heaven and for God to watch over him always." "We are so grateful we saw Namu only a few weeks ago," wrote one local family. "He was so beautiful and gentle." "Without our friend Namu, the waterfront will be a lonely place," added a mother in Seattle. "We hope you will consider getting another whale."[1]

The notes helped, but Ted Griffin hadn't been himself since his friend's death. The process of forming the first close human bond with a killer whale had produced an intense emotional high, and the animal's death sent him into a spiral of depression. Usually frenetic, the aquarium owner found himself listless and untethered from reality. "At first I told myself he would come back, as I had believed my father would after he died," he later wrote. "I had never faced the reality of death as a fact of life." Try as he might, Griffin couldn't pull himself together. "I wanted to shed my burden of guilt," he reflected. "I had brought Namu into the polluted water where the bacteria had killed him. My loved one died tragically, and indirectly by my own hand." As the weeks went by, his children became confused by their father's behavior, and Joan grew worried. Friends suggested that he try bonding with another whale, but they might as well have urged him to replace a lost spouse or child. Something in Ted Griffin had died with Namu.[2]

Nearly fifty years later, he sat with me at his dining room table and tried to convey this change. "After Namu died, I kept trying to find that connection," he explained. "I kept hoping for it with another animal, but I couldn't find it." So he turned his mind to business. "I figured, 'I've got these skills, I've got all this equipment and men. I might as well make a living doing this.'" But becoming a professional orca captor meant steeling himself to the implications of the trade. If he caught more whales, some were bound to die, in either the process of capture or the course of captivity. If he wanted to become a businessman, rather than a dreamer who owned a struggling aquarium, he couldn't let himself become attached to another orca as he had to Namu. At the time, it seemed the only way to go—he couldn't bear the pain again. "I wanted to be a different person. I wanted to change," he admitted. "I wanted to distance myself from that emotional connection with animals."

"Were you trying to become someone you were not?" I asked.

"Completely."[3]

Initially, it seemed it might work. The sale of Shamu to Sea World had won Griffin a reputation as the world's only purveyor of live killer whales, and by the fall of 1966 he was receiving inquiries from around the world. With his plans for a new marine park in Seattle stillborn, he turned his attention to orca capture. In the process, Namu Inc., initially founded to control marketing and merchandising for the world's only captive orca, became the whale-catching arm of the Seattle Marine Aquarium. Although the capture of Shamu had resulted in the death of the whale's mother, it had revealed a promising collection method, if it could be refined. Griffin threw himself into the task.

Over the following years, observers would call them "whale hunters," "sea cowboys," or simply "whalers," but in fact catching orcas had little in common with these descriptors. One did not follow tracks, shoot to kill, or drive herds from one pasture to the next. It certainly wasn't whaling in any recognizable sense. Whalers made their living by killing cetaceans and rendering their bodies into oil and meat. In contrast, what made orcas commercially valuable was their living bodies. As a result, Griffin and his colleagues had to become pioneers of what biologists Michael Bigg and Allen Wolman later dubbed "the live-capture killer whale fishery."[4]

Commercial fishing is more art than science. In addition to understanding currents and tides, fishermen must grasp the migration patterns, instincts, and foraging habits of the species they catch, which requires long experience at sea. After years of observing, chasing, and cavorting with killer whales, Griffin understood them better than anyone else on earth. In addition to his large spotting network throughout Washington State waters, he had a subtle grasp of orcas' behavior, as well as their physical abilities and limitations. He knew they tended to appear in areas where chinook salmon were present.[5] He knew pods with calves couldn't stay underwater as long, or travel as quickly, and that lone females would try to lure away pursuers. Experience had also taught him that despite their great power and agility, orcas in the Pacific Northwest seemed reluctant to tear through or jump over nets.

For technical expertise, he looked to Don Goldsberry. Although a competent manager of the aquarium, Goldsberry proved most valuable for his fishing experience. His years working on purse seine vessels in Puget Sound and Alaska had given him a keen understanding of lines, nets, and setting techniques, which had facilitated Namu Inc.'s first successful capture. "During the capture of Shamu," Griffin explained, "I recognized that Don had a skill set necessary for capturing whales which I did not possess." By late 1966, he had made Goldsberry an equal partner in the Seattle Marine Aquarium and Namu Inc.[6]

Goldsberry, in turn, proved a critical link to local fishermen, particularly Adam Ross and Peter Babich. With salmon runs declining in Puget Sound, these Gig Harbor men made most of their money fishing in Alaska in the summer, but in the off-season they were always looking for ways to make their skills pay. Both had an intimate knowledge of local waters, and Ross's capture of Shamu and her mother had shown that a purse seiner could do the job. Moreover, Goldsberry trusted both men. He had known Ross for many years, and Babich's father, Spiro, had given Goldsberry his first job on a fishing boat. Equally important, Ross and Babich found killer whale capture thrilling. Although Ross had joined the Shamu catch on a lark, he soon became obsessed with pursuing orcas, and Babich wasn't far behind. "It was *extremely* exciting," recalled Babich's son Randy, who joined his father on several captures as a teenager. "It was like cracking a new frontier." At the time, the capture and training of marine mammals had an aura of adventure not unlike space travel, and the lure proved irresistible. "Let's face it," Randy Babich told me. "They stopped fishing in mid-November every year—fall fishing in Puget Sound—and it's a long haul between November and June. This fit in so nicely. They could make a few extra bucks, plus all the limelight."[7]

Yet catching orcas was not fishing in the true sense. Unlike the salmon netted in the Salish Sea, killer whales would retain their value only if kept alive, and this was no small challenge. Normally, purse seiners use heavy nets, cinching them up from the bottom using a lead line threaded through metal rings. But like all marine mammals, killer whales need access to the surface: once entangled in nets, they will drown quickly. In fact, in July 1966—within days of Namu's death—a female orca had become wrapped in a gill net near Steveston, British Columbia, and drowned.[8] Such incidents taught Griffin and Goldsberry valuable lessons. "It appeared that the strength of the net was not as important as the way it was used to encircle them," Griffin noted. The initial setting of the net needed to be large enough to prevent the animals from panicking, and the circle had to be tightened very slowly. If a whale did become agitated, it needed to be able to break out. To ensure this, Griffin and Goldsberry commissioned a special capture net. At 3,600 feet, it was twice as long as a normal seine net and made of a light nylon mesh designed to tear if the animals collided with it.[9]

The partners also planned to use the net differently. Whereas a normal purse seine net would be cinched together after making a set, the whale capture net would need to be anchored to the seafloor. This placed great importance on the selection of a capture site. Although Puget Sound offered many shallow bays where killer whales could be cornered, it also posed numerous obstacles. In addition to ferries and growing commercial and recreational boat traffic, local waters were littered with reefs, snags, and discarded naval equipment. In addition, many areas had large tides that could collapse the nets, endangering whales and divers alike. Griffin and Goldsberry learned this the hard way when heavy currents in Bainbridge Island's Port Madison pulled their anchors off the bottom, nearly causing the net to drown a group of trapped whales. "For a few minutes the whales raced back and forth in a fast-narrowing corridor," recalled Griffin. "With no space remaining, they all broke through and escaped." His team had suffered similar mishaps earlier in Whidbey Island's Penn Cove and Eld Inlet near Olympia, twice losing potential mates for Namu.[10] If the whale catchers wanted to avoid disaster, they would need to choose their locations carefully.

As Griffin and Goldsberry worked to refine their capture method, environmental concerns were entering the local and national discussion. The thoughts had long been swirling in the postwar ether. Aldo Leopold's posthumous *Sand County Almanac* (1949) had helped revive the movement for wilderness preservation in the 1950s, and in 1962 Rachel Carson followed up her popular *The Sea around Us* (1951) with *Silent Spring*. Drawing attention to industrial pollutants such as DDT, Carson's book delivered an environmental wake-up call to suburban North America.[11] A year later, Canadian writer Farley Mowat published *Never Cry Wolf* (1963). A fictionalized account of the author's time in the Canadian Wildlife Service, it painted a vivid portrait of a wolf family and lambasted predator control policies. As Mowat asserted, in words that could easily be applied to orcas: "We have doomed the wolf not for what it is but for what we deliberately and mistakenly perceive it to be: the mythologized epitome of a savage, ruthless killer." Beyond altering popular perceptions of wolves, *Never Cry Wolf* proved a seminal step in what historian Tina Loo has called "an emerging urban sentimentality about predators."[12]

Rising environmental concerns found expression in government policy, particularly in the United States. In the wake of the first Clean Air Act (1963) and the Wilderness Act (1964), President Lyndon Johnson signed the Endangered Species Preservation Act in October 1966, a key step in the federal government assuming stewardship over species at risk. The following year, the Interior Department placed seventy-eight species on the endangered list. Among them were not only popular, aesthetically pleasing animals such as the ivory-billed woodpecker but also once-vilified predators such as the timber wolf and grizzly bear. The department also listed several marine mammal species, including the Florida manatee and Caribbean monk seal. Yet it left the fate of cetaceans in the hands of the International Whaling Commission.[13]

The IWC was undergoing its own change, much of it emanating from the Pacific Northwest. In 1960, University of Washington mathematician Douglas Chapman, who had taken part in statistical studies of the fur seal harvest on the Pribilof Islands, joined a special IWC committee. Tasked with estimating world whale populations and making harvest recommendations, the committee held a critical meeting in Seattle in December 1962, resulting in a recommendation of far lower international hunting quotas, particularly for fin whales. Although the major whaling nations of Japan, Norway, and the Soviet Union protested, Chapman and his colleagues provided compelling evidence that whale stocks were crashing. Their report reframed the thinking within the IWC and, in the words of historian D. Graham Burnett, opened the way for a "global backlash against commercial whaling after 1965."[14] The document likely contributed to the industry's final demise in the Pacific Northwest. With the IWC's banning of the harvest of humpbacks and blue whales in the North Pacific, Bioproducts Inc. shut down its short-lived whaling operations in Warrenton, Oregon, and in 1967 Coal Harbour—the last whaling station in British Columbia—closed after a disappointing harvest of only 496 whales.[15]

Meanwhile, another whale drama was playing out on the other side of the continent. On January 21, 1967, a fin whale had become trapped in a saltwater lagoon near the town of Burgeo, Newfoundland. Under normal circumstances, the animal's fate would likely have drawn little attention. After all, the province still boasted two thriving whaling stations. Long known for their commercial drives of pilot whales, local whalers also processed hundreds of fin whales each year, mostly for the Japanese meat market, and just off the coast, Norwegian ships harvested even more whales, in addition to participating in the annual harp seal hunt.[16]

But as fate would have it, author Farley Mowat was living in Burgeo, and he launched a campaign to save the trapped whale, who soon became known as Moby Joe. Claiming that the lagoon offered ideal conditions to study the animal in captivity, Mowat began toting buckets of herring to feed it. At first, it seemed he might form a relationship with the whale not unlike Griffin's with Namu. "It was something, after all, to be the nominal possessor of such a fantastic creature," Mowat admitted. "No other human being had ever had a fin whale for a 'pet.'"[17] Despite his efforts, fishermen and local residents shot at the trapped whale and drove motorboats over it for sport. When Mowat discovered an infection on the animal's back, a number of outside institutions rallied to help the trapped creature, among them the Vancouver Aquarium. On February 8, however, newspapers across North America reported the death of Moby Joe, with Mowat and others bemoaning the cruelties that humans had inflicted on the animal. That same day, Premier Joseph Smallwood eulogized the whale, declaring, "I'm sure all Newfoundland, all Canada, and even all North America will hear this news with regret."[18]

On the face of it, this was a curious episode. After all, commercial whaling still generated few protests in North America, and Smallwood himself strongly supported the industry in Newfoundland. But as Mowat had shown, the small-scale spectacle of one whale's fate could have a powerful impact on public opinion. It was a lesson Ted Griffin would learn the hard way.

A week after Moby Joe's death in Burgeo, officers at the Coast Guard station in Port Angeles, Washington, spotted orcas headed east past Ediz Hook and telephoned the Seattle Marine Aquarium. For Griffin, the timing was fortuitous. He had an order from Sea World for more killer whales, and promoter Bob O'Loughlin—the same man who had previously attempted several captures with Griffin—hoped to display an orca at his upcoming Boat and Sports Show in Portland. For the past month, Griffin had been following a group of killer whales in Puget Sound, but he had lost track of them a few days earlier.[19] Now a different pod was approaching, and the Coast Guard had provided advance notice. With orcas cruising at three to five knots, Griffin calculated the group would enter Puget Sound early the next morning. This left plenty of time to organize a capture, and because it was winter, Ross and Babich were still in Puget Sound.[20]

The next morning, Wednesday, February 15, Griffin boarded a chartered helicopter and located the whales east of Bainbridge Island. Soon after, the chase was on. As in previous operations, Griffin believed a buoy attached by the Greener harpoon rifle offered the only means of tracking the animals, yet he hesitated to use it. "Following the death of Shamu's mother," he explained, "I became concerned that the harpoon was more dangerous to whales than I had been led to believe," but he assumed it would pose less risk to a large bull. Such thinking made sense at the time. After all, the juvenile Moby Doll had survived a strike from a mounted harpoon cannon designed to kill much larger whales, and the death of Shamu's mother seemed the fluke result of her sudden roll. Griffin's opportunity came when a female and her calf surfaced beneath the helicopter, prompting an adult male to push them back underwater. "Did you see that?" Griffin yelled to the pilot. "He's protecting them!" The aquarium owner then fired his harpoon into the bull's right flank.[21]

In the following hours, Goldsberry and the fishermen used the three trailing buoys to track the pod. Now at the wheel of *Pegasus*, Griffin detonated sticks of dynamite in the water from a distance of half a mile. "It's a calculated risk," he later wrote. "The whales, highly sensitive to sound, should be little more than frightened at the unaccustomed noise." As the orcas fled west into Rich Passage, he noticed a passenger ferry headed directly for them. "Ferry *Illahee*—the *Pegasus*—Stop! You are running over my whales!" he screamed over the radio.

"The Ferry *Illahee*, back," responded the skipper. "I see them now; I'll give it a try." With that, the vessel turned sharply to starboard, drifting sideways among the panicked animals. Buoys from the harpoon line bounced against ferry's hull, and passengers rushed to the deck to watch the melee.[22]

By late afternoon, the team had cornered the whales in nearby Yukon Harbor. Sheltered from heavy currents with a gently sloping seafloor, it seemed a perfect capture site. As the *Chinook* took its position, Adam Ross's nephew, Danny, launched the vessel's skiff, quickly pulling out hundreds of feet of net before turning sharply toward the beach. By the time he had trapped the whales, it was 6:00 p.m., nearly dark. Goldsberry secured the nets with heavy anchors, while an elated Griffin circled his catch, in his words "consuming the whales with my eyes." With the orcas swimming counterclockwise in tight formation, it was difficult to get an exact count, but there appeared to be at least ten—maybe more. The next day, Pete Babich arrived in the *Pacific Maid* with steel antitorpedo nets to help secure the catch, and Gary Keffler and his divers made sure that the capture net was free of snags and no whales were entangled. Despite the murky water, Keffler caught repeated glimpses of spectral black-and-white bodies passing by.[23]

~

The corralled killer whales had a very different perspective. Namu Inc. had just captured fifteen members of what scientists have since labeled K pod, and they were confined and confused. It was likely the first time these orcas—or any of their ancestors—had been trapped in this manner, and they searched desperately for a way out. Young animals stayed close to mothers, while other whales perused the edges of the enclosure. On several occasions, large males rose vertically in the water, a behavior known as "spy-hopping," to gain a visual grasp of their predicament. On the first full day, the only moment of chaos came when a navy destroyer passed by, using sonar so powerful that Griffin felt its concussion through the hull of *Pegasus*. "The shrieking of the smaller animals is something I haven't heard before," noted Griffin. "One after another the whales breach, some clearing the water by several feet." But even with this scare, none tried to escape through or over the nets.[24]

This reluctance puzzled observers, as the whales needed only to jump or even slip over the cork line. How could an animal apparently so intelligent fail to solve a problem so simple? One factor may have been cultural conservatism. As researchers have noted, orcas often have difficulty responding to new situations.[25] Their pod cohesion probably contributed to the challenge. Even if one or two animals felt brazen enough to slip over the net, the desire to stay close to family likely deterred them—as it had Namu two years earlier.

But the most important factor was surely their reliance on sonar. Killer whales depend on echolocation to map their surroundings. But that extraordinary navigation system doesn't work in the air. To be sure, orcas often breach when playing or chasing prey, but their senses are attuned to life underwater. Although the nylon and steel nets produced different echoes than a solid surface, the whales perceived the barrier surrounding them acoustically, which they then verified visually. As they surfaced and spy-hopped, however, they saw no obstacle on the surface. The resulting sensation must have been confusing, even surreal. Despite the perception of confinement underwater, they could see an escape route right over the net. But these animals trusted sound over sight, and they quite literally couldn't believe their eyes.

Although this was not the first orca capture in Puget Sound, it was the first made in full view of the public. Located just south of Bainbridge Island, at the intersection of three major ferry routes, the operation immediately drew the attention of local residents and media. Most press coverage was positive. The

Seattle Times hailed Griffin as the city's "modern-day Ahab" and marveled at his "Great Whale Hunt."[26] "It looked like Ted Griffin was holding tryouts in Yukon Harbor yesterday to pick the cast for his trained whale show," wrote a reporter for the *Bremerton Sun*, who marveled at the trapped animals' spectacular "arching leaps."[27] But the expedition raised questions that Griffin's earlier activities had not. Because Canadian fishermen had accidentally caught Namu, there had been little controversy, other than the presence of the whale's "family" during the journey south, and the deaths of the two orcas in the fall of 1965 had stirred only brief outrage. But as Griffin soon discovered, while locals would pay to see a killer whale perform at Pier 56, they were less eager to witness a capture right in front of them.

Although no laws required him to do so, Griffin planned to release most of the animals. Yet no one had ever attempted to sort and separate killer whales, and it didn't go well at first. Late at night on February 22, after the team towed the pen closer to the old Harper Ferry dock, one of the young orcas became tangled in the steel net. By that time, the divers had gone home, and Griffin and Goldsberry could only watch as the little animal drowned with its helpless mother swimming nearby. Eager to examine the carcass, the federal Marine Mammal Biological Laboratory in Seattle sent a vessel to collect it.[28]

In the following days, Griffin made certain he had divers available, but the danger to the animals remained. One night, Stanford's Tom Poulter, who was recording the captured animals, heard what he recognized as distress calls, prompting the team to rush to the scene. As Goldsberry freed a female orca from the net, Griffin did the same for her tiny calf. "The whale shudder[ed] with fright and bob[bed] his head," Griffin later wrote.[29] In the following days, two more whales drowned—one of whom was the harpooned bull. Once again, Griffin donated the carcasses to the lab, and he packed the Greener harpoon rifle away, never to use it again.

Meanwhile, criticism had started to mount. "The slaughtering of whales for commercial uses is carefully controlled by international treaties," declared the Reverend Peter Raible of the University Unitarian Church. "Evidently, though, any carnival promoter can track and abuse these animals without control."[30] On Saturday, March 4, a small group of teenage boys picketed the aquarium with placards reading "Stop the Whale Killing," and the protests grew when photos emerged of one of the dead whales hanging from the boom of the Marine Mammal Biological Laboratory's research vessel. For some, the image confirmed their suspicions that the whale captors were concerned only with profit. "Don and I tried to keep our heads down," Griffin recalled, "but there was no place for us to hide and our good-will factor vanished."[31]

By this time, Namu Inc. had transported five whales to Pier 56 and released the remaining animals at Yukon Harbor. In the aquarium's new display pool, staffers administered antibiotics and worked to get the newly arrived orcas to eat. Nearly bankrupted by Namu's demand for fresh chinook salmon, Griffin hoped to feed the whales herring. But like captive whales before them, the new animals refused dead fish. On the suggestion of Dave Kenney, a visiting Sea World veterinarian, Griffin's team pried open the young whales' mouths and pumped in a mixture of herring and heavy dairy cream. Although this revived them, they still refused dead herring, leading Griffin to a creative solution. "As an experiment we added a thousand live bait herring to the tank with the whales," he explained. "The now very hungry animals chased and fed on the herring, and the following day we added some dead, previously frozen bait herring." Soon all five were conditioned to eat dead herring as their primary food.[32]

Griffin sold two of the orcas to Sea World. In the arrangement, the company agreed to pay out the remainder of Shamu's lease, as well as buy the new whales, soon to be named Ramu and Kilroy.[33] It was a bargain for the San Diego company, which was rapidly building its identity around killer whale shows. Griffin agreed to rent one of the three remaining whales from the catch to Bob O'Loughlin's Boat and Sports Show, which was about to move from Portland to Vancouver, British Columbia.[34] Two years earlier, Griffin had taken a Canadian-caught orca out of British Columbia. Now he found himself transporting a US-caught whale north across the border.

Griffin arrived at Vancouver's Pacific National Exhibition grounds on Saturday, March 8, with a small orca whom O'Loughlin was advertising as "Walter the Whale." Eager for the animal to make a good impression in the tiny above-ground pool the show had provided, Griffin warned Vancouverites that Walter might act a bit skittish. In addition to fatigue the animal felt from the journey, he reminded reporters that killer whales were "quite gregarious" and that the youngster "probably misses the others." The following day, he coaxed *Vancouver Sun* reporter Bob Purcell to take a swim with Walter. From that point, Griffin recalled, "the public turnout and response to see the whale was astounding."[35]

Indeed, Walter stole the show. Each hour, hundreds of sightseers filed into the tent, where Griffin gave a brief presentation about killer whales, assuring listeners that the species was far friendlier than most believed. Visitors could then approach the pool, where the bravest could offer the whale herring and even rub his rubbery skin. The event's high point came on March 16, when Griffin arranged a phone call from Walter to his two podmates at the Seattle Marine Aquarium, now named Kandu and Katy. Broadcast live on Jack Webster's popular show on CNKW radio and covered by newspapers in Vancouver, Victoria, and Seattle, the exchange cast the species in a new light for many listeners, who could now imagine orcas' relationships with one another. "Once they started 'talking' it brought tears to my eyes," Griffin recalled. It also brought larger crowds. By the end of the ten-day show, one hundred thousand people had attended.[36]

Among those who paid for a glimpse was Mark Perry, the same young man who had sneaked onto the Jericho Army Base to see Moby Doll three years earlier. "It was a big long lineup, and I think we paid a buck or two," Perry recounted. After enduring the wait and climbing the steps to the tent, he braced for the excitement of seeing the whale. But he was disappointed. The water in the pool was dirty, making it hard to get a good look. "I don't know what was in it, but it was turbid and kind of crappy-looking," he later told me. "All you could see was the fin, like Moby Doll." To be sure, it was fascinating to see an orca that close, Perry acknowledged, "but I also thought, at the time, 'this is really sad.' "[37]

For Griffin, the most powerful moment came outside the limelight. After one routine presentation, he spotted a man lingering near the back of the crowd watching Walter frolic as the other sightseers filed out. "Are all killer whales like this one?" the visitor finally asked. "I always thought they would kill a man if given the chance, and they eat a lot of salmon, you know, cause trouble for fishermen." As the stranger looked back to the young orca, Griffin studied the man's strong, weathered hands, concluding he was a fisherman. Then, with little warning, the aquarium owner found himself hearing a confession. "A while back I came across some killer whales up north," the man said in a quiet voice. They were in a shallow bay, he explained, chirping and playing. "Just shot them dead," he muttered, holding back tears, "thought they were good for nothing."[38]

It was a searing moment for Griffin, and it highlighted his complex relationship with old Northwest and new. On the one hand, the meeting held a promise of redemption. In the past two years, his capture operations had caused the deaths of four orcas, and he faced rising criticism in Seattle. Yet this brief encounter strengthened his belief that he could convince Northwesterners, especially fishermen, to stop harming the species. On the other hand, fishermen had made Griffin's orca dreams possible. Two Canadian gill-netters sold him the whale that made him famous, and the equipment and techniques of Puget Sound purse seiners enabled him to launch the world's first orca capture business. At heart, Griffin remained an animal lover, still fascinated with killer whales. But in ways imperceptible at first, the aquarium owner was starting to think like a fisherman. Once impelled by curiosity and an obsessive desire to connect, he had begun to see orcas as resources, as commodities to be captured and sold, even as he helped others see them as so much more.

Through most of the boat show, the Vancouver Aquarium remained aloof, showing only limited interest in Walter. Biologist Gil Hewlett, a veteran of Namu's voyage, sometimes came for the whale's feedings, and curator Vince Penfold also visited. But neither man broached the possibility of buying the animal. Indeed, when Griffin made his appearance on Jack Webster's radio show for the "conversation" between Walter and his podmates, he expressed surprise that the aquarium didn't want to keep the whale in Vancouver. Two years earlier, Murray Newman had stewed over the loss of Namu to Seattle and threatened to cross into Puget Sound to capture an orca. Now Griffin had brought one to him, and in contrast to the summer of 1964, when the aquarium lacked accommodations for Moby Doll, the facility had just completed its new pool, sponsored by the British Columbia Telephone Company (BC Tel). At sixty feet long and 125,000 gallons, the BC Tel Pool was small by modern standards, but it could accommodate a young whale like Walter.[39]

In reality, Newman wasn't about to let Griffin leave the country with another orca. Although he played coy, the aquarium director worked feverishly behind the scenes to get approval for the purchase, and Griffin had unknowingly helped. Not only were members of the Vancouver Aquarium Society aware of the stir Namu had caused in Seattle, but the boat show made Walter a Vancouver celebrity. Moreover, Griffin agreed to sell at a reasonable price—the same $20,000 Newman had quoted to Marineland for Moby Doll and balked at for Namu. The society allocated the funds just in the nick of time. As Griffin prepared for the trip back to Seattle, Hewlett and Penfold arrived to buy Walter.[40]

It was turning out to be a banner year for the Vancouver Aquarium. The institution had made headlines the previous month for its attempts to help Moby Joe in Newfoundland.[41] Now its purchase of a captive orca splashed across local newspapers. Walter entered his new pool at 10:00 a.m. on March 20, 1967, and soon after, Penfold joined him for a swim.[42] Determined to avoid their previous embarrassment with Moby Doll, aquarium officials wanted to verify the whale's sex. As it turned out, Walter was a female, probably about six years old. Soon after, Newman and his staff renamed her Skana—the Haida nation's term for killer whale.

For his part, Griffin returned to a city enjoying the antics of Katy and Kandu. It had been eighteen months since Namu's death, and the prospect of seeing killer whales again at Pier 56 stirred public excitement. Schoolchildren crowded around the pool to get a look at the "baby" killer whales. Griffin himself hoped to form a close connection with one of animals, but he couldn't bring himself to try. Perhaps he feared the prospect of bonding with and losing another whale. Perhaps the new pool was too painful a reminder of Namu's death in polluted Elliott Bay. Perhaps he was now more interested in catching orcas than in connecting with them. For whatever reason, he backed away from performing with the animals, even as he found himself besieged with applications from men and women eager to become killer whale trainers—a job that hadn't existed before he entered the water with Namu. Among them was Jerry Watmore, a former "sea maid" at Sea World who became the world's first female orca trainer. Arriving at the Seattle Marine Aquarium in 1967, she quickly bonded with Kandu and performed regularly with the young whale over the following year. "That brief one-to-one relationship with her was a joy of my life, and my whole life has been to better understand her world," recalled Watmore (now marine ecologist Jerry McCormick-Ray) fifty years later. "I am forever grateful to Don and Ted for hiring me."[43]

But the ranks of Griffin's critics continued to grow, among them some scientists. Having helped block Griffin's plan for a marine park at Seattle Center, Dixy Lee Ray now denounced his "constant harassment" of local orcas. "Certainly, I'm in favor of the careful study of all wildlife," she explained, "but utilization of animals for study must require handling in the best possible way for their comfort and continued survival." For his part, Victor Scheffer questioned whether Griffin had "the right to frighten and harry these magnificent creatures." "Some hazard to them is acceptable when the opportunity for scientific study is considered," he noted, but "when it is done for profit, that's something else again."[44]

It was a peculiar stance for the respected biologist to take. As Scheffer later admitted in his memoir, he himself had sold animals to zoos and aquariums, including a "modest, bring-'em back alive business" trapping and selling mountain beavers for twenty-five dollars apiece.[45] Moreover, his research at the Marine Mammal Biological Laboratory directly supported the killing of

FIGURE 7.1 Kandu greets visitor at Seattle Marine Aquarium, late 1960s. Courtesy of Ted Griffin.

marine mammals for profit. Since beginning his work on the Pribilof Islands in 1940, Scheffer had facilitated the slaughter and processing of more than a million fur seals, and he was currently working closely with the Fouke Fur Company, which held the fur seal contract from the US government. "Our sole assignment was to . . . support efforts to harvest as many sealskins as possible each year," Scheffer acknowledged. "There was no room for sentiment in that task."[46] Moreover, his colleagues at the lab worked closely with US whaling companies, and as Scheffer surely knew, even if most Seattle residents did not, their research involved killing orcas. By the time of the February 1967 capture at Yukon Harbor, the lab had slain at least eight killer whales for Dale Rice's study, and it would kill two more that summer.[47]

But Scheffer was no fool. He sensed a change in the political winds. Killer whales were becoming a symbol of the Pacific Northwest's shifting environmental values, and it was in the lab's interest to distance itself from Namu Inc. Few residents of Puget Sound realized that the work of Scheffer and his colleagues hinged on the killing of marine mammals. Two years earlier, an internal report completed by the US Fish and Wildlife Service had noted that the Marine Mammal Biological Laboratory was "well shielded from the public by its present location on a military base and by the nature of the resources with which it is concerned." The lab's acceptance of the dead orcas from the capture threatened to bring public scrutiny, and Scheffer pushed to keep the focus on Namu Inc.[48] The strategy worked. After consulting with Scheffer, the local KIRO television station released an editorial on March 13 dismissing Griffin's claims that orca capture contributed to scientific research. Noting the two whales recently sold to Sea World, the station concluded that Namu Inc.'s sole purpose was "to make music on the cash register."[49]

Yet Griffin had his defenders. Tom Poulter, for one, found KIRO's claims ludicrous. "I have personally made thirty-three miles of magnetic tape recordings of Ted Griffin's killer whales," he informed station executives—part of a quarter-million-dollar research program that had already yielded six scholarly publications on the biosonar and communication system of the species. If scientists such as himself were forced to pay for the capture and maintenance of a killer whale, Poulter explained, "there would be no funds left with which to do research."[50] But the pithiest response came from Griffin's wife, Joan, who defended her husband in the May issue of a local magazine, *Puget Soundings*. "One of the greatest tributes to Ted's recent activities is, in fact, the criticism of those who bemoan the fate of the killer whale," she wrote. "Two years ago no one would have cared."[51]

8

Skana and the Hippie

ON SEPTEMBER 21, 1967, Vancouver columnist Himie Koshevoy of the *Province* newspaper witnessed an unexpected Cold War encounter. Soviet minister of fisheries Alexander Ishkov had come to see the Vancouver Aquarium, and Murray Newman invited the reporter along. Ishkov had visited in 1956, when construction of the aquarium was still underway, and he was so impressed that he carried a copy of its plans back to Moscow. Eleven years later, he had returned for a grand tour, and Newman was happy to oblige. He showed his Soviet guest around the exhibits, proudly noting that each year forty thousand schoolchildren visited the aquarium, "gaining knowledge of their coastal environment."[1]

Like most visitors that year, Ishkov was especially eager to see Skana, Vancouver's captive killer whale. According to Koshevoy, what ensued between communist fishing minister and US-caught orca amounted to "a Little Yalta." With "squeals of delight," Skana showed off her acrobatic feats, earning a handful of herring for each one. Although Ishkov may not have grasped the significance of Skana "profiting through her labors," Koshevoy quipped, the Soviet official clearly enjoyed the performance. When his hosts suggested Ishkov try feeding Skana himself, however, he hesitated. "You could almost see the thoughts racing," mused Koshevoy. "Was she a potential aggressor? Could he deter an attack? The first-strike ability was clearly on the whale's side." Finally, coaxed by trainer Terry McLeod, Ishkov made his démarche, and Skana accepted. As Ishkov smiled and rubbed the whale's head, it was clear the crisis had passed. The interspecies summit closed, Koshevoy noted, with "mutual expressions of goodwill on all sides."[2]

It was a waggish depiction of the visit, to be sure. Although the Soviet official didn't want to look skittish in front of his North American hosts,

FIGURE 8.1 Soviet fisheries minister Alexander Ishkov feeds Skana as (left to right) Murray Newman, Terry McLeod, and Vince Penfold look on, September 1967. Courtesy of Terry McLeod.

the Cold War was likely far from his mind as he dropped herring into Skana's maw. Ishkov shared Newman's interest in fish and marine mammals, and he had recently outlawed the killing of small cetaceans in Soviet waters. Yet he was undeniably part of the political world, and he represented a

Soviet empire that was asserting its interests around the globe. As fisheries minister, he directed a whaling fleet that was decimating whale populations in the Pacific and hunting ever closer to North American shores. In fact, it was the Soviet hunt off Vancouver Island over the past three years that led to the closure of the Coal Harbour whaling station soon after Ishkov's visit. In that same period, Moscow had stepped up aid to the communist government of Hanoi in response to US military escalation in Vietnam. That conflict, in turn, had a profound impact on the Pacific Northwest. By 1967, US soldiers and Boeing-built B-52 bombers were flowing steadily from Washington State to Vietnam while antiwar and countercultural movements challenged the assumptions and hierarchies of American society. Those currents flowed over the Canadian border, making southern British Columbia, and particularly Vancouver, a haven for draft resisters, hippies, and activists of all sorts.

That Thursday morning, Newman and Ishkov could pose for poolside photos and pretend none of this mattered. Nestled in Stanley Park, the Vancouver Aquarium seemed isolated from geopolitics. Yet little did Newman know that a newly hired, freethinking scientist would soon upend his stately institution—and eventually cause trouble for Ishkov as well. It all started with Skana.

Upon her arrival, Skana's training had fallen to Terry McLeod. Although just twenty-two years old, the six-year veteran of the aquarium had collected and cared for many of the facility's animals. Three years earlier, he had worked closely with Moby Doll. At the time, Newman and neuroscientist Pat McGeer had captured headlines, but it was McLeod who spent the most time with the orca at the Jericho Army Base. Recognizing McLeod's talents, Newman sent the young man to Sea Life Park on Oahu to apprentice in marine mammal training. By the fall of 1965, McLeod was chief trainer, focusing most of his efforts on the aquarium's new Pacific white-sided dolphins, Diana and Splasher.[3]

McLeod approached Skana as he had the dolphins, training her to perform a range of behaviors on command and gradually integrating her feeding into scheduled show times. But she was clearly different from her tankmates. Skana was the largest animal McLeod had ever worked with, and this made the experience of diving with her unique. As he recounted in the aquarium's November 1967 newsletter, Skana often wanted to "hold hands" with people. "She accomplishes this," wrote McLeod, "by squeezing the diver's hand between her flipper and her body, and then swims with him slowly about the pool."[4] But the young whale could be temperamental, demanding attention and devising means to keep trainers in the pool. "Skana had her own personality, no question about it," McLeod told me decades later. "She had her good days and bad days."[5] For the next three years, it was he who had the closest relationship with her.

A close second was Mark Perry. After sneaking in to see Moby Doll at the Jericho Army Base and paying for a glimpse of Skana (Walter) at the boat show, Perry had taken a job as a floor boy at the Vancouver Aquarium in the spring of 1967 and spent most of the "Summer of Love" sweeping up popcorn.[6] For him, the opportunity to work with Skana was transformative. "I had already been at the aquarium, so I knew what was going on," he explained, "but to be up close and personal with an animal like that, and literally inches away from forty-four conical-shaped teeth—it really did have a profound effect on me, on the way I thought about animals." To that point, only McLeod had trained Skana, but the young orca quickly accepted Perry. "It was a rapport," he reflected. "Just looking in her eyes, you could see the intelligence and almost feel what she was thinking about me." Perry's time with the young orca proved intoxicating. "I'm a twenty-year-old guy, and I'm a whale trainer, and there are audiences that are applauding and cheering every time we do a show," he told me. "It was all new, and I was caught up in the excitement."[7]

Perry wasn't alone. Skana brought heady times to Vancouver. From the moment she took her first turn in the pool, the young whale was the aquarium's prime attraction and an invaluable asset to the city's growing tourist industry. That first spring, Stanley Park saw traffic like it never had before, with tourists and locals alike packing the aquarium. Then came the summer crush. In August alone, 119,746 people visited, bringing the year's total to 527,536—an 80 percent increase over the same period in 1966.[8] Scrambling to handle the crowds, the aquarium hired a hundred volunteer docents, while Newman eyed possibilities for expansion.[9]

Young killer whales grow fast, and Skana was no exception. The BC Tel Pool made for intimate performances, with the orca and her sidekicks delivering splashes and thrills to spectators crowded around the guardrails. But it was not built for a killer whale. "The tank was a small environment," noted Perry. "There was enough room for Skana to get up to speed, but all she could do was swim in circles."[10] Newman was aware of the problem. Despite the aquarium's recent expansion, the addition of Skana forced its director to

FIGURE 8.2 Skana performs at Vancouver Aquarium with Moby Doll sculpture visible in foyer. Courtesy of Terry McLeod.

think on a new scale. "You can't remain small and keep killer whales, at least not in a responsible way," he later explained:

> The major reason is a physical one: they need as much space as you can give them, and we were always trying to find ways to extend our killer whale habitat. Another reason is financial: they cost a lot to house and to feed, and you need a larger organization to do it. And finally, killer whales attract crowds of people, and you need an infrastructure to deal with them.[11]

Such factors help explain the rise of Sea World and other highly capitalized marine parks, but they also hint at the challenges confronting the Vancouver Aquarium. Thanks to Skana, business was booming, with nearly a million visitors projected for 1968. But the aquarium's main attraction required investment if she were to remain in Stanley Park. For Newman and his supporters, these questions went hand in hand. In the eyes of Vancouver Aquarium Society president Ralph Shaw, the institution had no option but to grow. "The million figure is beyond our physical facilities," he explained, "and Skana the whale is outgrowing her pool"—which soon became painfully clear.[12]

~

January 22, 1968, started as a normal Monday morning in Stanley Park. Just after 9:00 a.m., visitors began filtering into the aquarium. The bulk of them were schoolchildren, and they wanted to see Skana. The young whale had by now attained legendary status among Vancouver youngsters. With time before the ten o'clock performance, aquarium docent Sue Justice led a visiting class of eighth-graders into the underwater viewing area, where they watched Skana's graceful form behind an observation window. Justice then ushered the students upstairs, where they gathered with anticipation around the pool. Dressed in wetsuits, McLeod and Perry prepared for the show, carrying buckets of herring to the training platform. Skana was growing excited, chirping and cruising around the pool. She leapt once, twice, and then collided with the observation window just feet from where the students had been standing moments earlier. The impact shattered the three-quarter-inch imported British plate glass, and as water poured out, the resulting suction pulled Skana's head into the window frame. The whale found herself wedged into the two-foot-by-two-foot opening and held fast by jagged glass. "There was a terrific crash and suddenly blood and water came gushing through," described a witness who was standing near the window. Then, as a *Vancouver Sun* reporter put it, "Visitors watched in horror as the whale churned the water for about five minutes."[13]

It was a race against time. Young killer whales can hold their breath for up to fifteen minutes if they have properly charged their bodies with oxygen. But Skana had expected to come right back up—she didn't have that long. As blood darkened the water around her, McLeod and Perry alerted other staffers and raced to a construction crew renovating a nearby pool. Working together, trainers and construction workers looped a line around the animal's tail and used a crane to extract her from the shattered window. Pulled against the glass still embedded in the frame, Skana must have experienced intense pain. But she stopped thrashing the moment she felt the tug of the line, perhaps trusting her trainers were trying to help. With the whale freed, the staff shooed away visitors and drained the pool. Deeply shaken, her trainers knelt beside her lacerated head, afraid to look. "I was in tears calling the veterinarian," McLeod recalled. "I thought she had lost her eyes."[14] Miraculously, Skana suffered no serious injuries, but she bore the scars of that day for the rest of her life.

There was no shortage of commentary on the incident. The local SPCA expressed concern that Skana had suffered "psychological" damage and might never perform again.[15] For his part, Newman declared that the crash proved that the aquarium needed to expand. Noting that the young female had grown two feet and gained nearly one thousand pounds in less than a year, he even suggested that the accident stemmed from sexual frustration. "It might be she was thinking about a mate and family of her own when she crashed into the window," he speculated. "After all, it's not very nice to be cooped up here all the time." To resolve the issue, he proposed acquiring a male orca and building a new pool where Skana and her captive-bred children could reside.[16] It was an ambitious plan, but before he could proceed very far with it, Newman had other problems on his mind.

Paul Spong came from a different world than Murray Newman. He was born in 1939 in New Zealand—a nation with a long history of whaling. Although the industry's heyday had long since passed, New Zealand whaling continued into the 1960s. In 1958, when Spong was nineteen, crowds had gathered in the

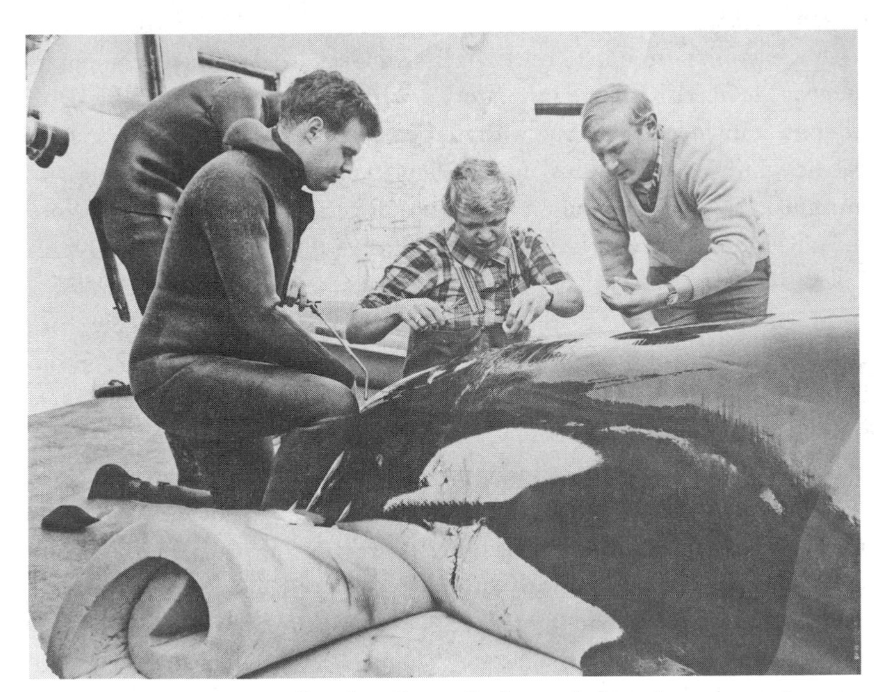

FIGURE 8.3 Aquarium staff tend to Skana after her crash through an observation window, January 1968. Courtesy of Terry McLeod.

capital of Wellington to welcome the Soviet whaling fleet, and local politicians called for renewed efforts to harvest whales off the nation's shores.[17] But the event made little impression on the young Spong. Although raised on the Bay of Plenty, within sight of nearby Whale Island, he initially showed no interest in cetaceans. After earning his bachelor's degree at the University of Auckland, he briefly studied law at the University of Canterbury in Christchurch before entering the doctoral program in physiological psychology at the University of California, Los Angeles (UCLA) in 1963.

Spong enjoyed Southern California. On campus, he proved a gifted researcher, using cutting-edge computers to analyze brainwaves. In the process, he became accustomed to utilizing animals as test subjects, implanting electrodes in their brains to map cerebral functions. In his mind, this was science, and he didn't question the abuse of lab animals or the superiority of *Homo sapiens*. Outside the lab, however, Spong immersed himself in Southern California's thriving counterculture. At UCLA, and on campuses across the country, students were questioning authority, as well as broader assumptions of social hierarchy. By early 1967, even as he wrote up his dissertation, Spong joined in protests against the Vietnam War. It was in the midst of this flux that he received word of an opportunity to study cetaceans in Vancouver.[18]

It was a split position, conceived by Pat McGeer. In addition to an appointment at McGeer's Kinsmen Laboratory of Neurological Science on the UBC campus, it included a contract with the Vancouver Aquarium to research its Pacific white-sided dolphins and possibly its new killer whale. Spong was intrigued, but he didn't know anything about cetaceans. In the weeks before the job visit, he read everything he could find on the subject, including John Lilly's *Man and Dolphin* (1961) and his more ethereal *Mind of the Dolphin* (1967). But it was a conversation with former Marineland of the Pacific curator Ken Norris that brought inspiration. Despite extensive research on the biosonar of dolphins, Norris told Spong, very little was known about their sight. Inspired by this suggestion, Spong boarded the plane for Vancouver with a proposal to study the visual acuity of the aquarium's cetaceans.[19]

Sporting a new haircut and trimmed beard, Spong made every effort to appear professional during his visit, but he felt out of place. His corduroy suit and velvet vest clashed with the UBC faculty's conservative attire, and he found McGeer curt and intimidating. On a visit to the aquarium, he also struggled to connect with the distant Newman. Staff biologist Gil Hewlett seemed intrigued by the research proposal, but Spong left Vancouver skeptical of his chances. To his surprise, he soon received a call from McGeer offering him the position, and four months later he and his wife, Linda, were on their way to Vancouver.[20]

The city to which they moved was changing rapidly. Over the previous year, an influx of US draft resisters had boosted the antiwar movement in southern British Columbia. Meanwhile, a growing number of young Vancouverites espoused elements of the counterculture associated with California's Bay Area. In March 1967, just a week after Skana arrived, thousands had gathered in Stanley Park for the city's first "Human Be-In"—inspired by a similar event in San Francisco. Two months later, the first issue of the countercultural newspaper *Georgia Straight* appeared.[21] "Vancouver was Haight-Ashbury North," recalled Mark Perry. "There was huge opposition from the mayor and the police against the young people trying to do their thing. It was a tumultuous time."[22] The city's countercultural epicenter was the Kitsilano neighborhood, where the Spongs made their new home.

When Paul Spong strolled into the Vancouver Aquarium in August 1967, many of the staff were stunned by his appearance. Gone were the corduroy suit and trimmed hair and beard. The young scientist looked more like a street musician than a researcher. "He was kind of a freaky-looking guy," laughed Perry, who couldn't believe Newman had hired him. But the trainer soon grew to appreciate Spong's incisive mind. "He was the kind of guy who could look at you, and he's got you figured out right now," Perry explained. Terry McLeod was less impressed. "Dr. Spong was a true hippie," he recalled, "and he had some strange reflections about him."[23] Still, McLeod had faith in Newman. If the director had hired the scientist, there must be a reason.

With two jobs to balance, Spong needed help running his research program, and he found it in a twenty-three-year-old UBC undergraduate named Don White. White had grown up in the blue-collar town of Prince George, British Columbia, where his father worked in the logging industry. After graduating from high school, he had hopped a ship to England in October 1962, crossing the Atlantic amid the Cuban Missile Crisis. "There was a fairly strong suspicion that by the time we hit Southampton, it wasn't going to be there," he told me. After several years bouncing between jobs, White enrolled at UBC. Drawn to behavioral psychology and intrigued by the writings of John Lilly, he visited the aquarium to ask about research possibilities, and Vince Penfold suggested he contact Spong.[24]

Like aquarium staffers, White was surprised by Spong's countercultural demeanor. He had expected a clean-cut scientist; instead, White laughed, "the guy I met looked not dissimilar to [countercultural poet] Allen Ginsberg." But such an appearance "wasn't aberrant" for the late 1960s, White emphasized. "It just declared a cultural position." After a brief conversation, the scientist hired White to run his research at the aquarium. The visual acuity test Spong devised was simple. It required an animal to distinguish between two panels—one with two lines and one with a single line. If the test subject pushed the correct lever identifying the two-line panel, the animal would receive a partial herring as a reward. Once the test subject grasped this concept, researchers could reduce the distance between the lines to test the limits of the subject's vision. Spong and White began with the dolphins. Although it took time for the two animals to accept separation and understand the trial's objective, they were cooperating by December 1967. Soon after, Newman instructed Spong to shift his focus to Skana.[25]

❧

Initially, Spong's direct involvement was limited. In addition to his duties at the Kinsmen Lab, he spent much of the spring and summer of 1968 in Pender Harbour, studying killer whales the aquarium had purchased from local fishermen, who had captured them there. As a result, the research at the aquarium itself fell to White, who ran trial after trial on Skana. Despite her species' reputation for intelligence, the captive orca couldn't seem to get it. Unlike the dolphins, she struggled to grasp the experiment's goal, and White worried that data collection might never begin. Finally, on July 25, 1968, after two thousand attempts, she made a breakthrough. "About halfway through the testing session," White told reporters, "Skana started to charge over, push the right button, swim around and around the pool and charge back again. She was very pleased with herself." Relieved, White could now adjust the distance between the lines. For the next several weeks, Skana cooperated with test after test, her accuracy rising to 90 percent.[26]

Then she balked. "I went out one morning and dropped the thing to get started, and she comes over and looks at me," White explained. Instead of cooperating, Skana began vocalizing loudly while slapping her pectoral fins on the water. "I could see that she was agitated," White recalled. "So I walked away and came back in twenty minutes." And that's when it happened: "She suddenly switched and started giving me zero percent."[27]

❧

This was the moment I'd been waiting for. For nearly an hour, the thoughtful White had easily related events from more than four decades earlier, making connections I hadn't even considered. Now we had come to the fulcrum of the story, the moment when Skana began answering "all wrong."

"On purpose?" I asked.

"Well, I won't say that," White replied. "All wrong." But that fact had enormous implications. After all, he explained to me, "Zero percent is just as significant as 100 percent, and there's no way for behaviorism to explain that." In contrast, he noted, cognitive psychology could explain it easily. "Just say that she got bored, she got pissed off," and she decided to cease cooperating. "For me personally, that was transformative," White asserted, "because now I'm sitting there, and I've got an organism in front of me whose behavior I can explain by assuming it has similar thought processes to my own, so that's step one."

"It's a moment of empathy for you, right?" I added.

"Well, that's step two," he corrected. "Step two is asking yourself, using that same thought process, 'What does it feel like being in this tank?' And that changed things for me."[28]

White related the incident to Spong, but the scientist was skeptical. He knew of no record of captive animals behaving in a similar fashion. Upon returning from Pender Harbour, he and White ran dozens more trials, all with similar outcomes. Skana's responses were precisely and consistently wrong. Spong knew such results were nearly impossible statistically if Skana had been choosing randomly—"like flipping a coin and getting eighty-three heads in a row," as he later put it.[29] At first frustrated with the whale's resistance, he, like White, began searching for the reasons behind her actions. Then the subject became the researcher.

It happened one warm day in late August 1968. As Spong sat at the BC Tel Pool contemplating the circling orca, Skana suddenly opened her mouth, grazing the scientist's bare feet. Stunned, Spong yelled and yanked his feet from the water. After recovering from his shock, he eased them back in, only to have the passing whale rake them once more. The two repeated this dance, eleven times in all. Then Spong found the courage to leave his feet in, and with that Skana ended the trial. "So you're the scientist now, eh, Skana?" Spong joked. "Who's experimenting on whom?"[30]

It was a revelatory moment for Spong, spiritually as well as intellectually. McLeod and Perry performed daily in the water with Skana, and White often swam with her. In contrast, Spong had spent little time with her, and he had remained fearful of killer whales. But this close encounter with a captive orca changed all that. Whatever Skana's actual intentions, Spong came to believe that her actions held a higher purpose. "I thought she did that deliberately to change my attitude toward her," he later reflected. "I consider that a great gift, as I've never felt fear around another whale again."[31]

In the wake of the incident, differences between Spong and his assistant began to emerge. Skana's behavior had affected White deeply as well. For him, empathy for the captive orca not only influenced his thinking about animals but proved a springboard for espousing a broader conception of human equality. "If you start extending the idea that there are rights and values for other organisms, then how can you not do so with other people?" he explained. "And if you do [extend rights and values] to other people, how can you not do so with other organisms?" Ultimately, however, White remained committed to understanding other species on their own terms rather than projecting his needs on them. Spong's response was different. "Paul was more in the counterculture than I was—there was no question about that," White noted. "He embraced a lot of its values, and I thought it was another social movement that was no more to be trusted than the war movement." In the end, he concluded, "Paul was more looking for spiritual importance and meaning."[32] Indeed, Spong had begun to re-examine the assumptions on which his scientific career was based.

As Spong pondered these questions, events in Vancouver took a radical turn. In October 1968, he and Linda attended a speech on the UBC campus by American Yippie leader Jerry Rubin, after which hundreds of students occupied the Faculty Club, declaring it a site of "authoritarianism." The activists failed to incite police repression, but word of their action spread quickly, drawing many from Kitsilano to the UBC campus.[33] Although Spong did not join in the protest, he could not help but connect such events to his own journey, particularly as his views clashed with the conservative culture of the Vancouver Aquarium.

Meanwhile, Murray Newman continued his efforts to acquire cetaceans for display. In August 1968, he mounted an expensive but unsuccessful expedition to capture a narwhal near Pond Inlet, a small Inuit community on Baffin Island in the Arctic. Weeks later, a native hunter shot a female narwhal and captured her calf, but neither he nor local Canadian officials could hail the Vancouver Aquarium by radio, and two days later he put the youngster "out of its misery."[34] No sooner had Newman learned of this missed opportunity than he received news that one of the orcas in Pender Harbour had died during a medical examination—the second whale the aquarium had lost there in five months. Then, on February 13, 1969, someone released the last remaining whale.[35] In this context, Newman had little patience for the questions posed by his countercultural researcher.

Yet Spong continued on his quest. A key moment came in early 1969, when he visited Pat McGeer on the UBC campus. McGeer was about to take public office, and he had something he wanted to give Spong. Reaching into the lab's specimen cooler, he handed Spong the frozen brain of Moby Doll. The young scientist was dumbfounded. Not only was the organ much larger than a human brain, but it presented a higher density of convolutions—a feature often used to distinguish human intelligence. Moreover, the neocortex was immense, particularly the portion devoted to the processing of sound. For the first time, Spong grasped the fundamentally acoustic nature of killer whales and in the process realized what a limited view of them his visual acuity tests had provided. Indeed, for a scientist attuned to cerebral structure, the brain of Moby Doll was an epiphany. What secrets did this large and complex organ hold? What sort of creature was *Orcinus orca*?[36]

The mystery deepened with a strange incident at the aquarium. On the evening of March 14, the dolphin Splasher was injured, apparently during a game of tag that Skana often played with her tankmates. At first, Splasher seemed to recover, but three days later trainers found him dead at the bottom of the pool. Although veterinarians concluded that the dolphin had died of internal injuries, they discovered light tooth marks from Skana on his body. To McLeod and other staffers, the evidence seemed clear. At some point during the night, Splasher had sunk to the bottom of the pool, and Skana had grasped him gently in her teeth and lifted him to the surface to breathe. If this were the case, the event suggested an interspecies empathy that further complicated popular and scientific views of killer whales. The behavior certainly puzzled Newman. "This is very unusual for a predatory animal to do for one of its natural victims," he declared.[37] Ultimately, the director framed the incident as further evidence of the need for expansion, and he called for the building of a new million-dollar pool. For his part, Spong believed the incident revealed something deeper about orcas.

The young scientist was spending more time at the aquarium, especially after closing. He thought Skana needed company, and he and Linda took great interest in a male orca calf the aquarium had brought from Pender Harbour. Having recently become a father himself, Spong worried the little whale was lonely, and he dubbed him Tung-Jen (Fellowship)—a name inspired by his reading of the I Ching, or the Chinese Book of Changes. Convinced the two whales were acoustically deprived, Spong brought various sound makers to the aquarium and invited Kitsilano street musicians to perform for them. But this didn't sit well with Skana's trainers or with Newman, who worried about the animals' rest and mental health.[38]

By spring 1969, the director had similar concerns about Spong, who seemed haggard and out of control. McLeod had certainly tired of Spong's antics. Although the trainer found the "down-to-earth" White easy to deal with, he thought Spong was "off in some fantasy world." The tipping point came when McLeod found wine glasses at the bottom of Skana's pool.[39] When confronted, Spong explained that he had accidentally dropped them the previous day. Regardless of the explanation, Newman was furious. Two years earlier, Skana's podmate Katy—the popular calf at the Seattle Marine Aquarium—had died after ingesting a stick tossed into her pool by a careless visitor.[40] Spong had no training in animal husbandry, and in the eyes of McLeod and others, he had become reckless.

On Tuesday, June 3, 1969, Spong arrived at the UBC campus to give a presentation on his research. The lecture hall was full of attendees, including Newman, McGeer, and Penfold, as well as numerous faculty, students, and journalists. Spong began with a detailed discussion of his research, noting that killer whales seemed to have visual acuity similar to that of a cat. But he went on to emphasize that, as fundamentally acoustic creatures, orcas inevitably suffered when confined to pools. Skana, he asserted, was "starved for stimulation," and he suggested that Tung-Jen (the calf) had likely suffered mental damage from his isolation—an isolation that Spong himself initially insisted on. Then came the kicker: "*Orcinus orca* in the wild, in the company of family, is a decidedly different creature than the *Orcinus orca* that we observe in the aquarium," Spong declared. It might be possible to contain killer whales in "semicaptive" conditions and train them for "release and recall," but the conditions at the Vancouver Aquarium were unacceptable. "We should put killer whales back into the ocean as quickly as possible," he concluded. "There they can live like killer whales."[41]

It was an astonishing act of professional courage. As he stood at the lectern, Spong must have realized he was threatening the aquarium's greatest economic asset and thereby his access to the whales who had come to mean so much to him. He had a clear view of Newman in the audience, and his presentation underscored the differences between the two men. Ethereal hippie though he was, Spong's attention was on his research and its implications, while Newman, as director of the aquarium, understandably had his eye on the bottom line. As historian Frank Zelko puts it, "By demanding that the whales be freed, Spong was questioning the aquarium's raison d'etre."[42] Newman stormed out of the hall, and the following day he canceled the scientist's contract with the aquarium and prohibited him from contact with the two orcas.

But Spong wasn't finished. On Monday, June 16, 1969, he appeared at the entrance to the aquarium demanding to play music and swim nude with Skana. When the staff refused him entry, Spong piled musical instruments and an amplifier under a tree outside and held impromptu interviews with reporters. "I wanted the whole bloody city to know she loves electronic music," he announced. "She'll flip her skull when she hears it." When asked why he was fired, Spong expressed disbelief. His research was going well, he explained, and he had discussed plans with Newman to expand it in the coming months. "The trouble is I look like a freak with this beard and everything," Spong declared. "I have a lot of freaky ideas and freaky looking friends but they play fantastic music." In Spong's eyes, it was the threat of the counterculture itself that led to his dismissal. "I've always liked Dr. Newman," he explained, "but he's just a bit dull."[43]

For Mark Perry, who had come to admire the unconventional scientist, Spong's breakdown was hard to watch. "He camped outside the aquarium," Perry recalled, "and they wouldn't let him in. They banned him, took away his keys and everything." Perry had developed his own doubts about orca captivity, and Spong's vigil was a daily reminder of that inner tension. "I'd go into work, and there were times when I felt, 'Jeez, I'm on the wrong side here. I should be sitting with Paul.' " But his bond with Skana brought him back. "I wasn't doing it for the aquarium," he told me. "I had a connection with that animal, and that's why I kept going to work every day." On several occasions, he tried to engage Spong in conversation, but he soon gave up. "He was usually pretty out of it," Perry noted, "and he was hard to talk to, even harder than usual."[44]

Soon after, Spong's department at UBC had him committed to the campus psychiatric ward, where reporters from *Georgia Straight* somehow gained entry and convinced the scientist to sit for an interview. After recording his long, incoherent rant, the magazine decided to publish it in full. Spong admitted, among other things, to taking the hallucinatory drug mescaline before visiting Skana at night. "It helps me work. Helps me tune into the killer whale space," he declared. "They didn't understand, man, but that's ok. They'll come round, y'see, cause the I Ching says so." The interview also included more damaging passages. "I was thinking of destroying the Vancouver Public Aquarium, and letting the whale go," Spong announced. "I was just beginning to get into Skana's space, just beginning to feel what the whale needed, what the whale wanted, what the whale was, just beginning to feel it, man. And they fired me."[45]

By the time of Spong's release, he was a Kitsilano folk hero but an academic pariah. The aquarium wanted nothing to do with him, and UBC was so eager to distance itself from his ravings that it offered him $4,000 in severance money—essentially a bribe to go away.[46] Then, just as he began searching for a place to study wild orcas, he remembered that fishermen in British Columbia were still catching them.

9

The Whales of Pender Harbour

BY EARLY 1968, Cecil Reid Jr. had given some thought to orcas. A gill net fisherman based in Pender Harbour, the thirty-one-year-old Reid—"Sonny" to his friends—had seen many killer whales over the years. As a boy growing up in the 1940s, he heard locals grumble about blackfish, and he watched family members take shots at the animals as they passed by. "My grandfather lived out around the corner from Irvine's Landing," he recalled, "and when the whales showed up, they would get the guns out and start shooting them."[1] Yet Reid knew live killer whales had become lucrative commodities, and when his father suggested catching one, he decided to give it a try. It was winter, however, and there weren't many orcas around. Then, to his surprise, they came to him. In the late afternoon of Wednesday, February 21, a pod wandered into Pender Harbour, passing Reid's waterfront home on Garden Bay. Momentarily stunned, Reid raced down to his boat, *Instigator One*. "I just happened to have my San Juan net still on the drum—which is a lot deeper and touched bottom," he later recounted. "So when they came into Garden Bay the first time, I just set my net across."[2]

The whales eluded his first attempt, but they lingered in the harbor, and the following morning Reid convinced other fishermen to help him, including several of his brothers and members of the local Cameron and Gooldrup families. In all, nine fishermen worked to seal off Garden Bay, and as the sun set over Irvine's Landing, Reid felt certain they had trapped at least three orcas. But that night, one of the nets tore loose, and in the morning only one whale remained. Disappointed, Reid and his partners secured the animal—a fifteen-footer they believed to be male. Like those caught previously, the trapped orca hesitated to challenge the frail net surrounding him, much to the fishermen's relief. Within minutes, the animal was swimming placidly in its makeshift enclosure. "Maybe it likes it here," mused Reid.[3]

For the Pender Harbour fishermen, the timing of the catch could hardly have been better. Just a month earlier, Skana had crashed through the observation window, and Vancouver Aquarium officials declared their interest in buying the captured whale as a mate for her. "They'll be welcome to [it]—at a price," responded Reid. "We're certainly not going to give it to them—we've worked too darned hard for that." Within a day, he and his partners had accepted $5,000 for the animal—a fourth of what Newman had paid Griffin for Skana.[4] Asked why they sold their prize for so little, Reid pointed to patriotism: the aquarium had repeatedly lost out to US buyers, and the men of Pender Harbour wanted this whale to stay in Canada. It was the politic thing to say, but in truth the group was already eying future captures. "We'll be ready for them if they come," Reid declared.[5] Like their counterparts in Gig Harbor, Washington, he and his friends had caught blackfish fever.

Before the whales came to town, life was quiet in Pender Harbour. An intricate maze of bays and inlets fifty miles northwest of Vancouver, the area had once been home to a dense indigenous population. White newcomers made a living in trade before turning to logging and fishing in the early twentieth century. The close-knit families of Irvine's Landing, Garden Bay, and Madeira Park had seen their share of ups and downs. Salmon runs had dropped steadily through the 1930s before fortunes rebounded with the World War II fishing boom. Born in 1936, Sonny Reid grew up immersed in this extractive economy. "We were all fishermen or loggers here," he recalled. Commercial fishermen like his father, Cecil Sr., caught lingcod and various species of salmon in British Columbia waters. By the 1950s, Reid and his brother Eddie were jigging herring, which they sold to hand-line cod fishermen and to the growing number of sport anglers who came to catch the chinook salmon found in Pender Harbour year-round. Soon their uncle Bill Cameron supplanted them. Cameron began seining herring and selling them to reduction plants in Vancouver, while another Cameron brother built artificial "ponds" in Garden Bay, where he kept live herring that were packaged and sold as fresh bait.[6]

Despite these niche markets, local fishermen faced challenges. By the mid-1960s, the population of small sharks known as dogfish was spiking, and with it came a decline in herring and young salmon. Still worse, the Department of Fisheries reduced the number of commercial fishing days while forcing permit holders to pay for a dogfish control program.[7] Squeezed on all sides, the fishermen of Pender Harbour searched for ways to boost their income, and some noticed the high-profile stories of orca capture. Like many Northwesterners

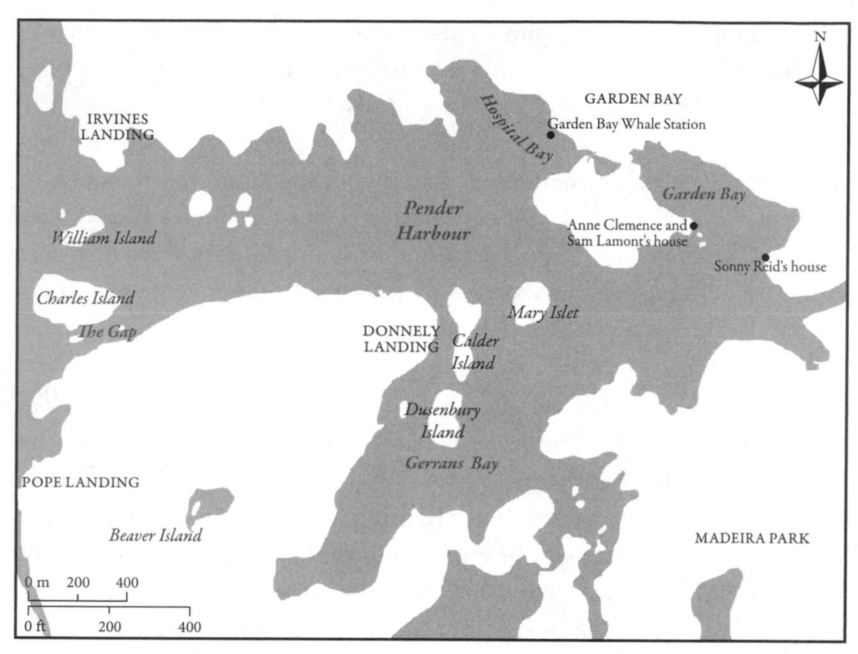

FIGURE 9.1 Pender Harbour in 1968.

on both sides of the border, Reid had paid only passing attention to Moby Doll's brief captivity in Vancouver. "We heard more about Namu," he told me. "That's where I got my idea."[8] Killer whales were common visitors to Pender Harbour, and nothing prevented Reid from catching them. The Canadian government had no official "blackfish season" and placed no limits on the number of orcas that could be taken. In 1966, the Department of Fisheries did begin requiring permits for killer whale capture, but it doled them out readily, as fisherman Sam Maki happily discovered.

A purse seine skipper based in Sointula, Maki had been fishing for salmon near Port Hardy in late July 1967 when several killer whales approached. After swimming alongside the net for several minutes, a small calf slid over the cork line—essentially capturing itself. Realizing he had struck gold, Maki towed his net slowly toward shore and transferred the orca into a makeshift pen. At twelve feet, the young whale seemed ideal for captivity. Having just hired Paul Spong at the Vancouver Aquarium, Murray Newman made a push for the animal but lost out to Marineland of the Pacific. While the Southern California oceanarium was thrilled with the acquisition—its first captive orca since 1961—Newman publicly fumed.[9] Once again he had lost out to a US buyer. Where was Canadian patriotism, asked the American-born director? The exchange was surely on the mind of Sonny Reid when he caught the killer whale in front of his home in February 1968 and sold it to Newman on the cheap.

The purchase of Reid's whale stirred excitement at the Vancouver Aquarium, particularly when Spong and Vince Penfold traveled to Pender Harbour and declared the animal was almost certainly male. "It could very well be Skana's fiancé," announced Newman. Noting that no one had ever bred orcas in captivity, he predicted it would be "a fantastic zoological first."[10] But the aquarium had no room for the animal in Stanley Park. Already rebuffed by Vancouver taxpayers, Newman hoped to find a large donor to build a new pool for Skana and her mate. For the time being, however, he leased one of the Cameron family's herring ponds, where the whale was currently being held, and he sent Terry McLeod to begin the animal's training. Living aboard Reid's *Instigator One*, McLeod worked to get the captive orca eating, a task made easier by the pen's proximity to local fishermen. Even before the sale, Bill Cameron had begun offering the whale herring, and McLeod soon integrated the local fishing industry into the whale's training. Each day, he would drive an aluminum boat to a herring pond, scoop out live fish, and drop them into a fine mesh enclosure for electrocution. Then he offered the freshly killed herring to the whale.[11]

Meanwhile, Newman's ambition soared. Perhaps, he thought, the aquarium could establish a permanent presence in Pender Harbour. In addition to providing a natural setting to study captive whales, Garden Bay offered opportunities for additional captures. As he explained to one reporter, "We are considering keeping the whale up there for some time and perhaps developing a research and collection station."[12] This option grew more attractive when Newman received disappointing news about the aquarium's new whale. In early March, his staff undertook the first medical examination of the animal. Following the four-hour procedure, Penfold reluctantly announced that the staff had once again missexed a killer whale.[13] Concluding the animal was female, they named her Cecilia—in honor of Cecil Reid. Convinced the whale could not mate with Skana, Newman decided to keep her in Pender Harbour.

This was just fine with Reid and his friends, who were enjoying their collaborative relationship with the Vancouver Aquarium. Having sold the orca, they now began offering other live specimens. Reid provided two large octopuses and a wolf eel to the aquarium, while Cameron stocked its Hall of Fishes with herring.[14] But the men weren't done catching killer whales.

In the days after Cecilia's capture, locals noticed orcas lingering near the entrance to Pender Harbour. Sometimes the animals entered the harbor, vocalizing in the direction of the penned whale, but they never went far from the entrance. Now known as A5 pod, these were northern resident killer whales, far from their usual range, and their reasons for coming south are unclear.[15] Perhaps they were hunting chinook, which are common in Pender Harbour even in winter. Regardless of the explanation, the whales didn't want to leave their podmate, and their loyalty made them vulnerable. Accustomed to trading in baitfish, Reid and his partners now used the aquarium's captive orca to lure the others into a trap. But the whales seemed to sense that danger lurked in Pender Harbour. Despite efforts to herd them in, they lingered out of reach, unwilling to enter but reluctant to leave, and by mid-April, the pod had vanished.

By this point, Reid and Bert Gooldrup's blackfish obsession had stirred talk in Pender Harbour. The two men bought up old mesh and sewed capture nets, keenly awaiting reports of orca sightings. "You'll forget to catch real fish," warned friends, and they were right. "I'd go out salmon fishing and someone would report whales, and I'd come roaring into home, pull my salmon net off, put the whale net on," admitted Reid. "Of course, it didn't amount to nothing, so I lost a lot of money."[16] He even became the target of a practical joke. On Friday, April 26, 1968, Reid and Gooldrup were scheduled to attend a formal banquet for their bowling league at the Jolly Roger Pub in Secret Cove, some ten miles away. Friends planned to interrupt the party with an announcement that killer whales had entered the harbor, anticipating that the men would tear off in excitement. Reid had gotten wind of the prank, however, and didn't intend to fall for it.[17] Yet as fate would have it, the A5 whales returned to Pender Harbour that very evening.

~

Anne Clemence had paid little attention to orcas up to that point in her life. Originally from Kent in southeastern England, she had lived through the Battle of Britain as a child and later trained as a nurse in London. In the early 1960s, she moved to Pender Harbour, where she met local log salvager Sam Lamont. She was troubled by the Vancouver Aquarium's harpooning of Moby Doll, but like other locals, she followed news stories about Namu and knew that Reid and his partners planned to catch more whales. She and Lamont were living in a little cottage on Garden Bay in April 1968 when the whales came. "I heard this slap, slap, slap outside in the bay and went out, and it was killer whales!" she later told me. "We couldn't tell how many there were because they seemed to be milling around quite a bit, but Sam immediately thought of Sonny Reid." Unable to reach Reid at home, Lamont rang up the fisherman's mother, who in turn phoned the Jolly Roger Pub. With the other partygoers assuming it was part of the arranged hoax, Reid and Gooldrup raced back to Garden Bay, where they met Bill Cameron. There, the three fishermen, still in suits and dress shoes, worked to cordon off the entrance, but the whales repeatedly eluded them. "It was still going on after dark, and of course the boats had floodlights so it was all quite exciting and quite dramatic," Clemence told me. Meanwhile, one of her neighbors had phoned the Canadian Broadcasting Corporation (CBC), which immediately began reporting the event. "I am watching the capture, and it is coming out on the news," laughed Clemence. "That is something to be said for the CBC!" In the end, Reid and his partners corralled seven whales in Garden Bay.[18]

The following day—Saturday, April 27, 1968—a thousand people flocked to Pender Harbour for a look at the orcas, including representatives from oceanariums throughout North America. Among them were Newman, John Prescott of Marineland of the Pacific, and Don Goldsberry. The fishermen had hit the jackpot, and this time Reid wasn't feeling so patriotic. "We know what we want for the whales," he told reporters. "It's up to the aquariums to meet our price."[19] Unlike Griffin, Reid and his partners intended to sell every whale they had caught. Determined to get his share, Newman instructed Penfold to begin talks, but the fishermen found Penfold, a former prison guard, arrogant. Privately, Reid told McLeod to keep Penfold away or the Vancouver Aquarium wouldn't get any of the whales. With the young trainer running interference, Newman closed the deal, purchasing two of the captured whales—a young calf likely not yet weaned, and a female believed to be its mother. Four others went to California. Marineland scooped up two, and two more went to Marine World—an oceanarium in Redwood City, California, owned by the American Broadcasting Company (ABC).[20]

With all the media attention focused on Pender Harbour, it was inevitable that the whales' departure would become a public spectacle. On May 9, Marineland towed its whales to Madeira Park for shipment, and Anne Clemence and Sam Lamont passed the evening on their boat listening to the sounds of the separated family. "It was awful," recounted Clemence. "They were calling to each other all night long."[21] The next morning, Marineland veterinarian Lanny Cornell supervised the animals' removal. Using a mobile crane, workers lowered each animal into a twenty-foot box resting on a flatbed truck. The process proved emotional for many observers. Over the previous days, locals had enjoyed watching the whales in their pens. With the finality of their departure driven home, even the captors found themselves affected. "I was actually quite sad about the whole thing," recalled Reid. "The whales were crying when you got one up on the dock and the other ones were crying in the water."[22]

With the animals loaded, the trucks raced for the Langdale terminal, where they caught the 8:30 a.m. ferry to Horseshoe Bay. An hour later, the whales were iced down on chartered jets and bound for Los Angeles. Observing the entire process was F. J. Jones, executive director of the British Columbia SPCA. Although Jones emphasized his organization's opposition to whale captivity, he noted that "we are satisfied the handling and shipping of the two California-bound whales could not have been carried out with greater care."[23] At Marineland, the two whales would become known as Orky II and Corky.

Marine World's whales didn't fare as well. The oceanarium's staffers were new to the business of orca captivity, and they had little understanding of how to handle the animals. "They had to cut a section out of the end of the box to get the tail in and the poor whale thrashed and was bleeding and throwing blood around," recalled Clemence, who found herself horrified by Marine World's treatment of the whales. Reid's partner Bill Cameron rode

FIGURE 9.2 Marineland of the Pacific staffers prepare whale for shipment, May 1968. Courtesy of Anne Clemence.

the ferry with the oceanarium's staff and was surprised to find that they packed insufficient ice to keep the animals cool.[24] But worst of all, neither the fishermen nor the Marine World staffers knew that the two whales, both mature females, were pregnant. Lifted by slings and transported by land, sea, and air to the Bay Area, the two expectant mothers delivered stillborn calves shortly after arrival. Soon after, one of the females, named Bonnie, died in her pool.[25]

Meanwhile, the Vancouver Aquarium tended to its whales in Pender Harbour. In order to convince them to eat, staffers moved the two new animals into the pen with Cecilia. In the process, Penfold realized, to his chagrin, that he and his colleagues had been wrong when they thought they had misidentified the whale's sex. Cecilia was, in fact, a male. Amid the confusion of buying and moving whales, the aquarium quietly renamed him Hyak. Newman also arranged to purchase the last remaining whale from the April catch—a twenty-foot male dubbed Irving.[26]

This final acquisition represented a milestone for the Vancouver Aquarium. As Penfold proudly declared, "We now have the most whales in captivity owned by any single aquarium in the world." This included the largest and smallest captive killer whales. "As a matter of fact," he boasted, "the new whale qualifies as the largest captive mammal in the world, too." Penfold admitted that Irving would likely remain in Pender Harbour. "We'll probably keep two or three of the whales there permanently for display." After all, he noted, "We haven't got anywhere to put them in Vancouver."[27] But not all was well with the animals. Although Newman assumed he had bought a mother and her calf, the little male didn't nurse and refused herring. For her part, the large female—named Natsidalia—didn't seem to be lactating. By mid-June, the calf had grown so weak that the aquarium staff decided to intervene, lifting the youngster from his pen and pumping heavy cream into his stomach. Once he had regained his strength, they flew him to the Vancouver Aquarium, where Spong later named him Tung-Jen.[28]

By this time, the appeal of keeping the other whales in Pender Harbour was obvious. Not only did the open-water pens afford the animals more space and clean water, but the area offered a natural setting to study captive orcas. The latter appealed to Spong, who spent much of the summer observing the whales and getting to know Reid and other locals. For her part, Anne Clemence found the antics of the young scientist amusing. "He had all the education, but he didn't have much practical knowledge," she observed. "He had transponders and microphones and things draped all around the pens because he was trying to record the vocalization, but he was always dropping them in the chuck, and a diver would have to come in or one of the fishermen would have to fish out these very expensive bits of equipment." All in all, locals didn't know quite what to think of Spong. "Fishermen are a practical sort, and they know what they are doing and they do it well," Clemence noted. "Dr. Spong was considered a little bit weird."[29]

The presence of captive whales brought changes to Pender Harbour. The animals quickly became a major attraction, and that summer more than ten thousand tourists visited.[30] To meet this demand, the aquarium built an informal display facility in nearby Hospital Bay. Newman officially launched the Garden Bay Whale Station on August 1, 1968. Although it had already been open to the public for a month, the announcement received broad media attention. On hand to celebrate were a hundred local residents and some fifty honored guests, including H. R. MacMillan and federal fisheries minister Jack Davis, who had recently announced a plan to reduce the British Columbian fishing fleet. Open from 9:00 a.m. to 5:00 p.m., the station offered four short performances per day and charged seventy-five cents for adults and twenty-five cents for children and students. Despite consistent crowds, admissions didn't even cover the cost of herring for the whales. But the shows hinted at Newman's long-term plans for Pender Harbour. In addition to developing a research facility at the whale station, he hoped to expand its exhibits to draw tourists.[31] In charge of the venture, he placed two young employees: Mark Perry and Graeme Ellis.

Having already worked with Skana, Perry was the natural choice to run the station when McLeod returned to Stanley Park. Over the following weeks, he bonded with the whales but sometimes experienced local hostility. One Saturday night, for example, he and Ellis found themselves barred from a skating rink by local youths. "It was a small community," Perry reflected. "We were city guys—big shots, I guess—and they wouldn't let us in." Like McLeod, he formed bonds with the fishermen who had caught the whales. "They were salt-of-the-earth guys trying to make a living," he observed. Yet Perry sensed tensions between the orca catchers and their neighbors, especially at the local pub. "I never saw any fights or anything," he told me, "but when Sonny or the others would come in . . . it would get quiet, and they'd be kind of shunned." On another occasion, Reid found himself barred from the pub when he attempted to enter with Spong—a victim of a "No Hippies Allowed" policy. "I think a lot of people were jealous, local people, about the money these guys had made," noted Perry, "and I think it affected them."[32]

Pender Harbour brought different revelations for Graeme Ellis. Just eighteen when hired as caretaker for the whales, he could hardly have imagined this summer job would set the course for his life. Born in the Vancouver area in 1950, Ellis spent much of his childhood in Campbell River. In those days, he joked, children in the area ran "almost feral," wandering the woods and frolicking in the water. But he knew of the fear killer whales generated among locals. "As a boy, I believed all that nonsense about orcas," he recalled. "Whenever my friends and I caught sight of killer whales, we threw rocks at them. Others used BB guns and later .22s."[33] But Ellis had never been close to one until he came to Garden Bay. Initially assigned to get Irving to eat, Ellis quickly befriended the animal. Having read of Ted Griffin and Namu in *National Geographic*, he decided to try swimming with the big orca. "You could see him coming through the murk there, and he opened his jaws as he came toward me," he recounted. "I just went out of the water like a shot. One moment I was in the water, the next moment I was standing on the logs." Ellis also watched Irving become angry when Spong, in a fit of enthusiasm, leapt onto the whale's back and tried to ride him. "Irving was a noble beast," explained Ellis. "He didn't go for that kind of thing."[34] Over the following weeks, Ellis trained the orcas for performances in Hospital Bay. He fed them, talked to them, nuzzled them. At first intrigued, then mesmerized, he found himself partially relieved when the whales started slipping away.

The first loss came near the end of summer. On August 26, 1968, Irving escaped through a tear in his net, apparently caused by marine growth.[35] In the following days, aquarium staffers scoured the area, hoping to lure him back with food. Spong and Don White drove up from Vancouver to help, bringing underwater speakers to make contact with the whale. Despite one brief sighting, however, no one saw Irving again. Months later, Natsidalia showed signs of distress.[36] In November, staffers lifted her from her pen to take blood samples, but she died in the sling. The subsequent necropsy revealed that she was a very old female and pegged the cause of death as a massive aneurysm—perhaps caused by the strain of her pod's capture.

And then there was one. Hyak, the first captured, was now the last remaining. Less than a year earlier, he had wandered with his pod into the area and found himself trapped and alone. When his family returned, seven more were netted, and one by one they were carried away. One of the pregnant females taken to Marine World may well have been his mother. And then Natsidalia, perhaps the matriarch of the group, had died within earshot. The other whales were gone, but Hyak didn't linger long in solitude. By early 1969, Perry and Ellis had returned to Vancouver, and their replacements failed to notice when someone unfastened the log boom, enabling Hyak to escape. Some suspected activists from Vancouver, but Reid was certain that local fisherman, jealous of the whale catchers, had done the deed.[37]

Many observers were appalled at the act, viewing it as vandalism or even theft. "A hunt is being pressed here today for a killer whale and the culprits who engineered his pen break," reported the *Vancouver Sun*. Vince Penfold declared it "a bit of skullduggery." Hyak had been "very well trained and knew a lot of tricks," he lamented, and his escape represented a critical loss. Initially, Penfold hoped hunger might draw the orca back. Despite help from local fishermen and the RCMP, the aquarium was unable to locate the escapee.[38] It was a devastating blow for Newman. Of the four orcas he had acquired from Reid and his partners, only the little calf Tung-Jen remained, and the aquarium had not a single whale left at its Garden Bay Whale Station. In the eyes of local boosters, the area had lost a key attraction. "There wasn't a day during the summer and fall when someone wouldn't show up and ask where the whales were located," former chamber of commerce president John Haddock noted sadly. "We had big things planned for Pender Harbour this year."[39]

In the following months, the Vancouver Aquarium dismantled the whale station, and things again grew quiet in the area. Perhaps the new Northwest would not come to Pender Harbour after all. That summer of 1969, as Spong's firing stirred public debate over captivity in Vancouver, Reid and his friends busied themselves with fishing. But after the last fall opening for chum salmon came and went, they again turned their minds to killer whales. That November, animal collectors from Florida hired Reid and Gooldrup to help them capture orcas. The two Americans came well equipped, with a device designed to clamp onto a whale's tail, as well as an assortment of firearms. "They were scary guys," recalled Reid. "They hated Indians, they hated black people, and they had pistols on them all the time." On one occasion, the Florida captors joined Reid and Gooldrup in a pub in Port Hardy. "It was mostly natives, and these guys had their guns ready to shoot one of them if something happened," he recounted. In the end, they failed to catch any orcas. "We were glad to get rid of them," declared Reid.[40]

The following month, foul weather brought Reid and Gooldrup good luck. On December 11, 1969, a nasty squall hit the coast, and a pod of orcas took shelter in Pender Harbour. The fishermen were ready for them, but it was a dicey operation. Facing heavy rain and howling winds, they struggled to get into position, and when they finally did, a school of dogfish slammed into their net, sinking the cork line. It seemed a cruel trick: the same small sharks that were cutting into herring and salmon catches had foiled an orca capture. Reid also noticed, to his horror, that his net held scores of chinook salmon—an illegal catch that he quietly dumped overboard. By 11:00 p.m., he and Gooldrup had managed to trap four whales near the Madeira Park dock. The fishermen stayed there through the night, tending to their nets in the howling storm. "You couldn't see the whales because of the sheets of rain and you couldn't hear them because of the noise of the engines," Reid told a reporter. The next morning, the weather calmed, and Bill Cameron joined in the effort, catching seven more whales who were lingering around the nets. By that time, the men were exhausted. As he stared at the stack of dead dogfish on his boat, Gooldrup expressed relief that they had caught something of value, and he knew the offers would soon come rolling in. "We're not calling anyone," he declared. "Once word gets around they will call us."[41]

But public views had changed since the last capture. Spong's clash with the aquarium had drawn attention to captive orcas, and many Vancouverites viewed those who caught them with scorn. Spong himself called on the fishermen to release the whales, and other activists turned to harassment. For Pender Harbour families immersed in the extractive economy, this was a new and frightening experience. As fishermen, Reid, Gooldrup, and Cameron viewed orcas as another commodity drawn from the sea. No one had denounced them for catching salmon or herring. From their perspective, they were finding ways to support their families, and now those same families were threatened. "I got a nasty phone call that if they didn't let the whales go, they'd kill my kids," Isabel Gooldrup recalled. "To me, that colored the whole tone of the thing. I didn't like it—any of it. It wasn't worth it to me."[42] The same happened to Sonny and Marie Reid.

∽

It was a rainy morning on the Sunshine Coast in July 2016 when the couple welcomed me into their home with coffee and pastries. Perched above Madeira Park, they enjoyed a sweeping view of their beloved Pender Harbour. These days, summer homes and pleasure craft dominated the area. But the

Reids, now in their early eighties, spoke in the clipped accent of an earlier time. Sonny grew up next to the dock at Irvine's Landing, where the ships of the Union Steamship Company once tied up. He began fishing at the age of ten and left school at fourteen to take a job as a whistle punk at a local timber camp, and he used his savings to build his first gill net boat. Marie was a local girl, equally at ease on the water. Married in 1957, the couple became lifelong fishing partners, and they enjoyed the freedom the sea afforded them. "I guess we were probably the first hippies in Pender Harbour," laughed Sonny. "We would just make enough money to pay our rent and for groceries, and then we wouldn't go out fishing anymore." Each May, they joined other gill-netters to catch chinook off the Bella Coola River, and they knew Willie Lechkobit and Lonnie McGarvey—the two men who caught Namu in the area in 1965.[43]

By February 1968, when Sonny caught his first killer whale, the couple was raising four little girls. "The kids were all excited and couldn't wait to get down to see it," Marie told me. "They were so proud, you know, 'My dad got it!' " With the second catch two months later, it seemed the glory would keep coming. Sonny received Pender Harbour's Man of the Month Award that April for bringing notoriety to the area. But then things started to change.

The first shock came when Sonny heard the whales' calls as they were separated. "I just felt bad because the mother was crying and the baby was crying," he explained.

"That is the part that really sticks with you?" I asked.

"Yeah, it was like part of you went with them, you know?" reflected Reid. "I can still hear them crying. It was like . . ."

"Breaking a family up," Marie interjected, and her husband nodded.

"Just like if I went and took your kids away," he added. "I might be going to look after them, but still, they are upset. They left you, so that is how I looked at it. It is just like somebody took one of my kids away."[44]

In fact, some callers threatened to do just that. Like the Gooldrups, the Reids received anonymous phone calls in December 1969 threatening to kill or kidnap their children if they didn't release the whales. "People would call and threaten my kids—my four little girls," Reid told me in a quavering voice. "They had to have police escorts to school."[45]

"When you are brought up in a small community like this, you don't really think of those things happening, of kids being kidnapped," added Marie. "You hear about it in the big city and that, but you don't think that it could happen around here."[46]

Yet it was the hostility of friends and family that proved most painful. Mark Perry had been right: many locals resented the fishermen's good fortune.

Amid the decline of salmon fishing, Reid, Gooldrup, and Cameron seemed to have found the golden ticket. "They thought we became millionaires," Reid recalled. "We'd walk into the pub, and you could hear a pin drop. We just maybe had a beer and left. We just felt terrible. Everybody hated us."[47] That jealousy drove some to target the whales themselves. Following the December 1969 catch, several locals threatened to kill the captured animals, and someone regularly took shots at them at night. Days later, when a female whale tore through the net and escaped during a transfer, some locals cheered the loss. Sonny felt hurt at the time, but he soon changed his mind. "Thank Christ she got away," he sighed, "'cause she was pregnant—had her baby out here in the bay."

It seemed the moment for a question I had been reluctant to ask: "Did you know that two of the whales from the second catch turned out to be pregnant and gave birth to stillborn calves?"

"No, we didn't know about it," responded Sonny, after a painful pause. "They never told us." And the room fell silent for a moment.

"We never talked about it much, you know, except if someone comes in and sits and just wants a conversation about them," reflected Marie. "We never actually owned up to it ourselves."

"No, we didn't," muttered Sonny.

"Would it be fair to say that this haunts you a little bit?" I asked.

"Yeah, it does in a way," said Sonny, and he offered a closing story. Years after the captures, around 1996, he and Marie had been fishing in Douglas Channel, near Kitimat. As they waited for a salmon opening, Sonny was jigging rock cod when killer whales suddenly appeared all around their boat, *Instigator One*. "We were surrounded by them, and they were just coming up and looking, you know, their eyes," he explained. "And it gave me an eerie feeling." He thought of the captures some three decades earlier:

> I had the same "Jimmy" diesel on the boat that I had before, and I guess the whales could tell the sound of it. I figured they hated the sound of that engine. "They're after me," I said. "They remember this boat, and they are going to get me." And we just left—didn't even set our net to fish, they scared me so bad.

∽

The December 1969 catch was the last Reid and his partners would conduct. One reason was that orcas avoided the area. Yet the enthusiasm of Reid and his partners had also waned. At first, the pursuit of killer whales was thrilling, but the personal costs had become too high. To be sure, they made a hefty profit from the last catch. Of the ten remaining animals, they released four and sold six.[48] Once again, most went to California: two to Marine World and three to Marineland of the Pacific.

The sixth whale undertook a different journey. Purchased by the new Marineland park near Nice, France, she was dubbed Calypso and loaded onto a plane for Europe. Because her pool was not yet completed, she took up temporary residence in Great Britain's Flamingo Park Zoo, which had recently acquired a male orca, Cuddles, from Ted Griffin.[49] Hoping the young female had reached reproductive age, veterinarians attempted the first artificial insemination of a killer whale in history, using semen collected from Cuddles.

It was a revealing moment in the saga of orca captivity. A northern resident killer whale had been caught in Pender Harbour and transported to North Yorkshire, where she was mated artificially (and unsuccessfully) with a southern resident killer whale caught in Washington State before being flown to the French Riviera. Clearly, times had changed. Killer whale capture may have begun in the Pacific Northwest, but it had become part of a rising international trade in marine mammals.

10

Supply and Demand

IN THE SUMMER of 1968, Richard O'Feldman must have wondered how he came to be playing the flute on the back of a killer whale. The curly-haired twenty-eight-year-old was no stranger to marine mammals. Growing up on Miami Beach in the 1940s, he had often seen bottlenose dolphins. "Back in those days, Biscayne Bay was teeming with them," he recalled, and his mother told him tales of dolphins rescuing downed pilots. Thirsting for adventure, fifteen-year-old O'Feldman lied about his age to join the National Guard and later enlisted in the navy. Over the next five years, he rode a US destroyer around the world, hearing his first dolphin calls in the ship's sonar room and training to become a navy diver. Not yet twenty-one when he left the service, he dabbled in treasure hunting off the Florida coast before finding work at the Miami Seaquarium.[1]

His first day on the job, O'Feldman joined the marine park's collection crew on an expedition to capture dolphins in Biscayne Bay. "In those days, you didn't need a permit," he explained. "You could do whatever you wanted." As a diver, his task was to search for entangled dolphins while keeping the net clear of coral snags. The collection method made casualties inevitable. "I would find dolphins wrapped up dead," he admitted. "We killed a lot." By 1962, O'Feldman had helped capture more than a hundred bottlenose dolphins. The Miami Seaquarium kept some for display, but it sold most to other marine parks. Among them were US buyers such as Marineland of the Pacific and Chicago's Shedd Aquarium, as well as a growing number of European dolphinariums. "Places were just opening," he noted, "and we were supplying them." Among the eager customers was his former employer, the US Navy, which had just launched its Marine Mammal Program. O'Feldman saw nothing wrong with captivity—"never questioned it at all," he told me. Never, that is, until he began working with the animals himself.[2]

He started with the dolphin shows at the Miami Seaquarium. Within months, MGM studios hired him as a trainer for the *Flipper* movies and television series, which did much of their filming at the marine park. By his midtwenties, O'Feldman was a high-profile and highly paid trainer of the most popular marine species in North America. But as in the case of Paul Spong, close contact with the animals brought misgivings. "I was living at the Seaquarium with these five dolphins for seven years . . . all day, every day," he explained, and as he got to know their personalities, he began to wonder about captivity.[3] Those concerns only grew in the spring of 1968, when the park welcomed its newest star: a young orca named Hugo.

Like Shamu's introduction to San Diego three years earlier, Hugo's arrival captivated Miami. Reporters raved about the city's first captive killer whale, and Mayor Chuck Hall proclaimed May 16, 1968, "Hail the Whale Day."[4] O'Feldman trained the youngster for performance, and like Spong, he added acoustic stimulation, strumming his guitar and even playing a flute during performances. Yet amid the cheering crowds, he grew uneasy. The pool was too small, even for a juvenile orca like Hugo, and the trainer worried about the animal's mental health. Eventually, O'Feldman would earn fame as an anticaptivity activist under the name Ric O'Barry. At the time, however, he gave little thought to Hugo's origins in the Pacific Northwest.[5]

Captured in Washington State in February 1968, the little whale was part of a new trend in the marine mammal trade. Until the mid-1960s, bottlenose dolphins had served as the oceanarium industry's prime attraction. Now orcas were becoming the stars, and marine park owners were shifting their gaze from Biscayne Bay to the Salish Sea. Just as the Seaquarium supplied many of the world's captive bottlenose dolphins, the Seattle Marine Aquarium seemed poised to corner the market on the industry's new signature species.

The relationship between captivity and commerce was hardly new. The modern zoological garden—or zoo—had developed in tandem with the wildlife trade. Beginning in the late 1840s, German fishmonger Carl Hagenbeck had bought seals and other marine animals from fishermen for public display. Taking over the trade, his son, Carl Hagenbeck Jr., drew on a network of collectors in colonial Africa and Asia, becoming the world's greatest purveyor of exotic wildlife and a pioneer in animal husbandry. Trained elephants, lions, and tigers from his collection were featured at the 1893 Chicago World's Fair as well as P. T. Barnum's traveling circuses and the amusement park on Coney Island. In 1907, Hagenbeck built his Animal Park near Hamburg, which became the model for modern zoos in North America. Over the following decades, display facilities around the world engaged in a vibrant trade in live animals, often exchanging locally collected species for those caught elsewhere.[6]

After World War II, North American oceanariums had followed a similar path. Most began by displaying local marine life, but as they expanded, they acquired less familiar species, which helped boost attendance. Among the masters of this public relations strategy was the Vancouver Aquarium's Murray Newman, who used tales of collecting expeditions in exotic locations to craft a persona as scientist-adventurer. As the industry grew, specimens, knowledge, and personnel flowed between the world's aquariums, and owners and curators viewed local megafauna as resources in this exchange.[7] Just as the Miami Seaquarium sold and traded bottlenose dolphins, Marineland of the Pacific became the main source for California sea lions and pilot whales, including those used by the US Navy. It was in this context that Ted Griffin approached the growing market for captive orcas.

By 1968, oceanariums were appearing throughout North America and Europe, and the blackfish boom was on. Just as elephants' immense size made them the signature display animal at twentieth-century zoos, killer whales were becoming essential attractions for major marine parks.[8] But with so few captured, different facilities scrambled for access to the striking animals. The purchase of Shamu in late 1965 gave Sea World a competitive edge over the more established Marineland of the Pacific, and when the upstart San Diego company acquired two more orcas from Griffin's Yukon Harbor catch in February 1967, attendance boomed. Sea World became a publicly traded company in 1968, and soon after it announced plans to open a new franchise in Ohio.[9] With Sea World growing and Marineland and others desperate to keep pace, it seemed a bullish market for captive killer whales.

Based in Seattle, with specialized gear and extensive experience, Namu Inc. entered 1968 well-positioned to meet this demand. But Ted Griffin faced an old problem. Orcas were difficult to track, even in the confined waters of Puget Sound, and pods showed a remarkable ability to confuse and elude pursuers. The Marine Mammal Biological Laboratory had helped resolve this problem with the Greener harpoon rifle, but Griffin had packed the weapon away after the death of the wounded male at Yukon Harbor. This left him in a bind: just as the market for orcas was spiking, he had renounced his only means of tracking them. It was no coincidence that a full year passed before Namu Inc. made another capture. Reports of sightings rolled in, and Griffin gave chase, but he struggled to track the pods, let alone trap them. Then, to his surprise, a dozen orcas trapped themselves.

The call came in the early morning of Monday, February 19, 1968. A group of killer whales was milling about in Vaughn Bay—a tiny harbor fifteen miles west of Tacoma. Frantic to keep the whales there, Griffin stationed skiffs to guard the narrow entrance and phoned Adam Ross and Peter Babich. It was winter, and he assumed the Gig Harbor fishermen would be available. "Once bitten by the 'bug,'" Griffin told me, "both Ross and Babich desperately wanted to get in on all the captures and would not leave town, except to fish in Alaska, for fear of missing out."[10] As it turned out, Ross was away in California, but Babich was home, and the eager fisherman quickly gathered his crew. But his *Pacific Maid* cruised at only nine knots, and for the next eight hours, it fought the ebb tide all the way to Vaughn Bay. Finally, in the late afternoon, the fifty-eight-foot seiner edged carefully past the sandbar at the mouth of the bay and sealed off the entrance. Soon after, Babich made an easy set, trapping all twelve animals.[11]

The next morning, the team was hard at work securing the nets and moving its steel antitorpedo screen into place. Among those helping was seventeen-year-old Randy Babich, who found himself stunned by the behavior of the whales. Like many, he assumed the large predators would be dangerous when cornered, but he saw no signs of aggression, even when divers entered the pen. "They were just as docile as kittens," he recalled. "That really amazed me."[12] Soon after, he and the other crew members were surprised to see the *Chinook* entering the bay. After learning of the operation, Adam Ross had boarded the first flight from Los Angeles to Seattle, gathered his crew in the middle of the night, and raced to the site.

Local residents initially reacted with enthusiasm. The capture operation was the most exciting thing to hit Vaughn Bay in living memory, and dozens gathered on the beach to watch the famed whale catchers. Teenagers in rowboats rained questions on Griffin. "How did you get the whales cornered in here?" "What are you going to do with them?" "Need any help?"

"Sure do," he replied, hiring the youngsters to ferry his crew back and forth to shore.[13] As the team constructed a pen and worked to sort the animals, more sightseers arrived. They came by land and sea, crowding the surrounding beaches for a look. One local man charged journalists for close-up views aboard his barge. Schools in the area bused whole classes of students to the bay to receive, as one journalist put it, "a 'live' lesson in marine biology."[14]

Among the students was eleven-year-old Kenneth Gormly, a fifth-grader at Vaughn Elementary School. Gormly watched the operation for several days

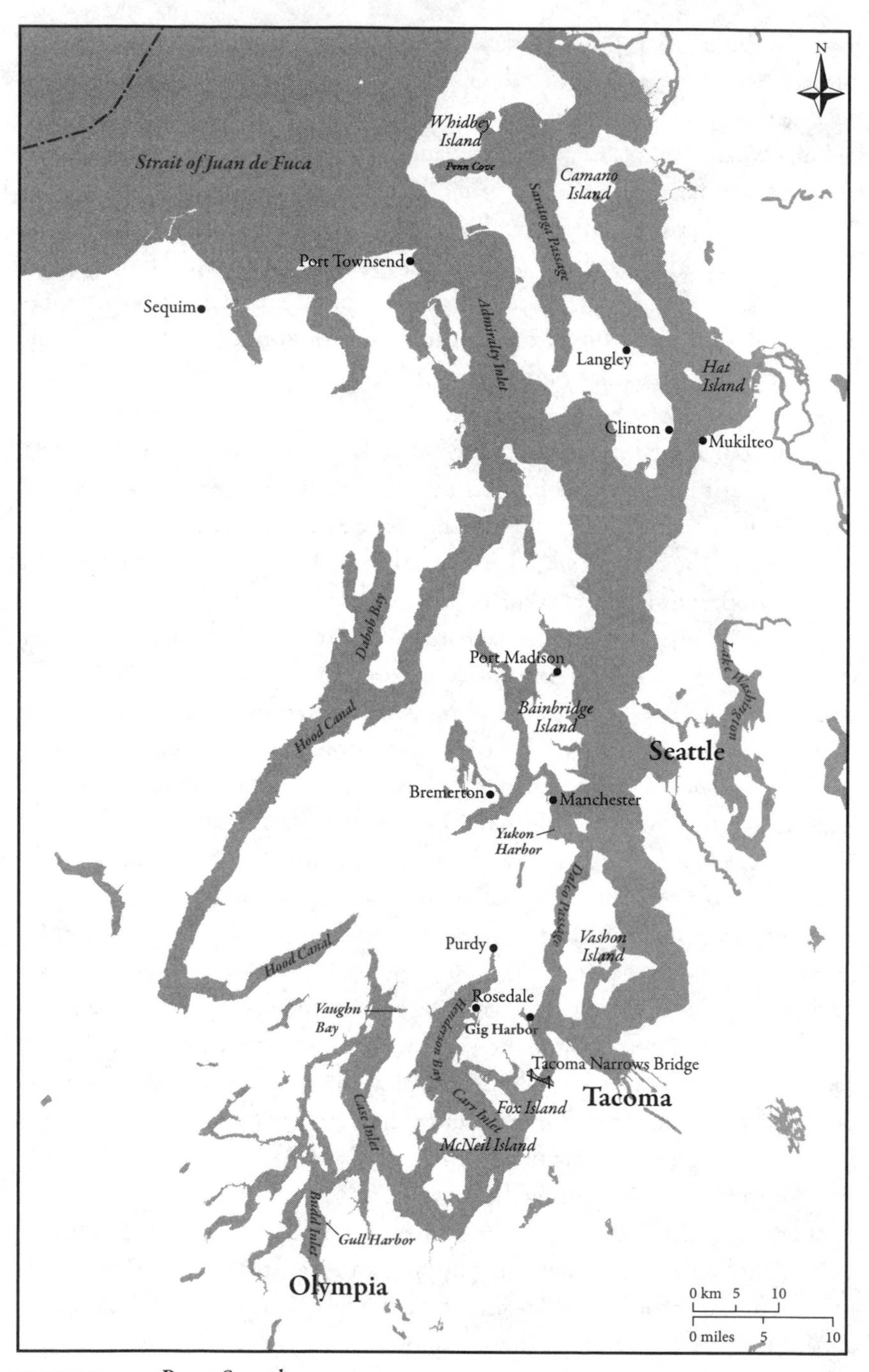

FIGURE 10.1 Puget Sound.

and wrote an eyewitness account published in a local newspaper. At first, the boy thrilled at the appearance of seiners and seaplanes in Vaughn Bay. Nothing like this had ever happened in his hometown. But as he observed the trapped animals closely, his feelings shifted. "The ones in the outer steel net were acting worried," Gormly wrote. "They were lying on their backs and flopping their fins, and jumping straight out of the water." When the Namu Inc. crew removed a young whale—the future Hugo—from the water, Gormly noticed that the animal's mother repeatedly tried to reach him. In the confusion, the divers held the youngster underwater, infuriating Griffin. "Let him blow! Let him blow!" Gormly heard the aquarium owner scream. Later, when the team lifted the juvenile orca from the water by sling, the boy wrote, "We could hear [him] crying for [his] mother."[15] Although Gormly could not know their meaning, the calls seemed to make a powerful impression on the boy.

The event was nearly fifty years old when I phoned Gormly in May 2016. Now a lawyer in Tacoma, he hadn't thought of the Vaughn Bay capture for a long time, and his only copy of his article had burned up in a fire years before. "We lived on the water, north side of bay, west of the sandspit," he explained. "Mom was a homemaker; dad worked for Weyerhauser." It was Gormly's mother who first told him of the whales, prompting him to drive his twelve-foot dinghy into the bay. "I did most of my observation from the boat," he recalled. "We got up right against the log boom." At first fascinated by the sight of orcas right in front of him, Gormly found himself disturbed as the captors began removing animals. "I could tell there was a connection between the whales," he noted, "and I remember thinking, 'This is really bothering me.'" It was their calls that left the deepest mark. "I remember vividly the noises they were making," Gormly told me. "I can still hear that sound in my mind."[16]

Griffin had different concerns. After loading the young male and a large female onto the seiners, he released the remaining ten orcas. But the freed whales showed no sign of leaving Vaughn Bay. What if they couldn't find their way past the sandbar? Or, worse, what if locals started shooting at them? After calling Ross and Babich back to the harbor, Griffin searched for a solution. Hoping the young whale's calls might lure the pod out of the bay, he instructed the crew of the *Chinook* to lower the animal back into the water. The result was electric. "All ten wild whales rush[ed] toward the bawling calf," Griffin recalled, "acting like relatives who [had] located a lost child on a Sunday outing." Suddenly, a female orca dashed between the divers and the youngster. "It's the mother! Get outa there!" shouted a crewman on the seiner, as the frightened divers pulled themselves into a skiff. "Would she have killed them?" Griffin found himself wondering.[17]

The ploy worked. Using the little whale as a lure, the team drew the pod past the sandbar and out of the harbor. But then the vessels faced a nasty squall. Heavy wind and rain pounded the seiners as the freed whales watched and listened. The *Chinook* pitched sharply, and the crew couldn't hoist the young male orca out of the water. Still worse, the rocking made it difficult to keep his blowhole clear. Griffin watched anxiously as the little whale arched his back, "trying to lift his head above the waves." Reacting quickly, Goldsberry secured a line to the *Pacific Maid*, keeping the *Chinook* off the beach and enabling its crew to raise the orca onto the deck. But relief for the capture team meant disappointment for the lingering whales. If the youngster's return to the water had raised hopes of his release, his return to the vessel ended them.[18]

No event better revealed Ted Griffin in all his complexity. Days earlier, he had arrived in Vaughn Bay seeking killer whales for sale. Despite the ravenous market, he had kept only two of the twelve whales and stayed to ensure that they left the bay safely. By any measure, it was a gutsy decision. In the face of a pounding storm, he had risked vessels, crew, and profit to help the pod. Without question, the fate of the region's orcas still moved him, even as he remained determined to act like a businessman. But his commitment to catching and selling killer whales concealed internal misgivings. Having watched the species closely for seven years, he was painfully aware of the social ties he was breaking. At that moment outside Vaughn Bay, they were impossible to ignore. As the vessels started for Seattle, the freed whales followed in a sad vigil for their lost podmates. They uttered "shrill cries and haunting shrieks," Griffin later wrote, "echoed by their companions aboard ship."[19]

The two whales he kept made history, becoming the first captive orcas to be displayed on the East Coast. After five weeks at Pier 56, the large female, dubbed Lupa, boarded a United Airlines DC-8 jet freighter for New York. At eighteen feet and nearly six thousand pounds, she was the largest animal ever to ride in an airplane, and she caused a stir when she touched down. Motorists watched in amazement as police cars escorted her down the Long Island Expressway to her pool at the New York Aquarium in Coney Island, where she lived for less than a year.[20] For his part, the little male whose calls had drawn his pod out of Vaughn Bay left Seattle in May 1968 and flew to the Miami Seaquarium, where he met Richard O'Feldman and became the beloved Hugo.[21]

⁂

Meanwhile, the orca market was beginning to shift. On the day Griffin's team departed Vaughn Bay, reports appeared of the first capture in Pender Harbour. At first, it seemed inconsequential. Sonny Reid and his friends had caught a single whale and sold it to the Vancouver Aquarium. But their capture of seven more in April 1968 was impossible to ignore. Goldsberry flew to Garden Bay for a look and briefly considered purchasing all of the animals. But Marineland of the Pacific and Marine World were eager to replicate Sea World's success, and they had quickly bought the whales. There was no telling where the orca captures north of the border would end. In July, a Vancouver fisherman named Aaro Palo accidentally netted eleven killer whales near Sointula. Unlike the Pender Harbour men, he released all but one, but the fate of that single whale spoke volumes. Palo sold the young male to Jerry Mitchell, a former Sea World employee and self-described "whale broker," who resold the animal to the Harderwyk Dolphinarium in the Netherlands. Dubbed Tula, he became the first captive killer whale in Europe. Although he didn't live long, his purchase underscored the growing market for orcas. And that wasn't necessarily good news for Griffin. Competition was rising, and despite his success at Vaughn Bay, he had not solved his tracking problem.[22] Luckily for him, the navy came to the rescue.

⁂

Following World War II, the Office for Naval Research had poured immense amounts of funding into oceanographic research, much of it aimed at monitoring the undersea world. This took on new urgency with the advent of Soviet nuclear submarines in the late 1950s. At the heart of this effort was an array of listening stations throughout the Atlantic and Pacific known as the Sound Surveillance System (SOSUS).[23] At the same time, naval researchers came to view marine mammals as potential assets. They not only studied the biosonar of dolphins to improve acoustic technology but also sought to utilize them in a variety of military tasks. With close ties to marine parks, the navy's Marine Mammal Program utilized species that had already proved to be successful display animals. Initially, it focused on sea lions and bottlenose dolphins, but as pilot and killer whales came into captivity, the navy considered these species as well.[24] In 1966, program officials had hosted Griffin at Point Mugu and expressed interest in orcas. "I was interviewed by military types, mostly ASW [anti-submarine warfare] people," he later told me. "They wanted to know if a whale would be able to assist them in their activities."[25]

Griffin subsequently developed informal ties with naval stations in Puget Sound. He gave military researchers access to captive killer whales, and the navy alerted him to the arrival of pods in the area, which it detected through its listening array. In the spring of 1968, navy officials took this assistance a step further, sending Griffin a portable hydrophone with electronic listening equipment. Although the aquarium owner later recalled no specific quid pro quo arrangement, it seems reasonable to connect the gift to the navy's interest in acquiring orcas.[26] Libertarian though he was, Griffin again accepted the US government's assistance in tracking killer whales, and he had the device with him that autumn when he made another catch.

❧

It was early October 1968, and Griffin had been following a gathering of orcas for a full day as they moved south through Puget Sound. The navy hydrophone worked like a dream, offering new insights into the animals he had long pursued. Cruising in his new twenty-six-foot runabout *Namu*, Griffin picked up constant chatter as the whales frolicked and occasionally approached the boat. As the pod rounded Restoration Point and headed west toward Bremerton, however, the *Chinook* was waiting. At first, it didn't seem a good spot for a capture. Goldsberry warned that the currents of Rich Passage were too strong. But Adam Ross was now a seasoned orca catcher, and he managed to corral the whales off Manchester. The team secured the nets to a nearby dock, which housed an experimental salmon farm. It was Namu Inc.'s largest catch to date, some twenty-five whales in all.[27]

Yet it almost ended before it began. When the *Chinook* launched its skiff, webbing had snagged on the seiner's fantail, shearing off a portion of the capture net. As Griffin cruised the open-water side of the set, he realized the cork line had no net attached to it—leaving a gap the length of two football fields. None of the whales had found it yet, but it wouldn't take long.[28] It was for such occasions that Griffin kept Gary Keffler and his divers on call. Decades later, Griffin would declare himself lucky to have had Keffler working with him. "Because of Gary's foresight and vigilance," he observed, "we never lost a diver nor to my knowledge was anyone injured."[29] But on this occasion, the master diver came close.

❧

Keffler was not one to scare easily. He had faced many underwater challenges, and this one seemed simple. After locating the web in a hundred feet of water, he and a partner began the job of mending it. But as Keffler wrapped up the task, the currents picked up, opening a vertical tear in the net. As the sections split away, he grabbed them, struggling to hold the net together. At that moment, a large male orca approached, and he didn't seem in a pleasant mood. Keffler had been in the water with these animals countless times, but at that moment he couldn't help but reflect on the name of the species: *killer* whale. "He could have gone right through me," Keffler told me. For a tense moment, the massive predator studied the diver. "He looked at me, and then he turned his head to the other eye and looked again. And then he went *whomp! Whomp! Whomp!* He chomped three times at me!" Keffler recounted, clapping his hands together like snapping jaws. "And I thought 'Oh, shit!,' but he was just testing me."[30]

From the orcas' perspective, the capture may have come at an awkward time. Although Griffin and Goldsberry didn't realize it, they had trapped portions of two pods, J and L, while they were in the process of socializing. Indeed, when two whales escaped through the net, the capture team watched in stunned silence as the animals copulated nearby. "Perhaps the trauma of capture is not as stressful for the whales as I once believed," Griffin noted hopefully.[31]

By this time, the operation had caught the US government's attention. Griffin's team had captured the whales just off the Manchester Naval Fuel Depot, and officials ordered the vessels to vacate the area. But higher-ups intervened. After all, the navy hoped to acquire some of these orcas. As Sam Ridgway, the Marine Mammal Program's chief veterinarian, later told me, "I was at a whale conference at the Stanford Research Institution when Tom Poulter came into the room and announced that Ted Griffin had caught twenty or so whales and some were available for research." Ridgway and his colleagues hurried to Puget Sound to take a look at the animals. Appreciative of the navy's assistance, and eager to support orca research, Griffin sold two whales to the program at cost for just $6,000 apiece.[32]

After selecting three more animals to fill orders, Griffin released the rest. Standing next to him as the freed orcas swam away was an incredulous Jerry Mitchell. The San Diego whale broker had offered to buy the remaining whales, and Griffin's refusal baffled him. "I could have sold them all. You're crazy!" he shouted. "You've tossed away a quarter of a million dollars."

"Just how many do you think there are anyway?" Griffin responded.

"Hundreds, thousands, I don't know, more than enough," Mitchell replied. "Those Canadian fishermen, who caught a pod in Pender Harbour, kept every last whale."

FIGURE 10.2 Orcas captured off Manchester, Washington, October 1968. Courtesy of Sam Ridgway.

"Yes, I've heard about them," Griffin replied, "but I don't conduct my business that way."[33]

In retrospect, Griffin had exercised noteworthy restraint. Demand for captive killer whales was becoming ravenous, and no law restricted their capture or sale. Yet of the nearly forty orcas he had captured in 1968, he released all but seven. Why? At the time, no research existed on the demographics of Pacific Northwest killer whales, nor did scientists know much about their migrations or social organization. Many observers estimated their numbers in the thousands, and research at the Marine Mammal Biological Laboratory was still limited to killing and dissection. But Griffin had begun to worry about the orca population, particularly with the catches in British Columbia. "There will always be competition with any business, and Don and I continued to have more orders for whales than we could or were willing to supply," he later explained. "My concern was the size of the whale population, the birth rate and what a sustained yield might look like."[34]

Although no scientist had yet determined these figures for orcas, Griffin's thinking represented a significant departure from the usual approach to animal capture. In earlier years, even collectors such as Frank Buck, famous for his "bring 'em back alive" expeditions to Asia in the 1930s and 1940s, engaged in an astonishing amount of carnage. In order to obtain young tigers, hippos, or elephants, wildlife collectors routinely killed the animals' mothers.[35] Marine parks, too, preferred young animals, which were less aggressive and required less food and space than older ones—hence the desire for smaller killer whales. But Griffin and Goldsberry had sought the means to obtain young orcas without injuring or killing others—a particular challenge when dealing with large marine mammals. By these parameters, they had succeeded. In their two captures of 1968, they had trapped large groups of orcas, isolated the desirable animals, and released the rest—all without the death of a single whale.

But market pressures continued to mount. In October, Griffin delivered Cuddles to the Flamingo Park Zoo in Yorkshire, and afterward he and Joan undertook a whirlwind tour of oceanariums in Europe. Eager to obtain killer whales, curators in France, Germany, and the Netherlands pressed Griffin with orders. Meanwhile, Goldsberry was discovering the same demand in the US Southwest. In the process of delivering a juvenile orca to the Texas Sea-Arama in Galveston, he received inquiries from other owners in the region. "The way I see it," he told Griffin, "we'll have orders for maybe 20 animals in the next year or so."

"That sounds great," Griffin replied, "but we've agreed to average no more than five or six whales a year." It only seemed prudent with the catches in British Columbia and growing calls for regulation in Washington State.[36]

~

The times were changing faster than the partners knew. In January 1969, the explosion of an oil rig near Santa Barbara gave a jolt to the environmental movement.[37] The following month, US senator Henry "Scoop" Jackson of Washington sponsored the National Environmental Policy Act. In signing the bill, President Richard Nixon declared that his administration was "determined that the decade of the seventies will be known as the time when this country regained a productive and enjoyable harmony between man and his environment." Soon after, Nixon signed the Endangered Species Act of 1969. Conservation advocates welcomed the legislation but noted that it did not require federal action until a species was on the brink of extinction, and many called for greater protection of commercially exploited wildlife. In the process, public discussion became more critical of enterprises that profited from animals.[38] These political currents reached Puget Sound, where many people had already come to question the motives behind orca capture.

By early 1969, a growing number of voices were pressing state officials to intervene. At first, Griffin and Goldsberry were receptive, assuming the task would fall to the Washington State Department of Fisheries. In addition to being familiar with Puget Sound, it was accustomed to regulating the commercial use of marine resources. But the Department of Fisheries refused, suggesting instead that the Game Department take up the task. The partners were appalled. The Game Department focused on recreational killing—hunting and fishing—not live capture, and it rarely working with commercial enterprises. Their anxiety only grew during legislative committee hearings in March 1969, when the head of the Game Department, Burton L. Lauckhart declared that his agency would prohibit anyone from capturing marine mammals for profit. Following testimony by Griffin and Goldsberry, however, the committee decided to table the bill. For the moment, at least, the partners had staved off state regulation.[39]

Despite this scare, the future of killer whale capture appeared bright. In addition to demand throughout North America and Europe, Namu Inc. had received its first order from Asia. Japanese entrepreneurs were building a new oceanarium in Kamogawa, near Tokyo, and they wanted an orca for their grand opening. Confident of future growth, Griffin and Goldsberry invested heavily in their operation, building a $30,000 dry dock to facilitate the transfer of killer whales and planning construction of their own high-speed capture boat. They also agreed to a contract with the Marine Mammal Biological Laboratory, which planned to mark whales caught and released by Namu Inc. As the *Seattle Times* explained, the lab's scientists would use the company's dry dock "for tagging and the attachment of sonic devices" and then monitor the freed whales "for research purposes."[40]

But the following months brought frustration for the whale catchers. In April, they chased a pod of orcas into Carr Inlet, only to find the waterway crowded with fishing boats. While most were catching herring, Peter Babich's *Pacific Maid* was trying to net the whales. Goldsberry was furious, urging Adam Ross to foil Babich's efforts. For Griffin, the sight of the two Gig Harbor fishermen jockeying to set on the same animals was surreal, but he understood Babich's motivation. For years, Griffin and Goldsberry had won most of the money and glory from captures while paying the fishermen fees for their help. In contrast, Sonny Reid and his partners in Pender Harbour had caught the whales and kept the profit. "Though I don't look forward to competing with Pete," Griffin observed at the time, "I recognize he has the same right to pursue them as I do."[41] In the end, both sides were disappointed. Babich failed to net any whales, and Namu Inc. kept only one of the eleven it caught.

The year's second catch proved heartbreaking. In October 1969, Namu Inc. trapped seven whales in Whidbey Island's Penn Cove and attempted to use its new dry dock to transfer them. Rather than soothing the animals, the confined space made them skittish. As the team attempted to isolate one female, she panicked, getting tangled in the nets. Griffin was frantic, demanding that the divers free the animal before she drowned, but Keffler concluded it was too dangerous. A fierce argument ensued between Goldsberry and Griffin. "Don always said the divers' lives were more important than anything," explained Keffler. "Ted never said that, ever." When push came to shove, the diver noted, "he was worried about the whales."[42]

Griffin lost the argument and was forced to wait as the orca drowned. It had been nearly three years since his team had lost a whale, and the experience was gut-wrenching. After releasing the other six orcas, Griffin offered the carcass to the Marine Mammal Biological Laboratory. Over the past seven years, its scientists had slain at least ten orcas, and it now held a contract with Namu Inc. for marking captured whales. But with opposition to capture rising, the lab's director refused to accept the body, declaring, "We don't want to be your dumping grounds." Bowing to the inevitable, Griffin towed the body east to the Everett Rendering Works. "Traveling in the darkness," he later wrote, "I feel the weight of the 7,000 pound whale as though chained to my neck."[43]

As 1969 drew to a close, Griffin and Goldsberry took stock of their enterprise. Despite utilizing expensive new gear, they had struggled to fill outstanding orders. Facing mounting debt, they decided to sell the Seattle Marine Aquarium's last remaining orca—Kandu—to Sea World.[44] It seemed a sensible move: the adolescent female was growing too large for the aquarium's pool. But the decision underscored the year's disappointment—particularly in light of the third capture in Pender Harbour that December. It also revealed how much the aquarium's purpose had changed. Conceived by Griffin as the means to obtain and befriend an orca, it was becoming a capture-and-holding facility for larger, better-capitalized marine parks. For his part, the boy who had yearned to swim with killer whales was now obsessed with catching and selling them.

11

The White Whale

WHEN BOB WRIGHT awoke on Sunday March 1, 1970, he didn't feel like getting in a boat. He had attended a wedding reception late into the previous night, and the morning in Victoria had broken cold and blustery. But he had promised to show his whale-catching operation to Don White, Paul Spong's former research assistant. Wright already had an orca at his new oceanarium, Sealand of the Pacific, but he was keen to try his hand at capture, and he especially hoped to trap an albino killer whale often seen in local waters. When White and a friend arrived for the excursion, however, Wright wasn't feeling very eager. "Bob is totally hung over, but he is feeling responsible," White recalled. "He has told me to come, so he feels like we've got to do it." Along with trainer Graeme Ellis, the three men piled onto Wright's twenty-foot Bertram runabout and started for Pedder Bay. As the boat rounded Trial Island and cruised west past Victoria, the sea became choppy and Wright grew queasier. But minutes later, as they approached Race Rocks, he forgot all about his hangover.

"Fuck!" he yelled. "It's the white whale!"[1]

Sure enough, a group of orcas with what appeared to be an albino member was passing Bentinck Island and heading straight for Pedder Bay. The sighting was lucky, but the timing awful. Wright wasn't set for a capture that day. His seine nets were in storage, and at first he couldn't hail any of his Sealand staff. Determined not to let this opportunity pass, he gunned the Bertram into the bay and made straight for the *Lakewood*—a charter fishing boat he had rigged for orca catching. As Wright gathered his crew on the vessel, the excitement was palpable. "We were playing macho whale hunters," White reflected, "and Bob Wright was our Captain Ahab."[2] With only one light net on board, the operation would have to be perfect, and everyone watched anxiously as the whales lingered near the mouth of Pedder Bay. Finally, as the sun began to set, the orcas entered. Once they swam past the *Lakewood*'s hiding place, the vessel motored across and netted off the bay.

The trap was sprung, but there was no guarantee it would hold. During an earlier attempt near Sealand, a group of orcas had easily torn through the flimsy net—"went through it like it was a spider web," Wright later recalled.[3] Desperate to prevent another escape, he stationed aluminum boats along the cork line, instructing crewmen to strike their hulls with paddles and drop small explosives known as "seal bombs" if the animals approached. Chilled divers checked and rechecked the nets, while Wright buzzed the *Lakewood* back and forth. Working through the night, the sixteen-man team built a wall of sound. Yet in the darkness, no one could be sure how many animals remained. Finally, around seven o'clock in the morning, the sun rose to reveal five orcas behind the net, among them the white whale. Soon after, two purse seine vessels arrived with heavy nets to secure the catch. Only then did the crew stop to reflect. Some felt remorse as they watched the trapped animals. "They were swimming back and forth making this distress call you could hear through the boat," Ellis recalled. "It just didn't feel right at all."[4] But Wright was elated. An upstart newspaperman who had stumbled into the oceanarium business, he longed to put Sealand on the map and to compete with Ted Griffin in the live orca trade. And he had just captured the most valuable display animal in history.

❧

Mysterious creatures were nothing new to the Pacific Northwest. Indigenous peoples had long spoken of Sasquatch stalking the region's forests, and fishermen and visitors alike reported sightings of Cadborosaurus—a sort of Loch Ness Monster of the Salish Sea. In 1937, the Naden Harbor whaling station on Haida Gwaii even found an mysterious skeleton in the belly of a sperm whale, which many claimed belonged to "Caddy." Haida Gwaii itself was known for its rare white ravens, which may have inspired the native story of Raven—the clever white bird who brought light to the world and in the process charred all ravens black.

Most famous of all were the region's rare, cream-colored black bears (now called "spirit bears"). Although indigenous peoples had long known of their existence, the subspecies came to widespread attention only in 1924, when Canadian officials seized a female cub from a smuggler at the US border and passed it to Francis Kermode, director of the Royal BC Museum in Victoria. For the next twenty-four years, this Kermode bear lived in a cage at the city's beloved Beacon Hill Park. Although SPCA members occasionally called for the animal's release, locals took pride in their unique attraction and mourned the bear's death in late 1948.[5] By that time, however, another ghostly vision had captured the public imagination.

A white blackfish. Rumors had circulated for years, but its existence was confirmed in August 1946, when boaters near Race Rocks observed what looked like an albino killer whale. Kermode's successor at the museum, Clifford Carl, viewed the animal as a unique opportunity. Up to that point, scientists considered killer whales' black-and-white forms virtually indistinguishable. The bodies of adult males and females differed, to be sure, but individuals defied identification, making it impossible to chart the whales' social ties and migration patterns. Did they organize themselves around a dominant bull? Did they form temporary groups or stay together for life? Did they have a defined range or wander the ocean? No one knew. Carl realized a white killer whale would be easy to spot and possibly track, if he had help. He set to work building the world's first orca-spotting network, circulating requests to lightkeepers, fishermen, Coast Guard vessels, and Department of Fisheries officials to report sightings of the white whale, and he issued pleas not to shoot the rare animal, whom he named Alice the Albino.[6]

In the following years, Alice and her companions were seen frequently in the waters around Vancouver Island. They were almost certainly mammal-eating Bigg's killer whales. In July 1947, the keeper at the Scarlett Point lighthouse near Port Hardy spotted the white orca attacking a small fin whale.[7] The following month, Alice and several other orcas visited Witty's Lagoon near Victoria, likely scouting for harbor seals. Although their appearance thrilled Sunday picnickers, the animals frightened the occupants of a nearby sailboat. Three years later, in October 1950, a Department of Fisheries vessel spotted Alice off Trial Island. The excited Carl loaded cameras and film into a skiff and raced to the scene, but the whales had gone. Still, the skipper of the boat had gotten a good look. He reported that Alice was starkly white and large—perhaps nineteen or twenty feet—and she appeared to be traveling with a calf of her own, a second white whale.[8]

Although few had seen her, Alice subtly affected local views of orcas, at least around Victoria. To be sure, many fishermen continued to view blackfish as competitors and threats. But the white whale's unique body, coupled with news stories and Carl's public statements, cast her in a different light. In 1955, when reports appeared that she was approaching after a long absence, newspapers in Victoria hailed the event. As *Victoria Colonist* reporter Humphry Davy observed, Alice was "the Pacific Ocean's most famous whale" and "probably the only whale in the world known as an individual." Three years later, when the white whale and her podmates frolicked in a kelp bed near the city's

Clover Point, dozens of spectators gathered to watch. Killer whales may have been wolves of the seas, but this one had established herself, in Davy's words, as "fun-loving Alice, the Pacific Coast's beloved albino whale."[9]

Clifford Carl had mixed feelings about this public affection for Alice. On one hand, it led to more reports of her movements and behavior. On the other hand, he worried that sightseers might come too close to orcas, which he still considered dangerous predators. In late February 1960, for example, he noted a report from the tugboat *P. F. Stone*, whose crew had tried to save a large sea lion from Alice and several other whales. After realizing the animal was already dead, the men lowered it overboard, only to be stunned by the ensuing violence. As soon as the carcass touched the water, reported the *Nanaimo Free Press*, the whales tore it to pieces "with a terrific lashing of tails."[10] Regardless of Alice's popularity, the awesome killing power of her species was undeniable, and with the rise of scuba diving in the early 1960s, many wondered if blackfish might attack divers, either by mistake or on purpose. Carl certainly thought it was possible. "There are no authenticated cases of killer whales eating humans," he declared in July 1964, "but there are numerous reports of attempted attacks, and some mighty suspicious circumstances."[11]

By this time, public attention had turned to Moby Doll, with newspapers asking the museum director to explain the captive whale's docile behavior.

FIGURE 11.1 Spectators watch white orca off Clover Point in Victoria, British Columbia, January 1958. Photo by Tim Sinclair. Courtesy of Royal BC Museum and Archives.

Yet even as the harpooned orca garnered headlines, Carl continued to receive reports of a white whale—off Salt Spring Island, Ten-Mile Point, Race Rocks.[12] Alice, or another whale who looked like her, was busy. So was Bob Wright.

Wright hadn't planned to get into the killer whale business. Born in Regina, Saskatchewan, in 1930 and raised in Edmonton, Alberta, he grew up far from the sea and went to work young. At sixteen, he took a job at the *Edmonton Bulletin* and two years later he moved to Victoria, where for the next fourteen years he worked for the city's two newspapers—first the *Victoria Times*, then the *Victoria Colonist*. Yet his real passion was the outdoors, particularly fishing. As a young man, he won several fishing derbies, and by 1960 he was recognized as one of the best anglers in the province.[13] But Wright realized that the region was changing, and he saw its potential as a tourist destination, particularly for sport fishing.

In 1962, just as Ted Griffin was building his aquarium in Seattle, Wright made an equally bold move. With just $600 to his name, he secured a thirty-year lease from the nearby municipality of Oak Bay to build a marina complex. The site was home to a boathouse that had once belonged to the famed Mount Baker Hotel—long since burned down. Over the decades, the boathouse had become a gathering place for Oak Bay's fishermen, artists, and unattended children.[14] By the early 1960s, however, it was badly decayed, and the city council leapt at Wright's proposal to build a new facility. Taking advantage of a federally funded breakwater and municipal subsidies—together totaling some $500,000—Wright opened his new marina and restaurant in April 1964. It was "the finest marine facility in western Canada," he boasted, "and possibly the finest in all the Pacific Northwest."[15]

Yet Wright needed an attraction to draw visitors. Victoria had yet to become a major tourist destination, and the Oak Bay marina lay more than three miles from the provincial capital's inner harbor. To solve this problem, he partnered with local entrepreneur Charlie White, who opened the Pacific Undersea Gardens, an innovative floating aquarium adjacent to the marina. As visitors rolled in, White's ambitions soared, and just three months after opening, he made his public bid to buy Moby Doll from the Vancouver Aquarium.[16] "It would be a tremendous attraction for Victoria," he declared. "Just think of it—the only killer whale in captivity and performing at that."[17] Wright supported the effort, and both men were disappointed when Murray Newman rebuffed the offer. The following year, Wright and White could only watch the publicity stirred by Namu in Seattle and wonder what might have been.

Nevertheless, their arrangement proved lucrative. With the Pacific Undersea Gardens drawing steady visitors, Wright's restaurant and marina flourished, and he expanded into fishing charters. In 1965, he acquired the *Lakewood*—a former navy torpedo recovery vessel—and began offering fishing excursions in the area. Business grew, and two years later he opened a second marina in Pedder Bay, thirteen miles to the southwest. Lobbying the Department of Fisheries to ban commercial fishing from the area, Wright was already eying a charter fishing empire. But in 1968, he learned that Charlie White was secretly planning to move the Undersea Gardens to Victoria's inner harbor. Worried his marina would suffer without a tourist attraction, Wright decided to build, as he later put it, "a better mousetrap." He began construction immediately, and by the time White towed the Undersea Gardens away, Wright was ready to replace it with another floating oceanarium: Sealand of the Pacific.[18] At first, White wasn't worried. After all, the Undersea Gardens was an established attraction moving to a choice location. But Sealand had something he didn't: a killer whale.

Wright had been calling Ted Griffin for weeks by the time Namu Inc. made its large catch off Manchester in October 1968. Of the five whales Griffin and Goldsberry chose to keep, all but one was spoken for. He was a young, healthy male about five years old and thirteen feet long—a bit smaller than Moby Doll. Wright immediately flew to Seattle for a look. Although Sealand wasn't yet completed, he made a deposit on the whale, at the time dubbed Junior. Taking careful notes, he queried Griffin and Goldsberry about killer whales—feeding, care, life expectancy, pool size. "Bob was a very likeable person and was very open about his passion to have a whale to display," Griffin observed. But Griffin also noted that his Victoria counterpart was eager to catch orcas himself. Although Wright had not yet made an attempt, Griffin predicted that "his ego will not be satisfied until he captures a whale."[19]

For the time being, however, Wright had other matters on his mind. With Sealand nearing completion and a newly purchased orca, he needed a trainer. Griffin gave him advice on that score as well. Either a young woman or man could do the job, he told Wright, but a good whale trainer should be intelligent and patient and above all should have a deep love of animals and their well-being.[20] As luck would have it, just such a person came knocking on Wright's door.

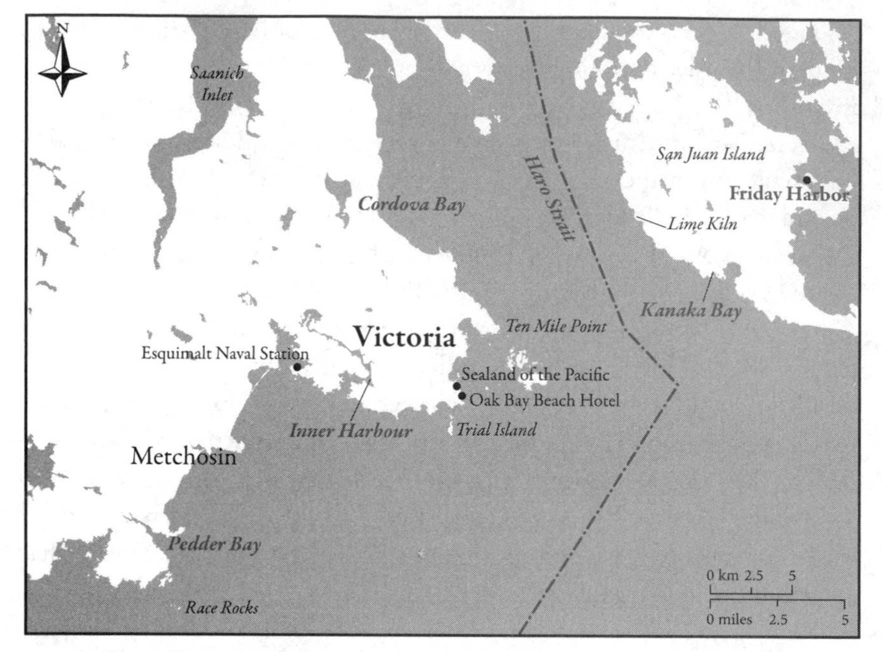

FIGURE 11.2 Victoria in 1970.

By early 1969, Graeme Ellis had devoted nearly a year of his life to captive killer whales. Though just nineteen, his time in Pender Harbour had made him one of the world's few experienced orca trainers. He had worked especially closely with Hyak, the first whale captured there in February 1968, and he had been visiting family in Vancouver a year later when someone cut the whale loose. "I was really bummed," he later admitted. "Not because I lost my job, which was $279 a month, but I just had a real attachment to that animal. It is like someone decided to take your dog away from you." Still keen to work with killer whales, Ellis learned that Wright had just acquired one. He immediately visited Sealand to offer his services. "They hired me on the spot," Ellis laughed. "I guess they didn't have anyone."[21]

When Junior arrived from Seattle in April 1969, however, Ellis was nervous. "I really hadn't trained one from scratch," he told me. But the young male, now renamed Haida, made his job easy. "He had been kept with a trained animal. They usually laid down the fundamentals in Seattle." Within a week, trainer and whale began offering shows, and the crowds came to Oak Bay. Rejecting construction of an expensive pool, Wright decided to keep his prize in an open-water pen. Overall, Haida seemed happy and healthy, and his performances drew visitors by the thousands. At the end of the tourist season, Wright decided the young whale needed a mate, and when Sonny Reid and his partners made their third catch in December 1969, Wright sent Ellis to take a look.[22]

Soon after, Wright agreed to pay $20,000 for a young female. But as the fishermen prepared to separate the animal from her podmates, he abruptly declared that he would pay only $5,000 if she measured more than seventeen feet. Annoyed at the aggressive entrepreneur, Reid and his partners decided to sell the whale—the future Calypso—to European buyers, thereby infuriating Wright. "You should have heard him on the phone," recalled Reid. "He was screaming 'I'll break you, I'll bankrupt you guys.' "[23] In public, however, Wright told a different tale. "They sold it right from under us to a British outfit," he claimed. "We got tired of being treated as if we were from the sticks and decided to do something about it."[24] If Ted Griffin and Sonny Reid could catch orcas, Wright reasoned, so could he. And with luck, he might even net the white whale.

❧

Wright knew Alice's story well. He had moved to Victoria shortly after her first appearance, and his time as a newsman convinced him of the animal's public appeal. He had also come to know Clifford Carl, who shared his killer whale research with the young businessman. Wright hoped historical sightings of Alice might offer clues to her present whereabouts, and he soon became obsessed with the white whale. "Ever since I started working for Wright, he'd rant about catching an albino orca," recalled Ellis.[25] Haida was a marvelous draw for locals, to be sure, but Wright knew that Sealand lacked name recognition. If people elsewhere wanted to see killer whales, they were more likely to go to San Diego, Vancouver, or Seattle—not Victoria. If he could catch the white whale, however, he would have a unique attraction.

Wright's approach to capture was informal. After receiving a permit from the Department of Fisheries, he put Ellis to work building a net made of light gill-net mesh. Because it wasn't sport fishing season, Wright outfitted the *Lakewood* as a capture boat and hired local skipper Tor Miller to run it. Over the following weeks, the crew scoured nearby waters for killer whales. Most forays proved fruitless, and when the team did spot orcas, it had to contend with maneuvers that the same whales had developed to elude Ted Griffin over the years. "They were very clever," Miller recalled. "You'd see one making a fuss, and you'd think 'oh, gosh, the pod's gone there.' And there'd be just one, and the rest of them would turn around and go back the other way, and they'd be a mile behind you when they'd surface again." In February 1970, the team finally managed to corral about fifteen whales off Ten Mile

Point, just north of Sealand. But Wright was overanxious. As heavy currents tugged at the net, he revved the engines of his small skiff, trying to turn the whales away. Instead, the cornered pod charged through the net as the crew of the *Lakewood* watched in stunned silence. "They went like homesick angels," quipped Miller, "and they took 500 fathoms of lead line and webbing down the pass with 'em." More than forty years later, Ellis still found humor in the memory of Wright's shredded net. "They just ripped the shit out of it," he chuckled. "So then Bob wised up and went to Pedder Bay."[26]

In hindsight, it was an ideal location. A narrow, shallow inlet adjacent to Race Rocks, Pedder Bay was a favorite local fishing spot even before Wright built his marina. The bay's nooks and kelp beds drew salmon and seals, which in turn attracted orcas. Equally important, it was secluded. Victoria was a forty-minute drive away, and Wright controlled boat access through his marina. Moreover, there were few houses in the area. Bounded by the William Head federal penitentiary on one side and an army ammunition depot on the other, Pedder Bay lay outside the public eye. It was this capture site that Don White had come to see the day they caught the white whale.

I met White in the autumn of 2013. A youthful sixty-nine-year-old now working on a doctorate in psychology, he seemed far removed from his days studying orcas. But those events had left a deep mark on him. For more than an hour, he reflected on how his time with Skana had forced him to rethink his understanding of morality—toward animals and people alike. He detailed his research at the Vancouver Aquarium and the events that led to Paul Spong's firing. And he told me how he came to be at Pedder Bay that fateful Sunday in March 1970.

In the months after Spong left, White had completed their research project, working closely with the calf Spong had named Tung-Jen.[27] But soon after, Murray Newman suspended orca research at the aquarium. Like Ellis, White had become a whale man without whales. With the orca trade booming, however, there was no shortage of opportunity. "You have to realize, at this point, that I was probably one of the authorities on killer whales," he laughed. "Nobody knew anything about it, so you could be a world authority on it and know nothing."[28] Among his suitors was the US Navy, which had recently bought two orcas from Griffin. Instead, White approached Bob Wright, who gave him permission to study Haida. In the course of his visits to Sealand, White often found himself locked in conversations about the white whale. Wright showed him copies of Clifford Carl's research and talked incessantly about Alice. "This guy was obsessed with the white whale," White recalled. "He had a stack of offprints that was about a foot high about every sighting of this albino whale. It was like I was part of *Moby Dick* and here's this Ahab guy."[29] At the time, White shrugged it off, but on the morning of March 2, 1970, he found himself standing on the deck of the *Lakewood*, staring at a white orca and its pod behind nets in Pedder Bay.

Sealand's owner could only marvel at his luck. Bob Wright, champion sport fisherman, had just caught the ultimate derby prize. Local newspapers had reported on Alice for years, and Carl had spent a quarter-century on her trail. But it was a kid from Saskatchewan who netted the white whale. In Wright's mind, no one had taken him or Sealand seriously. Yet he had done something that even Puget Sound's famous whale catcher hadn't. "Half of Bob's excitement was that he could now show Ted Griffin up," recounted White. Almost immediately, Wright was on the phone with the Seattle Marine Aquarium. "He invited him up for exactly that reason," White explained. " 'You can catch all the black ones, Ted, but do you have any white ones?' He was almost as obsessed with Ted Griffin as he was with the white whale."[30]

Wright's comments to reporters were equally triumphant. "It is impossible to put a price on the white whale," he declared. "There is no yardstick to go by, but it has to be worth more than a quarter of a million dollars."[31] To be sure, the white orca was rare, and it was valuable, but it wasn't Alice. As Carl confirmed when he got a close look, this white whale was only eleven or twelve feet long—far too young to be Alice or the calf spotted with her eighteen years earlier.

Alice or not, protests of the capture began almost immediately. Some worried that outside buyers would scoop up the striking animal. Bruce Rogers, the lighthouse keeper at Albert Head, called on the federal government to take possession of the whale. If it remained private property, he warned, there was no guarantee the white orca would stay in Canada. "What's to prevent Wright from selling to the Americans?" he asked. "Money means nothing to them." Other critics denounced captivity more broadly. The Victoria Natural History Society demanded that fisheries minister Jack Davis divulge how many whale-hunting permits his office had issued, and local Victoria resident Dinah Kowalezk was even more outspoken. Killer whales were accustomed to "the freedom of the oceans," she asserted. "It's criminal to confine them in small areas like the pool at Sealand."[32]

Others dismissed such criticism. Carl, for one, declared live capture essential to research, and he lauded the care orcas received in oceanariums such as Sealand. "Whales in captivity are pampered and much better fed than other whales," he asserted. "Life in the wild for these animals isn't so simple."[33] University of Victoria chancellor Roderick Haig-Brown agreed. A well-known sport fishing writer, Haig-Brown was hosting an environmental teach-in, during which he urged citizens to become stewards of local waters. But he viewed orca capture and display as a part of this process. It was not long ago, he reminded listeners, that the federal government had considered shooting killer whales. In fact, Haig-Brown himself had attended the 1960 Department of Fisheries hearing in Campbell River that led to the mounting of the anti-orca machine gun above Seymour Narrows—a project he supported at the time.[34] "Bringing the killer whale into civilization has cured this nonsense," he observed. "A few killer whales on display are good for the killer whales. One or two are sacrificed for the good of them all."[35] Yet the most striking endorsement came from Ed Underwood, the Tsawout chief who had called for Namu's release five years earlier. Interviewed by Humphry Davy, the reporter who had written often on Alice over the years, Underwood declared that the rare orca was now Bob Wright's "whale spirit." "Good fortune will not only knock once, but again and again at his door," Underwood predicted, provided Sealand's owner took good care of the white whale.[36]

❧

The Sealand staff was doing its best. As he had at Pender Harbour, Ellis patiently offered herring to the captive whales. But as the days passed, none of the animals seemed interested. Although one reason was undoubtedly the trauma of capture, the principal factor was food culture. Wright and his crew had no idea they had caught a group of Bigg's killer whales. Just as the salmon-eating Moby Doll showed no desire for whale meat, these whales had no experience with, and hence no interest in, eating fish. But the Sealand staff doggedly tossed herring into the pens, hoping to entice them. In the process, Ellis and his colleagues realized that this was an odd bunch of orcas. In addition to shunning fish, nearly all of them had deformities. An older female, probably the white whale's mother, had a crushed lower jaw, earning her the nickname "Scarred-Jaw Cow." Another, named Florencia, sported a pointed rostrum, and a large male, likely Florencia's son, had an abnormally long lower jaw, resulting in the sobriquet "Charlie Chin." The Sealand team puzzled over the animals' appearance. Noting that nearby Bentinck Island had once been a leprosy asylum, veterinarian Alan Hoey declared the group "a leper colony of killer whales."[37]

More worrisome to Wright was the behavior of the white whale, soon to be named Chimo. Although curious and inquisitive, the youngster struggled to navigate her surroundings, particularly when separated from the other whales. Her swimming was erratic, and unlike the other four, she had difficulty avoiding the nets, becoming badly entangled on one occasion. To free her, Wright turned to diver Bruce Bott, an American expatriate who had gone AWOL from the US Army. "Bruce, you gotta cut her loose!" screamed Wright. "She's gonna drown!" As Bott desperately cut at the nets holding the white whale, Scarred-Jaw Cow approached. The large animal "glided watchfully alongside me and stopped two feet away," Bott later wrote, "as wisps of her daughter's blood from the net chafe swirled into our faces." He expected the mother to charge him at any moment. After all, he reflected, "if ever an attack on a human by an orca was warranted, this would have been the time." But to Bott's surprise, she simply watched as he cut Chimo free and pushed her to the surface.[38] In hindsight, it seems likely that the white whale's biosonar didn't function properly. In the wild, her podmates had ensured that she avoided danger and got enough to eat, but they couldn't do that for her now.

Three weeks after the capture, Wright decided he'd had enough. If the whales didn't start eating soon, he might lose his prize. After separating Charlie Chin from the four females, he decided to transfer the two youngest to Sealand. The first to go was the pod's one normal-looking member—a female later named Nootka. After depositing her into one side of Haida's partitioned pen, the Sealand team returned for Chimo. To supervise this delicate operation, Wright hired Bill Cameron, one of Sonny Reid's partners from Pender Harbour. Cameron's background as a herring fisherman greatly influenced his approach to moving captive killer whales. In stark contrast to the hoop-net system that Griffin and Goldsberry used, Cameron sought to reduce stress on the animals. "You have to keep cool," he later explained. "If there's one member on the team that gets excited you're in trouble." Regardless of this measured approach, the whales knew something was amiss. As Cameron directed staffers to shore up the nets, Scarred-Jaw Cow and Florencia made one last effort to shield the white whale. As one reporter observed, the two large females "seemed to know that they were to lose her."[39]

At 6:30 p.m. the seiner delivered Chimo to Sealand, where Ellis and the others carefully lowered her into the pen. Although happy to be reunited with her podmate, the white whale struggled to find her bearings. Swimming erratically, she collided repeatedly with the net. Initially, Haida seemed to pay little attention to the newcomers, but within hours he had accomplished what Wright's team had failed to do in more than three weeks: he convinced them to accept fish. To the surprise of all watching, the young male collected herring from trainers and offered them to the two females, who immediately began eating.[40]

The three remaining orcas in Pedder Bay did not fare so well. Wright had declared publicly that he planned to release Charlie Chin, who seemed too large for captivity, but he intended to sell Scarred-Jaw Cow and Florencia, and his most eager customers came from Texas. The city of Arlington had recently broken ground on its Seven Seas Marine Park, and boosters hoped to make a splash by acquiring killer whales for the grand opening. City leaders made deposits on the two animals but warned Wright that he would need to keep them in Pedder Bay until the pool in Arlington was completed.[41] This meant that Wright had to get the whales eating. To accomplish this, he turned to Don White.

By this time, White harbored deep misgivings about orca captivity, and he worried about the animals' health in Pedder Bay. Living in a houseboat next to the pen, he watched as the whales grew weaker each day. Although he regularly tossed buckets of herring into their pens, his dives revealed that they weren't eating anything. White didn't know that these orcas weren't fish eaters, but he knew that they were in trouble. "A mammal loses hunger symptoms very quickly," he explained to me. "So you've got to induce hunger." White pleaded with Wright to inject the whales with hydrocortisone and vitamin B12, but Sealand's owner refused.[42] By their seventy-fifth day in captivity, the animals were badly dehydrated, and they had lost so much of their blubber that observers could see the outlines of their ribs.

Then, on May 15, White awoke to a troubling scene. Scarred-Jaw Cow was swimming along the surface, vocalizing and slamming into the logs that held her pen together. "I pulled on the bottom half of my wetsuit," he recounted, "and I was going around with her trying to keep her from smashing into the logs." Ellis arrived soon after, and the two men watched helplessly as the whale began opening her mouth and gulping seawater. "She made one charge at the net and got halfway out up to her dorsal [fin] and got stuck," White explained. He and Ellis slashed frantically at the net, but the orca drowned right in front of them.[43]

Bob Wright ordered them to dispose of the carcass quickly and quietly. "The consideration was, do you weight it or slit its belly?" White later explained to author Erich Hoyt. "Dead whales wash up all the time. But if you actually slit its belly or attach weights to it, then that's evidence that it's one of yours."[44] That night, Sealand divers towed Scarred-Jaw Cow out of Pedder Bay, tied her to an anchor, and let her go. With little blubber left on her body, she sank like a stone. For Graeme Ellis, this was too much. "It was so wasteful," he reflected. "You have an opportunity to learn from a dead animal like that. Why did it die? Why wouldn't it eat?" But for Wright, there was "no interest at all. It was just 'get rid of the evidence,' " Ellis noted. "We towed that thing out and dumped it off Race Rocks." Disgusted, he resigned. "There is only one thing I'd like to do, Bob, before I leave," he told Wright. "I would love to let those whales go." At that, Ellis told me, Wright grew furious, threatening the young trainer and "wagging his finger like only Bob could."[45]

In the end, Scarred-Jaw Cow's death may have saved her two companions. Fearful of losing the other whales, Wright agreed to the injections White recommended and bought fresh salmon to offer them. Two days later, the orcas did something White would never forget.

By the time we came to this part of the story, our interview had run far over time, and I was preparing to leave. As I thanked White and headed for the door, he stopped me, as if something was gnawing at him. After the injections, he explained, the two whales were in the same pen, but they were listless. Florencia had spent weeks lingering on the surface, and her blowhole was "dried up and cracked, sunburnt." White and another attendant were offering salmon, with little hope, but on this occasion, Charlie Chin took interest. "He opens his mouth and he takes the salmon," White recounted, "and he is sitting on the surface and he's starting to vocalize." Then the big male approached Florencia. "I'm sorry," said White, as he paused to wipe away tears:

> He gets right in front of her and drops the fish, and she ducks her head and grabs it. Now she's got it hanging out at the edge of her mouth, and he grabs ahold of the other end of the fish, and they make almost one complete circuit around the pool vocalizing back and forth on the surface of the water and then they twist their heads and the salmon splits and they each eat half.

Charlie Chin retrieved another salmon for Florencia and then returned for one himself.

"What does that do to you?" I asked dimly.

"Well you see, eh? It became a part of me."[46]

In one sense, the behavior was not unusual. Scientists now know that prey sharing is an integral part of life for killer whales around the world. Like all cetaceans, they learn about food from their mothers. For orcas, this is the defining part of their culture, what separates mammal eaters like Florencia and Charlie Chin, mother and son, from fish eaters. But on this occasion, the son had taken the lead, prompting his mother to eat a food she likely found abhorrent and in the process saving her life. In White's eyes, it was an arresting display of selflessness. "You see a level of social interaction that you rarely see . . . even among humans," he reflected. "And we're standing there keeping these things in this pen?!"[47]

White spent the late summer and early fall nursing the whales back to health. He was heartened to see Florencia and Charlie Chin regain their strength, but he worried about their future. With the death of Scarred-Jaw Cow, he knew both would be headed to Seven Seas Marine Park, and he couldn't stomach the idea. "I had spent months living out there alone with those whales," he explained. "I began to feel it was wrong to keep them captive, to put them in a situation not of their own choosing, where, sooner or later, they are doomed."[48] In the meantime, he could not help but reflect on the fate of whales elsewhere.

FIGURE 11.3 Haida and Chimo perform at Sealand of the Pacific, summer 1972. In author's possession.

That September, the Vancouver Aquarium announced the deaths of three recently captured narwhal calves, and the carcass of a small orca, slashed and weighted, turned up in Whidbey Island's Penn Cove. Aboard his houseboat, White thought of the sunken body of Scarred-Jaw Cow off Race Rocks and wondered about the fate of Florencia and Charlie Chin. Then, on the night of October 27, someone lowered the nets and released them.[49]

Wright suspected White. The researcher claimed he was playing cards with friends at the time, but Wright urged authorities to question him. "I don't know that he ever accused me of releasing the whales," White explained, "but I refused to take myself off of the suspect list." When Wright arranged for the RCMP to administer a lie detector test, White balked. "Then bully Bob came out. He called me up and said 'If you don't come back and do this, I will hound you to the ends of the earth. I will follow you. You will never work with killer whales again.' "

Ultimately, the damage was less than Wright feared. Despite his loss of the whales, the capture at Pedder Bay gained him the reputation of orca catcher that he coveted. It also netted him a white whale, which drew crowds to Sealand.

For his part, Don White tried to make sense of his role in a business he had renounced. "I carry some guilt," he admitted, "because if I hadn't come back that damn Sunday, none of this would've happened."

"Do you really feel that way?" I asked.

"I have, sure, but it's not something I torment myself over."

"Well, you can't, right?"

"Right—it's absolutely serendipitous and unintentional. Shit happens."[50]

12

The Penn Cove Roundup

TED GRIFFIN AWOKE with a start, but he wasn't sure why. It was a warm night in August 1970, and all seemed calm and quiet. Water lapped against the boat's hull as the lights of Coupeville flickered a mile and a half away. Yet something wasn't right. The breathing of the whales behind the capture nets sounded clipped and nervous. "How long have they been blowing that way?" he asked the two men on watch.

"Blowing? What way?" they answered. "All night I guess."

Straining his eyes in the dark, Griffin scanned the enormous pen, anchored just off the old Standard Oil dock. Everything seemed to be in order—except on the north side. The marker lights there were too far apart. He roused Goldsberry, and the partners jumped into a skiff to investigate. When they reached the floating lights, Griffin stared down at a loose cork line, puzzled. The net looked split.

"Not split—cut!" yelled Goldsberry. "And in more than one place."

Griffin couldn't believe it. Suddenly the orcas' anxious breathing made sense. During the night, someone had slashed a section of the net. Large portions of loose mesh now drifted in the current, threatening to drown any whales nearby. Griffin and Goldsberry shouted for their crew, and in the following hours everyone worked feverishly in the dark—reattaching lines, mending mesh, anchoring nets. Had they reacted in time? Had the animals managed to avoid danger?

Griffin needed to find out. Donning his wetsuit, he slipped over the cork line and into Penn Cove's murky waters. At first, he was hopeful. All the whales seemed to be swimming near the surface. But a moment later, his eye caught a shimmer of white—perhaps a shark caught in the net? No, it was a tiny orca calf, no more than eight feet long. Ensnared in a floating portion of mesh, the little whale hung lifeless, head down. Other divers found two more, also calves. Initially, Griffin felt only nausea, but that soon gave way to rage. He wanted to lash out at those responsible. "Is this what it feels like to be on the other side?" he reflected. "Do those who oppose my capturing whales feel as I do now—angry, so angry they could kill?"[1]

For Griffin, it was an unimaginable disaster. In the three and a half years since the capture at Yukon Harbor, only one orca had died during his expeditions. Now three calves had drowned in a single night. Hours earlier, the capture seemed a triumph. After months of stalling eager buyers, Namu Inc. had made its largest catch ever—as many as ninety killer whales. Yet from the beginning, Griffin knew the operation was in trouble. There were too many animals to handle safely, even if the capture team was in lockstep, which it wasn't. Over the years, Griffin had worked closely with fishermen and adopted some of their methods. But his differences with them, including Goldsberry, were never more apparent than at Penn Cove. There Griffin made contested decisions to safeguard the killer whale population of the Pacific Northwest, even as critics denounced him as the greatest threat to the species' survival.

For the thousands who witnessed it, the Penn Cove Roundup was a pivotal moment in their lives. The sight of orcas crowded behind nets horrified most onlookers, and some claim to still hear the whales' cries nearly fifty years later. A few, such as the two men who cut the net, felt moved to action. But people throughout the Pacific Northwest took notice. In the past, Griffin's pursuit of killer whales had stirred controversy, but those who opposed it remained a vocal minority; Penn Cove tipped the political scale, transforming orca capture into a symbol of environmental destruction. Even without the calves' deaths, the event would have loomed large. The discovery of their bodies forever changed marine mammal policy.

Griffin and Goldsberry had begun the year with high hopes. In January 1970, they launched the *Orca*—a high-speed, thirty-two-foot "pocket" seiner. Goldsberry was thrilled. Over the years, he had bristled at his dependence on Adam Ross and Peter Babich. In the world of orca chasing, as in fishing, the skipper who made the catch had the bragging rights. Now Goldsberry had his own boat, custom-made for killer whale capture. In addition to a power block for retrieving a specially made net, the vessel boasted a four-hundred-horsepower engine, giving it a top speed of thirty knots—more than three times faster than Ross's *Chinook* and Babich's *Pacific Maid*. Unlike those purse seiner vessels, the *Orca* could keep pace with its namesake.[2]

Yet excitement soon gave way to frustration. In late January, Namu Inc. set up camp in Port Townsend, where the Strait of Juan de Fuca meets Puget Sound. With schools of herring gathering to spawn, Griffin knew the area was primed for the arrival of salmon and thus likely to attract orcas.[3] But the whales didn't come, and the partners looked elsewhere. In mid-February, they netted six orcas in Carr Inlet, but they kept only one, and two weeks later, Bob Wright made his surprise catch of five whales (including the white whale) near Victoria.[4] It was bad news for Namu Inc. Competition was heating up just as killer whales were becoming harder to find. By spring, Griffin and Goldsberry had fifteen unfilled orders from oceanariums around the world, and customers were growing impatient. The partners had spent a fortune on new equipment, and if they didn't start catching whales soon, they would be out of business.[5] But if they succeeded, they risked running afoul of the nation's rapidly shifting environmental politics.

Although concern with nature had been rising for years, 1970 marked a turning point. No longer a collection of fringe issues, the environment became a subject for mainstream politics. One register of this change was the growing use of the term "ecology." Coined a century earlier by a German scientist, it described the relationship of living things to their environment and each other. Biologists had used the concept since the 1930s, and universities began offering courses on the subject in the 1950s, but it was not until the late 1960s that ecology became tied to environmental politics.[6] Many today associate this shift with radical groups such as Greenpeace, but US activists and politicians in the Pacific Northwest had embraced ecological concerns long before that organization was founded in Vancouver, British Columbia. In Oregon, Republican governor Tom McCall (1967–1975) helped popularize concerns about pollution and livability, while in Washington Republican governor Dan Evans (1965–1977) created the nation's first Department of Ecology in March 1970. The following month, people throughout the region and world celebrated the first Earth Day on April 22.[7]

The event signified a great many things to a great many people, but to Ric O'Barry it meant ending cetacean captivity. Although the young trainer had enjoyed his time frolicking with dolphins and the orca Hugo at the Miami Seaquarium, he quit in early 1970. In his mind, the cost of captivity to the animals was simply too high. Months later, he was called back to care for Cathy—one of the *Flipper* dolphins he knew so well. Soon after he arrived, the sick animal sank to the bottom of her tank and died in his arms. Convinced Cathy had committed suicide, O'Barry decided to take action. Days later, he was on a flight to the Bahamas, where he planned to free dolphins he had previously delivered to a research station in Bimini. "It was just an act of passion," he later told me. "I was angry."[8] Yet the operation didn't go as planned. O'Barry nearly drowned while cutting the pen open, and the dolphin inside refused to escape. As dawn broke on Earth Day, the disappointed activist found himself sitting in the Bimini jail. But failed or not, O'Barry's mission made news around North America, and he used that publicity to launch the Dolphin Project. If humans were serious about loving nature and the earth, he argued, they needed to end cetacean captivity.[9]

Activists in the Pacific Northwest made similar connections. Among them was Bonnie Jean Schein, an adjunct instructor at Seattle Pacific College. After being approached by her former instructor Dixy Lee Ray, Schein visited the Marine Mammal Biological Laboratory, where she learned that Ray and Victor Scheffer had received a grant to study killer whales in Puget Sound. At their urging, and without realizing that the lab worked closely with the whaling and sealing industries, Schein published an open letter in the *West Seattle Herald*. Appearing one day after Earth Day, it denounced live orca capture. Surely it was better to observe whales in their "natural habitat," Schein declared, than to watch Ted Griffin "force feed a grief-stricken infant" at his aquarium, and she included a petition signed by ninety students calling for an end to the "unregulated killing and capture of killer whales and other marine mammals in the Puget Sound."[10] Some state officials agreed. Representative Hugh Kalich, chairman of the Interim Committee on Game, instructed the Game Department to gather information on the history and methods of killer whale trapping, and in late May he announced plans for a public hearing on marine mammal legislation.[11]

Griffin planned to cooperate with the state, but he had a lot on his mind. Although he hadn't told Goldsberry, he was considering stepping away from orca capture and focusing on display. In early April, he had submitted another proposal for the construction of a new oceanarium—Seattle Center Marine Life Park. Modeled on Sea World, it would be located adjacent to the Space Needle and boast a three-million-gallon whale pool.[12] The following month, he flew to Boston to deliver a lecture on orcas at MIT. Yet he made sure to meet with Game Department officials in Seattle to offer suggestions for regulation, urging, for example, that they only allow the removal of orcas between eleven and sixteen feet: smaller animals may still be nursing, he noted, and larger ones could suffer from transport. As for the method of capture, he invited officials to observe Namu Inc.'s next expedition.[13]

As summer approached, the debate heated up. On June 6, Kalich's committee held its public hearing in downtown Seattle. While Game Department officials proposed regulation of capture, animal rights advocates demanded immediate cessation. Bonnie Jean Schein urged the committee to remember that it had "higher responsibilities than just to those who stand to gain financially." Also in attendance was Victor Scheffer. Now retired from the US Fish and Wildlife Service, Scheffer had won fame for his book *The Year of the Whale* (1969), a fictionalized account of a year in the life of a baby sperm whale. In a thinly veiled reference to Griffin, the scientist declared at the hearing that there had been "unintended cruelties to killer whales by showmen in the Puget Sound area." As an example, he pointed to the death of the male orca from a harpoon wound in February 1967—failing to mention that his lab had provided the harpoon rifle.[14]

The following week, *Seattle Times* reporter Susanne Schwartz wrote a thoughtful column on the controversy. "Ten years ago the killer whale was a symbol of fear," she observed. "Today, the 'killer' is a popular aquarium performer . . . valued for his loyalty, quick mind and striking black-and-white good looks." So beloved had the once-feared predators become, in fact, that some Puget Sound residents were determined to stop all harassment of them, and therein lay a paradox. After all, Schwartz emphasized, "the target of the crusade to save killer whales is the one man most responsible for taming the once-feared creatures and teaching the world what is known about them today." Griffin was, she reminded readers, "the first man to tame, train, ride and perhaps love a killer whale." Yet Schwartz noted that critics were pushing hard for regulation. "The state legally owns all wildlife," Scheffer told her, and "the killer whale is important as a tourist or recreational asset." For his part, Don Goldsberry maintained that Namu Inc. would support regulation, but only after scientists had assessed the killer whale population. How could the state impose capture limits if it had no idea how many whales there were? "I killed the bill once," Goldsberry quipped. "I can kill another one."[15]

Little did he know that 1970 would in fact become the year of the whale.

~

For years, scientists had warned that the whale stocks of the world were crashing. By the summer of 1970, however, the ocean's great leviathans were becoming symbols of an imperiled global environment. The US government was quick to embrace the issue. Because there were only two small whaling companies in the country, US representatives could push for cuts in the international hunt with few domestic repercussions. But it wasn't just for show.

When the IWC refused to lower quotas in 1970, Washington, DC, took action. In July, the US Department of Interior announced plans to add eight species of great whales to the list of endangered species, which would prohibit all imports of products made from their bodies.[16]

Opposition to the proposal immediately appeared. The Commerce Department claimed that Interior had overstepped its bounds, and leaders in both the US military and defense industries protested the inclusion of sperm whales on the list. The president of Archer Daniels Midland reminded Senator Warren Magnuson that "sperm oil is vital to the national defense and to many important industrial and mechanical processes" and that Defense Department policy required that a strategic reserve of twenty-three million pounds be maintained. A former naval officer known for his economic and military boosterism in Washington State, Magnuson initially sympathized with these concerns. Other than helping Griffin import Namu five years earlier, he had given little thought to cetaceans up to that time, and he initially viewed the whaling question through the prism of the Cold War. But at that very moment, the southern resident killer whales were beginning a gathering that would convince Magnuson and many other Northwesterners of the need for marine mammal protection.[17]

Observers call it the "superpod," sometimes the "potlatch." Each summer, the three resident orca pods (J, K, and L) rendezvous in the Salish Sea. The purposes of the gathering are not entirely clear, though it almost certainly plays a role in mating and group cohesion. As the pods come together, whale watchers often witness the so-called greeting ceremony, in which separate groups line up facing each other before frolicking. While there appears to be no set time or location, the superpod frequently forms in summer around the San Juan Islands, at the height of the Fraser River chinook runs.[18] This was the case in mid-July 1970, when reports appeared of a massive group of killer whales in Haro Strait—at least a hundred, possibly more. Over the centuries, such gatherings had tightened the bonds of the southern residents. On this occasion, it made them vulnerable to disaster.

News of the whales' appearance traveled fast. From their base in nearby Oak Bay, Bob Wright and his Sealand staff watched the animals closely. Although he still had two whales penned in Pedder Bay, Wright hoped the mass of orcas might cross into Canadian waters, making another capture possible. Griffin was there, too. Eating and sleeping aboard the *Pegasus*, he followed the superpod for more than a week, repeatedly hailing Goldsberry on the radio. The whales might move into Puget Sound at any moment, he prodded—get the team ready. But Goldsberry was not in a whale-catching mood. Weeks earlier, he had run the *Orca* over a log boom in the fog, sinking the new vessel and nearly drowning in the process. The boat was in dry dock for repairs, and most Puget Sound fishermen—including Ross and Babich—were in Alaska. "We're out of business," Goldsberry complained.[19]

But Griffin stayed with the whales, hoping for an opportunity. Had the orcas remained in the deep waters of Haro Strait, they would have been fine, but they didn't know that, and in early August, the superpod started south toward Admiralty Inlet, lingering on the west side of Whidbey Island. Griffin kept up his vigil. Sleepless and sustained by Coca-Cola and chocolate bars, he teetered on the edge of delirium. Finally, on the evening of August 5, the whales headed east along the southern end of the island, and he followed. As Griffin rounded Possession Point, his mind was on the chase, but it couldn't help but drift to the past. Somewhere in the darkness off his port side was the Glendale dock, where he had taken his tumble into the water and first dreamed of killer whales. Was he still that little boy who loved animals? Or had he become someone different? Banishing such thoughts, he raced north. He was an orca catcher now, and he had to turn the whales back before they reached Deception Pass.[20]

The next day, Goldsberry was wrestling with his own dilemma. Namu Inc. had orcas to catch, but no vessel to catch them. After scouring the Tacoma docks for hours, he happened upon the seventy-two-foot seiner *La Touche*, skippered by Matthew Antone Vodanovich Jr. Even by the standards of local fishermen, Vodanovich was an imposing figure, standing six foot six and weighing nearly three hundred pounds. He listened to Goldsberry's proposal but showed little interest in catching whales. There were fishing days ahead, and he and his crew weren't about to miss them chasing after some damned blackfish. Desperate, Goldsberry offered him $2,000 per whale captured—double the usual fee—and Vodanovich agreed. That evening, *La Touche* stopped at Pier 56 to load the capture net and motored north at eleven knots.

The seiner reached Whidbey Island early Friday morning, August 7. By that time, Griffin had lost sight of the superpod, which seemed to have split apart. From the window of a chartered seaplane, he scanned the waters below. He noticed some whales milling to the south, but they looked like a diversion party. The bulk had fled elsewhere. Finally, through a break in low-hanging clouds, he spotted them.

"Where are they, Ted?" radioed Goldsberry from *La Touche*.

"They're in . . . en . . . ove. I found them," Griffin's voice crackled. "Everyone is there, maybe a hundred or so."

"I can barely make you out. Did you say Penn Cove?"

"Yes, Penn Cove. They're heading in."[21]

~

At first, the operation went smoothly. With Goldsberry directing him, Vodanovich made a large set, trapping some fifty orcas. Believing the capture complete, Griffin flew to Seattle for supplies. He then stopped on Bainbridge Island, where enterprising teenagers in Port Madison had corralled an orca calf—likely separated from its pod during the melee of the previous days. After locals helped Griffin load the little animal onto a flatbed truck, he transported it to the Seattle Marine Aquarium. Griffin didn't return to Penn Cove until that afternoon, and even from the floatplane he could see that something was amiss. In his absence, *La Touche* had made a second set, netting forty more whales. A beaming Goldsberry explained how it had happened. Following the first set, the whales outside the net had refused to leave their trapped companions, with most crowding along the cork line. Seeing an opportunity, he had cut the net in half and made a second set. Now they had ninety whales.[22]

Griffin could not have known that he was looking at all three pods of southern resident killer whales, but he knew his partner had caught too many animals. "We didn't have enough anchors to control the nets, and there was not enough experience in the world to handle this," he later told me. It was a turning point in his relationship with Goldsberry. "I'm a control freak. Suddenly I have two catches, and I can't imagine what is going through his mind." Convinced the animals were in danger, Griffin demanded that his partner open one of the nets. "Don and I did not come to blows," Griffin noted, "but I was hostile." At that moment, the partners' divergent priorities became painfully clear. "I knew I had to stay there because I couldn't trust Don to take care of the animals. He could be trusted to take care of the whales as inventory, but not as life."[23] For his part, Vodanovich was incredulous at Griffin's order. He had been promised $2,000 per whale captured, and he had captured ninety. Why on earth would he let half of them go? As Vodanovich fumed over the release, however, Griffin had more pressing problems.[24]

When he had directed *La Touche* to Penn Cove, Griffin had given little thought to the location. After all, Namu Inc. had netted whales there on two previous occasions. But those had taken place in winter. It was now

August, the height of pleasure-boating season. Even before the divers could check the nets, several vessels had approached so their passengers could watch. At first, most simply seemed excited to see the whales. But on the morning of Saturday, August 8, 1970, newspapers around Puget Sound reported the capture, and by the afternoon hundreds of people had arrived by land and sea to take a look. The scene quickly became chaotic. "The situation was out of control," Griffin recalled. "You could have walked across the water on boats."[25]

Regardless of one's views on animal captivity, the collection of wild specimens is, by most definitions, an act of violence. Young animals are torn from their habitats, context, and connections in order to serve human purposes. Because the spectacle of capture itself tends to offend public sensibilities, zoos and aquariums have normally collected specimens outside the public eye.[26] But Griffin gave little thought to this dynamic. One reason was that Puget Sound, populated though it was, offered the safest capture sites for both whales and workers. Another was that many of the locals Griffin encountered seemed excited to watch the captures. "I may have been so full of myself that I believed all others would see capture events as exciting and transformative," he later reflected. "I could not believe there were people hostile to my activity."[27] In hindsight, it was a crippling blind spot. So immersed was Griffin in his own adventure that he failed to grasp the impact captures might have on other people, or on the orcas themselves.

No one can be sure how the southern resident killer whales experienced the Penn Cove Roundup, but it must have been traumatic. Their joyous reunion in Haro Strait had become a harrowing chase into Puget Sound. Herded by speedboats and underwater explosions, they found themselves separated and crowded behind nets. The forty whales Griffin released stayed in the area, refusing to leave their companions. Those who remained trapped must have been anxious and confused. By this time, capture was a familiar experience for many of them. Portions of all three pods had been chased and trapped at least once before, and for some this must have seemed old hat. But the corralled whales included at least twelve youngsters, several of them less than a year old, who had never been behind nets before. For these calves, proximity to mother and family had provided safety and guidance in the wild, but their elders had no answers now. Adult whales swam frantically back and forth, spy-hopping for clues and calling out in a constant chatter. Usually, the sounds of family provided comfort. This was a panicked cacophony.

The animals' distress had a powerful impact on witnesses. Many recoiled at the sight of an increasingly iconic species crowded behind fishing nets. Although pleasure boaters on Puget Sound gave little thought to fish caught by vessels such as *La Touche,* the spectacle of trapped killer whales looked different and sounded worse. Unlike salmon, orcas can cry—or, rather, they can make sounds humans interpret as cries. Griffin and his crew were inured to these calls, but for those who had never witnessed a capture, the sounds were troubling. This was certainly true for Terry Newby.

❧

When I met him at a Tacoma cafe in September 2015, the seventy-five-year-old Newby looked every bit the grizzled ex-fishermen he claimed to be. But in 1970, he was a promising graduate student in marine biology at the University of Puget Sound. In the mid-1960s, he had undertaken cutting-edge research on harbor seals, in the process meeting Victor Scheffer and other scientists at the Marine Mammal Biological Laboratory. After completing his coursework, he was looking for a job, and he saw opportunity in the rising debate over killer whale capture. At the Game Department's request and with Griffin's permission, he traveled with three department officials to Penn Cove to observe the operation.

The Game Department men didn't impress Newby. "They were the most lamebrained bunch of guys you ever met in your life," he scoffed. "You should have seen them sitting up there with coolers full of beer on top of that seiner." As a young biologist keen to observe both orcas and their capture, he couldn't believe the Game Department had been assigned the task. "They didn't know nothing," he continued. "They were the same guys that would go kill wildlife and have their big feed every year." In contrast, Newby liked Griffin. "I saw a man running a business who had a genuine heart for the animals," he explained. "He was trying to do what he knew [how] to do—he was selectively culling the group to get the youngest, smallest animals he could for safe transport." But the operation disturbed Newby, nonetheless. "I felt excited and everything, but I was sad after the fact. I was even crying," he admitted. Like the witnesses in boats and on shore, it was the animals' calls that affected him most deeply.

Thirty at the time, Newby had already seen a lot in his life. In 1967–1968, he had completed a tour in Vietnam, leading a platoon in the Mekong Delta and fighting house to house in Saigon during the Tet Offensive. So when he arrived at Penn Cove, Newby had heard his share of screams. Yet the orcas' calls left a different sort of scar. "At night you hear it. The squealing," he told me. "To this day, if I hear killer whales, it is almost like a PTSD explosion in my head."[28] And some witnesses decided they had had enough.

Michael Park and Ron Bertsch meant well when they pushed away from shore that night. A home designer who had moved to Whidbey Island recently, Park was appalled at the capture, and along with Bertsch he had decided to free the animals. Rowing quietly, they steered for the pen's marker lights. Outside the net, other whales kept their vigil, and the two men marveled at the dark shapes passing beneath their boat. Using a bread knife, Park and Bertsch made a series of cuts through the cork line and mesh. But like the dolphin O'Barry had tried to release in Bimini, the whales failed to rush for freedom. Disappointed, the two men returned to shore convinced they had done their best.[29]

Maybe they had, but they knew little about nets and currents. Their cuts had detached sections of mesh from the anchors and lead line, and when the heavy flood tide hit Penn Cove, the crippled net began to drift and fold. The panic among the whales must have been severe. By now, they had acoustically mapped their enclosure and learned to navigate it. But the net was now collapsing unpredictably. The experienced older animals managed to avoid it, but the youngest weren't so lucky. The next morning, Griffin found the calves drowned in the nets. All three were between eight and ten feet long—still of nursing age and below the minimum length Griffin had declared viable for removal. They would have been set free a few days later.[30]

For many who witnessed the capture, the activists had made a forgivable error. "I would have supported what they did because it was all with good intentions," declared local John Stone, whose parents owned the Captain Whidbey Inn, where the Namu Inc. divers took their meals. "The fact that it caused the death of the whales is tragic . . . , but I don't blame Michael Park and Ron." Terry Newby disagreed. As he later told me, "I was pissing mad. What the heck are they doing—what are they thinking?" Forty-five years later, he still blamed the activists for the whales' deaths.[31]

Ted Griffin wasn't so sure.

It was April 2016 when he finally opened up to me about it. Now eighty years old, Griffin was still a guarded man, and he didn't express his feelings readily, even to those who knew him well. But we had just finished an interview on Seattle public radio, and his emotions were raw. Under the Space Needle, at

FIGURE 12.1 Ted Griffin in Penn Cove, August 1970. Photo by Wallie V. Funk. Courtesy of Center for Pacific Northwest Studies, Western Washington University.

the very spot he once planned to build an oceanarium, he told me about the moment his life changed—the moment he saw the dead calves. "I will never forget coming upon that scene," he said, looking away. "It haunts me to this day." Sure, it was tempting to blame the activists. "For Don, it was just that easy," Griffin explained. "He just said to himself, 'It's not our fault. Someone else did it.' And that was that." But that reasoning didn't work for Griffin. "I could say that, but I didn't *feel* it, not in my heart," he noted. "It doesn't matter if someone else cut the nets. Those are *my* nets with dead whales in them. I can't explain that away. They drowned because I caught them. *I* did that."[32]

Regardless of blame, Griffin and Goldsberry had three dead orca calves on their hands, and the Marine Mammal Biological Laboratory refused to take them. Although the lab still worked closely with the US whaling industry and had a contract with Namu Inc. for holding and marking whales, its researchers had no desire to be associated with the dead whales. Soon after, Newby witnessed a heated exchange between the two partners at the Captain Whidbey Inn. Griffin argued for coming clean. "Nobody will believe us," Goldsberry responded. "They don't believe anything we are doing." The only solution was to sink them, declared the ex-fisherman, who assured Griffin, "I can get rid of 'em and nobody will ever find 'em." Griffin relented, and soon after Newby joined Goldsberry to find anchors. Assigned by the Game Department to observe the capture, the young biologist wondered how he had gotten caught up in an effort to conceal the deaths of three young orcas.[33]

That night, after boaters and sightseers had gone home, Griffin, Goldsberry, and their most trusted divers towed the carcasses toward the entrance of Penn Cove and dragged them ashore. The bodies would need to be cut open to prevent gases from raising them to the surface, but no one wanted to do it. "They're dead whales. It makes no difference to a dead whale how you dispose of it," Griffin explained to me, "but the idea of doing it was horrible and watching it being done was almost unbearable." For one diver, it proved too much. After the team slashed and weighted the bodies, the man threw down his mask and sat on the bow of a nearby skiff, where Griffin joined him. "That was the worst part for me, seeing the pain on this man's face," he recalled. "It was not what he signed up for."[34]

Although no one other than Newby and the capture team knew about the dead whales, public scrutiny continued to mount. Among the keenest observers was Wallie Funk, owner-editor of Whidbey Island's two newspapers. Since covering Namu's arrival in Puget Sound, Funk had given little thought to killer whales, apart from seeing Skana perform at the Vancouver Aquarium. But the Penn Cove Roundup had brought the orca debate to his backyard, and on August 9 he hired John Stone to run him out to the capture floats in a skiff to take photos. When Stone returned to collect Funk an hour later, he found himself on a collision course with a large male orca—now known as J1. Afraid that any sudden move might threaten the animal, Stone held his course. "At about twenty feet off the bow of my little skiff, he goes under and about five feet off the stern he rises again," recalled Stone. "It was smooth as silk, but I was just frozen to my outboard motor." When Stone reached the pen, Funk was beside himself that he hadn't photographed the encounter. "God damn it, John! Did you see that?" he shouted. "And I'm out of film!"[35] Despite his disappointment, Funk had produced an extraordinary record of the capture, and decades later his photos still pulse with the stress of people and whales alike.

❧

The chaos only grew in the following days. On Monday morning, August 10, Matt Vodanovich abruptly declared he was going fishing. "You can't leave us without a boat!" Goldsberry protested. "We have to get our whales out of

FIGURE 12.2 Terry Newby rows among captured whales, August 1970. Photo by Wallie V. Funk. Courtesy of Center for Pacific Northwest Studies, Western Washington University.

the nets right now and turn the rest loose as soon as possible." Vodanovich didn't care. He wanted Namu Inc.'s gear off his vessel immediately. For the whale catchers, the rest of the week was a frantic blur. Griffin chartered a local tender boat, and the team worked to separate suitable animals.[36] Observers gathered to watch the operation, and many still recall the orcas' plaintive calls as they were removed for transport. "They let out screeches that just raised the hair on the back of your neck," recalled John Stone. "What sticks in my memory is my cat's response to it. It may be because it was a high-pitched screech . . . but she would just cower whenever a whale was pulled out of the water."[37]

The roundup ended with a final tragedy. On Sunday, August 16, a twenty-foot female tangled in the net and drowned before divers could free her. Witnesses swore she was trying to reach her calf. By that time, Namu Inc. had removed seven young whales—three females and four males, all between eleven and sixteen feet. Goldsberry wanted more. They had unfilled orders, nagging debt, and plenty of whales. Griffin refused, and on August 18 he opened the nets.[38]

∽

By that time, Northwesterners far and wide were debating orca capture. Some came to the whale catchers' defense. Although deploring the death of the large female, the *Bainbridge Review* reminded readers that Griffin, unlike commercial whalers, was "in business to catch whales alive."[39] Others were less forgiving, among them publisher James Scripps. A resident of Lopez Island, Scripps had flown his private plane over Penn Cove during the operation. He described the scene of the whales outside the nets trying to reach their trapped companions as "terrible" and "sickening" and called for legislation to protect orcas. State senator Peter Francis agreed, observing that "there seem to be no controls over the destruction of sea life in Puget Sound."[40] Days later, the new Progressive Animal Welfare Society organized a protest at the Seattle Marine Aquarium. It was a small affair, but Griffin noticed the signs: "Free the Whales," "Stop the Killing," "End Profits and Greed."[41] They were hard to take, but he had worse coming.

Matt Vodanovich was mad as hell. On Monday, August 24, the hulking fisherman stormed into Griffin and Goldsberry's office and slammed the door behind him. He had received his check for the capture, and he wasn't happy. As a fisherman, Vodanovich was accustomed to catching as many fish as he was allowed, and there was no law limiting orca capture. For the life of him, he couldn't understand why Griffin and Goldsberry had kept only seven whales, but that wasn't his problem. Goldsberry had promised $2,000 for every whale caught, and *La Touche* had netted ninety of them. Vodanovich figured that Namu Inc. owed him $180,000. Although Griffin and Goldsberry were formidable men, they knew they were in trouble. Both had received death threats and had taken to carrying concealed handguns, but they didn't want to use them. "Don and I thought we were going to have to kill him or he was going to kill us," Griffin recounted. Eventually, the aquarium owner coaxed the irate fisherman into leaving.[42]

That same day Griffin wrote a revealing letter to Vodanovich. He began by acknowledging the fisherman's frustration with the release of half the catch, and he conceded that no law prevented Namu Inc. from taking as many whales as it liked. Yet he stressed that "any true fisherman is also a conservationist in that he must do nothing to endanger the return and rebuilding of the marine resource which he harvests." Griffin acknowledged that he had orders for more whales than he had removed, but he emphasized, "I could not in all good conscience have taken every last one just in order to take care of customers."[43]

With Vodanovich at bay, Griffin turned to Olympia. He knew state legislation was coming, but what sort would it be? On September 3, 1970, he wrote to Representative Kalich urging regulation rather than prohibition.

"Marine mammals are a resource much the same as food fish," Griffin argued, and regulation should allow for "trapping in numbers which do not endanger the resource's ability to replenish itself." Along these lines, he offered to help fund research on the population and reproduction rate of killer whales. But he also pressed Kalich and his colleagues to think beyond state borders. Because killer whales "may range throughout the Puget Sound, British Columbia, Alaska, and California," he noted, all governments on the Pacific Coast needed to draft "similar marine mammal management regulations."[44]

Over the following months, Griffin focused on the care and delivery of the whales taken at Penn Cove. While two stayed in Seattle, the paths of the other five underscored just how global the orca trade had become. One went to Sea-Arama Marineworld in Galveston, Texas; another went to Safari Park in Great Britain. Japan's new Kamogawa Sea World took two. And on September 24, a young female orca went to the Miami Seaquarium, where she joined Hugo. Hoping to boost public interest with a hint of sexuality, the facility named her Lolita. The following month, Griffin delivered the fifth orca to the new Marineland of Australia.[45]

In his partner's absence, Goldsberry aggressively pushed to shape the capture debate. Rather than the state intervention that activists urged, he called for joint US-Canadian regulation. After all, Goldsberry observed, "they are international animals." He suggested an annual capture quota of ten orcas within the shared waters of western Washington and southern British Columbia. The area, he suggested, should stretch from Olympia to Neah Bay in Washington all the way north to Seymour Narrows. It was a remarkably prescient mapping of southern residents' primary range in what is now called the Salish Sea, but there was a problem. Goldsberry estimated an orca population of three hundred, with an annual birth rate of forty calves. If such figures were accurate, a yearly catch of ten animals would likely be sustainable. But neither Griffin, nor Goldsberry, nor any scientist knew the actual number of killer whales in the region, or how slowly the species reproduced. The real population of southern resident orcas numbered less than a hundred, and Namu Inc. had just removed a generation in Penn Cove.[46]

By the time Griffin returned from Australia, the debate was raging, and Victor Scheffer had entered the fray. The distinguished scientist had devoted most of his career to the commercial fur seal hunt, and that June he had defended the industry in a letter published in the *New York Times*. "The harvest of seal skins and byproducts is valued at $5 million a year," he wrote, asserting that the activists who opposed the slaughter "don't know enough."[47]

At home in Puget Sound, however, he posed as a champion of the new environmental values. "Ted Griffin has tried very hard to keep respectability and public sympathy," observed Scheffer, but at heart the aquarium owner was nothing more than a callous profiteer. "He gets $20,000 and more for each whale," asserted the respected scientist. "It's like the activities of the lumber barons of the last century—completely legal but immoral."[48]

Others came to Griffin's defense, including two of Scheffer's former colleagues at the Marine Mammal Biological Laboratory. Dale Rice noted that Namu Inc. had shown restraint by releasing most of the whales caught. After all, he emphasized, under the current laws "nobody could do anything about it if someone went out and caught every last one and ground them up for cat food if they felt like it."[49] For his part, Mark Keyes dismissed the criticism leveled by Scheffer and others while hinting at the lab's own history of killing orcas for research:

> Killer whales have been subjected to three forms of harassment: One, being shot at by fishermen and adventurers; two, being chased by biologists who wanted to kill them and examine their stomach contents; and three, live capture for display and study in oceanariums and research institutions. Whale protectionists are a bit late (and a bit derelict in their professed concern) to do anything about the first two forms of harassment, so I conclude their present protest is in response to the third.

Keyes also argued that those who called for observing the whales only in the wild overlooked their own privilege. "In 1969, 100,000 people, including 10,000 school children, saw Kandu the killer whale," he noted. "Few of these, I doubt, can afford a [waterfront] home or a boat on Puget Sound. Yet they have seen a killer whale, perhaps even touched one, and know of them far better than if Kandu, Shamu, Namu and all the rest had not been brought back alive."[50]

Yet all that reasoning lost its force when dead whales turned up in Penn Cove.

∞

The first calf had washed ashore in September. Namu Inc. denied responsibility, and attention quickly faded. The real scandal emerged two months later, on November 19—the very day Griffin was set to meet with Governor Dan Evans to discuss a bill regulating orca capture. As Griffin drove to Olympia, a local radio station reported the discovery of two more orca carcasses in Penn Cove. At 1:30 p.m., Evans welcomed him into his office, where Griffin made his pitch for regulation rather than prohibition of capture. Suddenly, an aide entered the room and passed the governor a slip of paper, informing him of the dead whales. "This matter is not going to go away," Griffin said, after which he walked out of the meeting.[51]

Soon after, photos began to circulate, and they had all the trappings of a crime scene—mutilated bodies, attempts at concealment, hints of sinister intent.[52] Someone had killed baby orcas and sunk them to hide the evidence. Many Whidbey Island residents were horrified. Living near the island's naval air station, locals were accustomed to upsetting news. "Our business was Vietnam," Wallie Funk's son Mark reminded me. "That year—my eighth-grade locker partner's dad was a POW," and Mark himself had seen dead whales washed up on shore. But this was different. "Those whales were white. They had anchors on their tails. Those carcasses were meant never to be found," he emphasized. "That was a searing experience for a lot of my friends who lived around Penn Cove," and it raised questions. "What did the whales

FIGURE 12.3 Local man recovers carcass of killer whale calf, November 1970. Photo by Wallie V. Funk. Courtesy of Center for Pacific Northwest Studies, Western Washington University.

do to deserve what was such a horrible death?" he asked, and "why were people hiding it?"[53] And could it be that Ted Griffin, the man Funk and other youngsters had once cheered, was responsible?

Reporters phoned the Seattle Marine Aquarium for comment, but Griffin remained silent, and Goldsberry wasn't admitting anything. "It could have been us or again it couldn't," he declared. "I'm checking into it."[54] Dixy Lee Ray weighed in forcefully. "The fact that someone tried to cover it up is the thing I find most reprehensible," she observed. "The honorable thing would be to donate the carcasses to study." Allen Wolman, a whale specialist at the Marine Mammal Biological Laboratory, agreed. On August 10, his lab had refused to take the bodies; now he bemoaned Griffin's failure to donate them as "a large loss to the laboratory."[55]

Soon after their discovery, the rotting carcasses arrived at the lab, but there was little Wolman and others could do but hold their noses and take measurements. A veteran of dissecting cetaceans on whaling ships, Wolman was unimpressed with the incisions made into the dead calves. "Whoever did it showed a complete lack of scientific information on whales," he declared. "A whaler would know there are cavities that have to be pierced before gasses can get out."[56] And joining him in the examination was none other than Terry Newby.

It must have been a surreal situation for the young scientist. He knew the circumstances of the animals' deaths only too well, and he had participated in the effort to conceal their bodies. Yet neither the capture nor the decision to deceive had been his, and he had no desire to tarnish his budding career with the scandal at Penn Cove. So he played dumb. "Whoever did it showed a complete disregard for scientific inquiry," Newby told a reporter, and in an article for *Pacific Search* magazine he went further: "If Washington had proper legislation for the control of hunting and capture of all our marine mammals, we might not have seen the tragedy of Whidbey Island."[57] And if the state needed a special observer for orca captures, Newby felt he was the man for the job.

From his perch on Vancouver Island, Bob Wright, too, called on the Washington State government to intervene. On November 26, 1970, just a week after the discovery of the dead whales in Penn Cove, Sealand's owner criticized his US rivals. "We were quite disturbed by their operation," he told reporters. "Their methods were deplored and completely unacceptable to us." He especially criticized Namu Inc.'s use of a "heavy, specially constructed whale net designed to try to contain the whales by force." If an orca attempted to tear through such a net, it would become entangled—likely the cause of the calves' deaths. In contrast, he claimed, Sealand used only lightweight mesh, which posed no threat if the whales "chose" to escape.[58]

It was an astonishing claim. Just that year, Wright had used heavy seine nets to confine orcas in Pedder Bay for eight months. Scarred-Jaw Cow had become tangled and drowned in one the previous May. Yet unlike the Penn Cove carcasses, that body never turned up. No one knew about it except the Sealand staffers who had sunk it off Race Rocks. "Our people are whale people," proclaimed Wright. "They're fascinated by them and are the last people who would want any harm to come to them."[59]

In one sense, the comments were understandable. Wright may have bought his first whale from Griffin, but Penn Cove put a spotlight on orca capture, and he had no desire to become tainted by his rival's disaster. But there were likely deeper motives. Claiming, inexplicably, that killer whale capture was "strictly regulated" in Canada, Wright declared that he supported legislation in Washington State and felt confident it would pass. "The team from the Seattle Aquarium is not allowed to hunt into Canada," he quipped, "and after the first of January, I don't think they will be able to hunt in the US either."[60]

That would be just fine with Wright. The market for live orcas was booming, and the Pedder Bay catch had established him as a supplier just as Namu Inc. came under public scrutiny. With any luck, activists and legislators in Washington State would drive his only competitor out of business.

13

Whaling in the New Northwest

DON GOLDSBERRY HAD been speaking for only a few minutes at the Game Commission's April 1972 hearing, and already Elizabeth Stanton Lay couldn't believe her ears. Branding killer whales with dry ice? Burning their skin with lasers? Confining them to pools for research and profit? What kind of men were these? After listening to representatives from the Audubon Society, Friends of the Earth, and the Washington Environment Council voice their opposition, the sixty-year-old Lay rose to speak. "I have never before heard such a frank statement of what seems to me a totally inhumane attitude toward living creatures," she declared. Marine mammals could do without the type of "research" Namu Inc. proposed. Whales were disappearing around the world, she reminded listeners, and the same could happen to orcas in Puget Sound. "When I was a very little girl, we used to see blackfish out in the bay, and we loved it," she recalled. Now locals rarely saw the great creatures, except when men like Goldsberry trapped them behind nets.[1]

Lay was never one to stand idly by. Named after Elizabeth Cady Stanton, organizer of the 1848 Seneca Falls Convention on women's rights, she would have made her namesake proud. Born in Tacoma in 1911, she had grown up in the nearby town of Rosedale on Henderson Bay and earned a history degree from Reed College in Portland, followed by a master's degree in political science from the University of Washington. She studied in Geneva, worked as a journalist in Washington, DC, and served in the new Federal Security Agency during World War II. From the mid-1940s to the mid-1950s, she worked as a historian for the US military, living in Paris, Frankfurt, and Seoul and producing a two-volume account of the Berlin Airlift. By the time of the Game Commission hearing, Lay had retired to Rosedale, where she played the organ at her Christian Science church, promoted forest preservation, and fought to stop orca capture.[2]

Her interest in the issue may have started with young Ken Gormly's 1968 account of the catch in Vaughn Bay. "Several years ago," Lay told listeners, "there was a newspaper report written by a small school child telling about the harpooning and the bay being full of blood." She had her facts wrong—no killer whale had been harpooned or badly wounded in that capture—but the image appalled her nonetheless. So when Namu Inc. came once again to Henderson Bay in March 1972, she and other locals had decided to act. They drew up a statement urging the state to ban the taking of marine mammals for "commerce, exhibition, sport or any other personal or commercial purpose" and especially to stop issuing permits for killer whale capture. She presented that document, with the signatures of eighteen neighbors, at the hearing. In doing so, she praised the Game Commission for its "compassion," expressing confidence that orcas were "in extremely good hands."[3]

It was not often that the Game Commission drew praise from animal lovers. After all, its primary task was to manage recreational hunting and fishing. But Lay and her fellow activists had come to protest capture, not killing. Perhaps some of them didn't grasp the commission's usual function. More likely, they were too focused on killer whales to see the paradox. Whatever their reasons, they were just as determined to save local orcas from capture as others were to save the world's whales from slaughter.

Lay's testimony came at a turning point in popular views of cetaceans. For decades, scientists had warned of collapsing great whale populations. By the early 1970s, however, a large number of North Americans and Europeans had come to perceive whales not as resources to be managed but as sentient beings to be protected.[4] Numerous factors contributed to this change, but none played a greater role than the display industry itself. In addition to giving scientists access to live cetaceans for the first time, oceanariums enabled millions to see orcas, bottlenose dolphins, and other small cetaceans close up. In 1971–1972, Sea World even held a gray whale calf named Gigi before releasing her into the wild.[5] To be sure, many visitors came away from such exhibits with misgivings about captivity, but most left more likely to view cetaceans as individuals and to care about the fate of whales in the wild. That fate was receiving more attention than ever before. Throughout the early 1970s, magazines such as *National Geographic* ran numerous stories on industrial whaling. Even more influential was the album *Songs of the Humpback Whale* (1970). Drawn partly from recordings made by the US Navy's SOSUS network and compiled by marine biologist Roger Payne, *Songs* became the best-selling natural history album of all time. Released in August 1970, it offered listeners a new perspective on cetaceans and convinced many of the need to save the great whales from extinction.[6]

That fall, the US government seemed to embrace the cause. In November 1970, Interior secretary Walter Hickel followed through on his department's proposal to add eight species of commercially harvested great whales to the list of endangered species. Soon after, the Commerce Department informed the last US whaling company that its permits would not be renewed. In hindsight, these seem surprisingly proactive measures, taken five years before Greenpeace launched its first antiwhaling voyage.[7] Yet they had little bearing on killer whales and other small cetaceans. Moreover, in contrast to Canada, where the federal government had jurisdiction over maritime space, state governments in the United States were responsible for the management of wildlife in their waters, apart from those species listed as endangered—and no one yet suggested that status for killer whales. If Washington State wanted to protect the species in its waters, it would have to pass its own legislation, and Governor Dan Evans planned to do just that.

When state senators and representatives convened in Olympia in January 1971 for the Forty-Second Congress of the state of Washington, they had a lot on their docket. Some were calling for revised gambling laws, others pushed for more open government, and everyone was worried about a new round of layoffs at Boeing.[8] But the legislators didn't forget about killer whales. On January 14, the House Committee on Natural Resources and Ecology began consideration of a bill for marine mammal management. Although it pertained to all marine mammals, everyone knew that orca capture had prompted it. For the next four months, interest groups wrangled over its purpose and scope. Activists demanded a ban on capture, scientists called for research funding, and Ted Griffin was determined to have his say.

Griffin's greatest fear was that the state would shut him down. Over the years, he and Goldsberry had invested heavily in Namu Inc. From his perspective, the company had improved its methods and carefully limited the number of animals removed—to the detriment of its bottom line. Now the state might prohibit further captures before the partners could even recoup expenses.[9] Over the following weeks, Griffin monitored the various bills, wrote to sympathetic legislators and scientists, and testified in Olympia. He declared himself willing to work with the state and again offered to fund killer whale research, but his libertarian views often bubbled out. "For the past six years I have assumed the responsibility of self-regulation for my business," he asserted in an April 20 letter to Lieutenant Governor John Cherberg. "When private business tries to conduct its affairs in a responsible manner and is accountable for its behavior, government intervention is undeserved."[10] Griffin had reason to expect sympathy—six years earlier, Cherberg had joined him on Pier 56 to welcome Namu to Seattle. But the aquarium owner underestimated how profoundly regional views of killer whales had changed. He also failed to see how his own decisions, including the concealment of the dead calves, ensured that many locals viewed his company as neither "responsible" nor "accountable."

In May, the legislature finally passed, and Governor Evans signed, House Bill 106, for the "Regulation of Endangered Fish and Wildlife, Deleterious Exotic Fish and Wildlife, Managed Marine Mammals and Killer Whales." No one was entirely happy with the new law. Although it created a framework for regulating orca capture, it included no funding for research. Despite Griffin and Goldsberry's pleas that the Department of Fisheries be placed in charge of enforcement, the responsibility went to the Game Department. It was a poor fit. Focused on the recreational killing of wildlife, Game Department officials planned to approach orcas as just another "game" species. But in the eyes of many Northwesterners, marine mammals were no longer pests to be eliminated or resources to be managed. They were ecological icons to be preserved at all costs.[11]

Many in Washington, DC, agreed. In March 1971, as legislators in Olympia debated their state bill, the US Senate began consideration of the Ocean Mammal Protection Act. Like legislation in Washington State, it reflected the growing popular view of marine mammals as symbols of the imperiled environment. Unlike its state counterpart in Olympia, however, the US Congress was more concerned with killing than capture. In addition to the open-ocean slaughter conducted by the Japanese and Soviet whaling fleets, legislators examined their own nation's violence toward marine mammals. Criticism of the government-sponsored fur seal harvest was rising, and scientists and citizens alike voiced concern over the hundreds of thousands of dolphins killed annually by US tuna seiners. Although this dolphin by-catch had been occurring for years, the American public was now predisposed to outrage. Over the past decade, millions had seen dolphins at marine parks and enjoyed the *Flipper* movies and television show. As a result, concern for dolphins and other marine mammals extended beyond the coast. In fact, the act's cosponsors, Democratic senators Fred Harris and David Pryor, hailed from Oklahoma and Arkansas, respectively.[12]

Scientists played a pivotal role in this process. In June 1971, whale specialists held an international conference at the Skyland Resort in Shenandoah National Park, where many called for a more ecological approach to population management. Nine days later, the IWC began its first-ever meeting in Washington, DC. Undersecretary of State U. Alexis Johnson opened the gathering with a call for sharp cuts to the international harvest, and when delegates balked, leaders in the US Congress decided to act. In the House, Michigan Democrat John Dingell consulted closely with marine mammal researchers Ken Norris, G. Carleton Ray, and William E. Shevill on potential legislation, and in the Senate Warren Magnuson played a central role. Over the course of a year, Washington State's senior US senator had undergone a remarkable transformation. The previous July, he had pressed the Commerce Department to reconsider protection of sperm whales; now he emerged as a leading advocate for cetaceans.[13] Although direct evidence is lacking, it seems plausible that the furor over the Penn Cove Roundup played some role in convincing him of the rising political importance of whales, particularly among his constituents in Washington State. Whatever the reason, Magnuson strongly supported the resolution put forward by Republican Pennsylvania senator Hugh Scott calling for a worldwide ten-year moratorium on the killing of all species of whales.

The resolution came at a profoundly divisive moment in the nation's history. Democratic hopefuls were eying the 1972 presidential nomination, and the chair of the Senate Foreign Relations Committee, William Fulbright, was a vocal critic of the Nixon administration. Yet despite partisan rancor, the resolution passed the Senate unanimously on June 29.[14] It was a stunning illustration of how far environmental values had come. As recently as the late 1960s, few members of Congress concerned themselves with the survival of whales. Now no US politician wanted to go on record opposing their protection. In July 1971, the House Subcommittee on International Organizations and Movements held hearings on the moratorium resolution. Most of the testimony was predictable. Environmental groups warned of ecological loss if the whales died off, while tuna fishermen and canners complained that protection of dolphins would cripple their industry. Yet public pressure overwhelmed all opposition. As marine mammologist and former curator of Marineland of the Pacific Ken Norris later wrote, the debate made clear that the dominant view was no longer "let us change the take of whales so their populations can prosper," but rather "we must take no whales at all, because they are incredible creatures whose lives we hardly understand." Cetaceans had become, he wrote, "a symbol of man's mindless destruction of nature."[15] In contrast to the discussion in Washington State, however, few legislators or witnesses at the hearings brought up the subject of killer whales—with the notable exception of George L. Small.

A professor at the City University of New York and author of *The Blue Whale* (1971), Small strongly supported the call for a whaling moratorium. "If worldwide protection is not now granted," he warned, "disaster will strike the whales as well as the human race itself." But his reasoning likely surprised many listeners. First, he argued that in the future, more and more people would need to eat whales. "If the Government of this Republic is truly concerned with the well-being of mankind," he declared, "it will raise its voice to protect this perpetual source of food." It was a comment profoundly out of step with shifting popular views. Second, Small warned that if the great whales were to vanish, people might find themselves on an ocean predator's menu. "There is one species of large whale that has no commercial value for man," he asserted, and that was the numerous and "carnivorous" killer whale. "When there are no more baby whales of large species for it to feed on, where will it find food?" he asked. "How long could the seals of the world sustain the killer whale population? When it eliminated the seals, what would it use for food?" He added ominously, "The beaches of the world have enough problems already."[16]

Small was making a point about unintended consequences, but his comments hinted at the lens through which many scientists still viewed *Orcinus orca*. From his perspective, killer whales were, above all, predators that ate other mammals, including his beloved blue whales. Who was to say they wouldn't attack humans, particularly as other prey grew scarce? Yet as many of his listeners surely knew, Small was wrong on both counts. By the time he testified, orcas had attained tremendous "commercial value" not only to whalers but to the marine park industry. Moreover, it was the species' reluctance to eat humans, even in captivity, that had warmed public opinion toward them. Indeed, even as Small warned of orcas feasting on beachgoers, activists in the Pacific Northwest worried the iconic animals would vanish just as people were getting to know them.

Game Department officials in Washington State had no idea what they had gotten themselves into. From the moment the state legislature passed Bill 106, activists shifted their focus from Namu Inc. to the department itself.

Although the Game Department would not officially assume its new duties until August 1971, Director Carl Crouse had already received an influx of letters urging him to refrain from issuing capture permits. Meanwhile, he and his staff scrambled to learn about marine mammals and the business of killer whale capture. Through the spring and early summer of 1971, they consulted frequently with Griffin and Goldsberry, and in this sense the legislation initially worked to the partners' advantage. By using Namu Inc. as a template, the Game Department raised the barriers for other entrepreneurs who might consider catching the species. Yet, as activists pointed out, departmental officials still had no concrete data on the region's killer whale population. And if they didn't know how many there were, how could they issue permits? To answer this question, Crouse and his staff looked north.

That summer, the Game Department participated in the world's first "killer whale census." Organized by Michael Bigg, an innovative young marine mammalogist for the Canadian Department of Fisheries, the census enlisted the help of hundreds of US and Canadian volunteers to make a one-day head count of orcas on the Pacific coast.[17] Conducted on July 26, 1971, the census resulted in a total of 549 killer whales spotted from Alaska to California. Most sightings occurred around Vancouver Island, with only 114 killer whales seen in Washington State waters—many of them reported by Griffin, who spent the day counting whales in Haro Strait. The numbers were lower than some observers expected, but that didn't necessarily point toward prohibition of capture.[18] No one knew if the killer whale population was static or fluctuating, resident or migratory.

The greater challenge was the identification of the individual animals. Without the ability to distinguish the orcas, scientists had no means to assess the species' migration patterns or social organization, and thus no way to determine if the same groups of whales were being captured and culled repeatedly in the shared waters of Washington and British Columbia. To resolve this problem, researchers considered a variety of means to mark orca bodies. One strong possibility was freeze branding. Developed by Washington State University professor R. Keith Farrell, the method used a mixture of dry ice and ethanol to inscribe marks on animals. Initially conceived as a more humane means of branding cattle, it had recently been used by the US Bureau of Land Management to help monitor wild mustangs. With herds of mustangs frequently crossing the US-Canada border, officials wanted a means of both quantifying and laying claim to the animals as US government property. Washington State Game Department officials now hoped the method might be used to manage the region's killer whales.[19]

Freeze branding initially enjoyed broad support. Griffin and Goldsberry endorsed it, as did Mike Bigg and activist-scientist Paul Spong in British Columbia. Of course, it was only possible to perform the procedure on captured animals. Under the Marine Mammal Biological Laboratory's contract with Namu Inc., federal scientist Mark Keyes had made the first attempt in February 1970, marking a whale caught and released in Carr Inlet and a second netted during the Penn Cove Roundup. The next opportunity came a year later, in the first capture supervised by the Washington State government.[20]

The Game Department officially assumed management of marine mammals on August 20, 1971. That same day, Carl Crouse issued Namu Inc. a permit for the capture of six orcas. The permit stipulated that the whales could be held in nets for no longer than ten days and that no animal shorter than eight feet or longer than sixteen feet could be removed. It also required the company to pay the state $1,000 for each whale kept. With at least ten outstanding orders from buyers in the United States and Europe, Griffin and Goldsberry immediately launched their operation.[21] To the chagrin of many, the chase again led them to Whidbey Island. Over the following days, they chased a large group of orcas into Holmes Harbor, where private boaters foiled one capture attempt. By the following morning, Namu Inc. had corralled fifteen whales. They were members of the southern resident L pod, and as in the previous year, they were trapped in Penn Cove.[22]

This time, it was a smaller, more controlled operation. After setting up its holding pen and platform, the Namu Inc. team anchored the nets, while divers ensured that no whales were entangled. Using a hoop net from a makeshift platform, Griffin then separated three whales for removal—all of whom went to Sea World.[23] Finally, the captors selected an adult female for freeze branding by Keyes and Farrell. The marks included "US" on her dorsal fin and "—1" on her left side. Game Department supervisor Garry Garrison later explained the operation as "an attempt to obtain knowledge of the whales' activities and migration routes."[24] Yet the branding marked the whale not only as a recognizable individual (in human eyes) but also implicitly as US property. By the evening of August 27, the remaining whales were released, and most of those who participated considered the operation an unqualified success. "I was amazed at the skill with which these animals were separated and held for marking," noted Farrell. "It was smoother than the rhino and elephant programs by far."[25] But that wasn't good enough for Don McGaffin.

FIGURE 13.1 Griffin casts a hoop net as Don Goldsberry looks on, Penn Cove, August 1971. Courtesy of Washington State Archives.

McGaffin was a combative television reporter, and he didn't scare easily. After serving in the Marine Corps during World War II and the Korean War, he had begun a career as a television journalist in California. Arriving in Seattle in 1968, he reported first for KOMO and then for KING, specializing in uncovering government corruption and malfeasance. As Washington Republican Steve Excell later put it, "Don McGaffin was the incisive, biting voice of the liberal left on KING-TV before Seattle was liberal and leftist."[26] And by the early 1970s, one of the liberal Left's great causes in the new Northwest was to stop orca capture.

With a summer home in Coupeville, McGaffin had had a front-row seat to the Penn Cove Roundup of the previous year, and when Namu Inc. made its second catch in August 1971, he was furious. As the capture crew set up its floats, he arrived in a skiff with his cameraman, demanding access to the pen. When Griffin and Goldsberry refused, McGaffin motored around the pen taking footage. "He could be a son of a bitch!" laughed local resident Mark Funk, whose father, Wallie, knew McGaffin well. "I have no doubt he got under the whale captors' skins."[27] Enraged, the partners took turns chasing McGaffin's boat away. In his subsequent news report, McGaffin claimed the partners had attempted to "swamp" his boat, but he also offered thoughtful commentary. The Game Department may be allowing captures, he noted, but "no one even knows how many killer whales there are." Although McGaffin

FIGURE 13.2 Namu Inc. crew subdues tethered whale, Penn Cove, August 1971. Courtesy of Washington State Archives.

positioned himself as a concerned observer, the clash with the whale catchers had clearly become personal. One evening at the Captain Whidbey Inn, McGaffin confronted one of Griffin's divers and knocked him unconscious.[28] And he wasn't the only one who felt like attacking the whale catchers.

By the fall of 1971, criticism of Namu Inc. had reached new heights but Don Goldsberry didn't give a damn. He happily embraced the image of whale rustler. "WANTED," read one of his flyers. "ORCINUS ORCA alias GRAMPUS alias KILLER WHALE"—featuring saloon-style front and profile images.[29] But the black hat didn't sit comfortably on his partner's head. Griffin had begun his pursuit of orcas out of a desire to connect, but somewhere along the line it had become about money, and he hadn't felt like himself in a long time. He dreamed of finding another whale like Namu and having the chance to bond with it far from the public glare. At the same time, he wanted recognition for what he had accomplished, for how he had cast new light on a once-feared predator. In the past five years, some twenty million people worldwide had seen orcas he had captured—only a tiny fraction of whom could have ever glimpsed the species in the wild.[30] But many Northwesterners wanted the captures stopped, and some part of him agreed.

His family had always brought him joy, and he turned to it now. By the fall of 1971, eight-year-old Jay could already handle the family's thirteen-foot Boston Whaler, and seven-year-old Jon wasn't far behind. Griffin took his sons on the water to show them his work, and sometimes they joined him in floatplanes to search for whales. "I wonder how the boys view my profession," he pondered. "Will they choose to join me? What influence do their teachers have?"[31]

The latter was a poignant question. Angry letters arrived regularly, some of them from schoolchildren. "You are the meanest, most cruel man on earth," read one. For a man who often visited schools to give talks about killer whales, these must have stung. But the phone calls from adults were worse. Most of the threats focused on him. "Griffin, I'm going to put a bullet between your eyes," promised one caller. Others targeted his family. "We know where your kids are," whispered another menacing voice, "and you'll never know when we're coming."[32]

Most—perhaps all—were crank calls. But Griffin had reasons to take them seriously. Decades earlier, a disgruntled mill worker had nearly killed his brother and infant half-sister in rural Oregon. Just ten years old and shaking with fear, Jim Griffin had faced the assailant down with a shotgun in an incident that became family lore.[33] Equally pertinent was the life story of Ted's close friend and fellow Bainbridge Island resident Jon Lindbergh. Son of the famous aviator Charles Lindbergh, Jon had grown up in the shadow of his older brother's kidnapping and murder. In short, Griffin couldn't easily dismiss the threats, and they contributed to erratic behavior that had his wife worried.

⁂

Forty years later, Joan—Joan Grant now—sat down to talk to me in a cafe in Poulsbo, Washington. Although the couple had separated in 1974, the pain was still evident on her face as she reflected on her life with Ted Griffin—a brilliant, driven man who seemed able to do anything. "He was really kind. He loved to help people," she told me, "and he was a good dad, a really good dad." But as the criticism mounted in the early 1970s, Griffin had changed. Unable to comprehend the new environmental politics restructuring his life, he retreated deeper into his libertarian worldview, becoming obsessed with the writings of Ayn Rand and attending local meetings of the John Birch Society. "I don't know what was the matter with him, but he just became different. This kind, wonderful man who could do everything, suddenly changed," she sighed, "and I blamed it on Ayn Rand just as much as anything else."

For Joan, it was the paranoid behavior that proved most alarming. "He built a special shed for the kids to wait for the school bus," she told me. "It even had communication to our house because he was afraid somebody was going to kidnap them!"

"Did he ever tell you about the phone calls threatening the kids?" I asked.

"No, he didn't," she replied, stunned. "He never mentioned it."[34]

As Griffin wrestled with his demons, activists nearby captured headlines. In mid-September, a ragtag organization in Vancouver, British Columbia, known as the "Don't Make a Wave Committee" launched a quixotic protest against US nuclear testing on Alaska's Amchitka Island. Made up of Canadian and expat US activists, the group hired a halibut fisherman to take it to the site. Although the activists failed to reach Amchitka or stop the detonation, the ensuing publicity spurred the Nixon administration's February 1972 announcement that it would forgo future tests in the area. By that time, the activists had returned to Vancouver and renamed themselves the Greenpeace Foundation.[35] It would be three years, however, before they embraced the antiwhaling cause that the Nixon administration was already promoting.

The following month, March 1972, Namu Inc. made another capture, netting nine whales in Henderson Bay. This was almost surely the event that spurred Elizabeth Stanton Lay to attend the Game Commission hearing a month later. Yet not all local residents opposed the capture. One local woman allowed Griffin to use her telephone and seemed enamored with the whale catchers. Another couple, the Johnsons, watched the entire operation from their waterfront home and marveled at the captors' patience and technique. Game Department officials, too, monitored the operation, this time bringing along Terry Newby. Now a doctoral student at the University of Washington, Newby had agreed to serve as a department consultant and was hoping to freeze-brand the netted whales in Henderson Bay. But Griffin and Goldsberry were still fuming over Newby's criticism of the Penn Cove Roundup, and they refused to let him aboard.[36]

The capture also caught the attention of Bob Wright, then in Washington State on a recruiting trip. After visiting the capture site, Sealand's owner assigned his newly hired curator, John Colby, to observe the operation. The twenty-four-year-old biologist was eager for the task, and he gleaned information however he could. After perching for three days in a tree taking meticulous notes on Namu Inc.'s methods, Colby invited Newby to dinner. The two men had much to discuss. Both were biologists, both alumni of the University of Puget Sound, and both fascinated with killer whales and their capture. Still smarting from his rough treatment by Griffin and Goldsberry, Newby didn't mind revealing trade secrets. Despite being on Washington State's payroll, he shared everything he knew about the partners' operation with Colby, a man who had just taken a job with their Canadian competitor.[37]

For Namu Inc., the March 1972 catch proved disappointing, yielding only one keeper—a thirteen-foot male, Canuck, who soon went to Sea World.[38] The company's permit was set to expire on March 31, and it had collected only four of the six whales allowed. Was it bad luck? Were orcas starting to avoid local waters? Or was capture depleting their numbers? Such questions confronted the Game Commission as it prepared for its April 11 hearing in Olympia, set to be devoted entirely to the question of killer whale capture.[39]

By the time of the hearing, the Game Commission was feeling the heat. In his opening statement, chairman Arthur Coffin reported that he had received 450 letters opposed to killer whale capture and only a handful in favor. Yet officials remained sympathetic to Namu Inc. Game Department supervisor Garry Garrison reminded the audience that "blackfish" had until recently been the subject of fear and violence. "They had no protection from man," he declared. "They were harassed, shot at, and killed at every opportunity." Capture had opened people's eyes and created research opportunities, but it had also prompted concern and regulation. He assured listeners that state management was working. In the past eight months, the Game Department had monitored two capture operations in which four whales were taken. "There has been no mortality," Garrison noted, "and during this time the permittees have abided by all Department regulations."[40]

The commission then turned to Goldsberry, who was representing Namu Inc. at the hearing. The first questions centered on the orca population. "In any one day," he estimated, there were probably a hundred killer whales in Washington waters. "I don't think they are always the same whales," he hastened to add. "Some will leave and others will come in and replace them." Was there a danger of taking too many out of one pod? asked a commission member. "No," replied Goldsberry. Although fishermen in British Columbia had sold entire pods, he and Griffin had carefully limited their catches and observed size limits that were now part of the state regulations.[41]

The hearing then turned to methods and objectives. When one commission member asked if critics had reason to consider Namu Inc.'s collection methods cruel, Goldsberry scoffed. "I could go out there and capture an animal in ten minutes instead of spending $25,000," he declared. "It would only cost me one bullet or a tranquilizer, and all I would have to do would be to kill a cow, pull her along the ship and the calf will come."[42] In fact, animal collectors had long used such methods to capture juvenile predators such as tigers and lions—but rarely in public view.

"What is the purpose for which you seek these permits?" asked another member.

"Strictly for our knowledge and for the animal to go to oceanariums," Goldsberry responded.

"Well, scientific purposes is one thing and taking the animals for profit is another. Now, what are you taking for scientific purposes and what for profit?"

"It is a combination of the two, sir," replied Goldsberry. "Yes, we are making profit, otherwise we wouldn't be there. But it is like the salmon or herring out in Puget Sound. There are commercial people after those animals if there is an abundance of them."

Yet, as Goldsberry surely knew, the sight of killer whale capture, unlike the netting of fish, stirred public outrage, and it soon became clear that the commission hoped to move Namu Inc.'s operations away from populated areas for precisely that reason.

"What would you think about restricting your area for this type of trapping, say north of Whidbey Island?" asked another member of the commission.

"That is like trying to sweep us underneath the rug," Goldsberry answered. "If we caught animals in that area, believe me you would receive just as many complaints as you do now."[43]

But the real fireworks came when the commission turned to a second applicant: Gig Harbor fisherman Peter Babich. Explaining that he had worked with Griffin and Goldsberry on numerous captures, Babich was now applying for his own permit to collect three orcas. After attempting to assess Babich's expertise, Crouse turned to the question likely on everyone's mind.

"I presume, Mr. Babich, that you have a place to keep the whales and a veterinarian?"

"Yes," he responded. "I have a letter here that I would like to submit from the Animal Clinic in Vancouver and the Vancouver Public Aquarium for the facilities, and my two associates who work at the Vancouver Public Aquarium."

The commission was flabbergasted. "Do I understand then that you might take these whales in our waters, under Game Department surveillance, and then take them into Canada?" asked one member.

"Yes. That is the closest tank available with all the necessary equipment."

"How would the whales be transported to Canada?"

"By truck, and they would be sold from that point."[44]

Still stunned by Babich's proposal, the commission opened the floor to public comment. Speaking for the local Audubon Society, Robert Ramsey questioned the notion of orcas and other animals as human possessions. "Man has always found it profitable to use natural resources which have cost him nothing to produce," he observed. "We are not in sympathy with the confining of such large creatures for the purpose of display." When Crouse asked if he opposed all captivity, Ramsey hedged, noting that whales "seem to be a more knowing type of creature and they merit special consideration"—a response that drew applause from the audience. Tacoma resident Helen Engle agreed, arguing that state policy should reflect the region's shifting values. "I think we have a natural resource here that has a tourist value and has an overwhelming value to the people of Puget Sound," she observed. "Is that animal for viewing and entertaining or more important in its natural habitat?" Various other speakers criticized Game Department officials for failing to assess the total killer whale population before issuing permits.[45]

By this time, commissioner Burton Lauckhart had had enough of these mouthy activists, who espoused entirely different views of animals and their proper use than he was accustomed to. "The Game Department operates totally on license money," he reminded his listeners. "You in this audience who are complaining about the lack of research, if you don't have a hunting and fishing license, you are contributing nothing to the management of the wildlife . . . in the state." Moreover, he asserted, the recent Canadian-led census indicated that "there is a good whale population along the coast of Washington and British Columbia" and that "we have no indication that the few animals that have been taken have in any way harmed the existing population."[46]

In the end, however, public pressure had an effect. While the commission rejected Babich's application on the grounds that it would be "basically a Canadian operation," it delayed its decision on Namu Inc.'s permits until its next hearing the following month. By the end of the meeting, one thing was clear: Game Department officials viewed the location of capture—not capture itself—as the primary issue. "We do not want any more whales taken for a while in Penn Cove or in that area," explained Crouse. "Public opinion up there is pretty stirred up."[47] In fact, the hearings had made clear that the debate was in many ways a struggle between the values of the old Northwest and the new. As *Seattle Times* reporter Lyle Burt put it, "Much of the opposition appeared to come from persons who live or have summer homes along southern Puget Sound beaches where Namu Inc., has done much of its hunting."[48]

Over the following weeks, Goldsberry assured Griffin that the state would issue the permits, but he did not warn his partner of the possible restriction on their range of operation. And as the two men waited for the next hearing, public pressure continued to mount. The *Seattle Times* published an editorial calling for the end of all killer whale trapping in Puget Sound, and on Sunday, May 21—a day before the Game Commission hearing—KING-TV broadcast a searing half-hour documentary by Don McGaffin entitled *Catch 33*. Aired in prime time at 8:30 p.m., it depicted Griffin and Goldsberry as driven by greed and posing a mortal threat to the orcas of Puget Sound.[49] For Griffin, such portrayals had become intolerable. Unlike the biologists at the Marine Mammal Biological Laboratory or the hunters the Game Commission regulated, he had never in his life set out to kill an animal. True, whales had died in his catches, but the greatest tragedy had come through the error of activists, and in the two years since, no whales had died in a catch. Yet public views were changing faster than he realized. When he had begun his quest, the live capture of a killer whale seemed the antithesis of commercial whaling; now many viewed them as one and the same.

It all swirled around in Griffin's mind as he took the long drive north to Mount Vernon to attend the Game Commission hearing on May 23, 1972. When he arrived, he took a seat quietly near the back of the room. At first, all seemed to go well for Namu Inc. The public attendees had their say, and then the commission announced its decision. Rather than the six whales requested by Namu Inc., it offered a permit for four. That seemed reasonable, thought Griffin, but then came a new condition: Namu Inc. would be required to give one of the four animals to the Game Department for research free of charge. As Griffin struggled to process this, the commission announced a second condition: capture would not be allowed south of Port Wilson and Point Partridge—in other words, it would be banned from Puget Sound. "It seems they think it's OK to catch whales so long as the public isn't watching," he thought.[50] And in truth, the decision was about public relations, not conservation.

Griffin was furious. He had repeatedly warned that barring capture from Puget Sound would increase the dangers to both orcas and men. Now, to appease his critics, the commission had done just that. "Up to that time, our whale capture activities had succeeded without any injuries to our crew, but we had many close calls," Griffin later explained. "Under the best of conditions, this was hazardous for whales and divers." In the rough and deep waters of the Strait of Juan de Fuca or around the San Juan Islands, a capture operation could be fatal for both.[51] But his political views also colored his reaction. In his eyes, Game Department officials represented an arbitrary state controlled by mediocre men out to crush an extraordinary individual. "It is the loss of one of my highest values, my freedom, which they threaten," he later wrote. "The freedom to live and enjoy life to the fullest without compromising my values or sacrificing them to the needs of society."[52]

Obsessed with his personal adventure, Griffin had failed to appreciate the broader shift in environmental values to which he had contributed, particularly through the transformed popular and scientific views of *Orcinus orca*. In 1962, his dream of capturing and connecting with a *killer* whale had seemed visionary, even quixotic. People around the world considered orcas dangerous pests, and fishermen and scientists in the Pacific Northwest routinely killed them. By 1972, however, his adversaries were no longer fishermen who shot orcas; they were activists who feared he might drive the animals from the waters that both they and he loved.

Yet Griffin's fall went deeper than that. For years, he had told himself that he would bond with another killer whale, if only he could find an animal like Namu. It never happened—not because he didn't catch the "right" orca, but because he was a changed man. No longer driven to connect with killer whales, he had acted, in his own words, "like a businessman." Critics' barbs about his being out for profit may have stung so much because deep down he feared they were true. He wasn't in it for the love of animals anymore. Most of that died with Namu, and the disaster at Penn Cove had sealed it. "When you pursue your hobby as a business," he later told me, "one side or the other fails."[53]

More than any other figure in the world, Griffin had changed the public perception of orcas, but he proved heedless of the consequences. His libertarian worldview caused him to focus on government interference rather than grasp the reasons behind rising public concern. "As I could never have thought others would not see me as I saw myself," he admitted, "I did not try to persuade anyone as to my intentions or try to justify my activities."[54] A firm believer in the free-market economy, he had failed to engage in the marketplace of ideas. As a result, his detractors could never imagine that such a man loved animals as much as they did, and Griffin could never accept that other animal lovers might honestly believe he posed a threat to the species he had done so much to promote. In the end, this man who had lived so public a life over the previous decade made a quiet exit in the spring of 1972. Annoyed at the new restrictions and torn by inner tension, he sold his share of the aquarium and Namu Inc. to Goldsberry and moved his family to eastern Washington. "I wanted to get away from myself," he later explained.[55] Good riddance, muttered his critics.

❧

Friends and admirers were sad to see him go, among them Gary Keffler. Over the previous decade, the champion diver had shared many adventures with Griffin. They had begun collecting wolf eels and octopuses and ended up capturing the ocean's greatest predator, which became the world's most popular display animal. Forty years later, Keffler put it in perspective:

> I think it was a good thing that they stopped Ted, that they stopped us. Knowing what we know now, we would have wiped the whales out no matter how careful we were. But we didn't know—nobody did. If Ted had known what we know now, I think he would have stopped himself.[56]

In fact, Griffin had stopped himself, but his former partner was about to double down.

14

Big Government and Big Business

JEFF FOSTER ARRIVED at Pier 56 in the summer of 1971 eager to get started. Although just fifteen, the Bellevue native already had extensive experience with wildlife. His father was head veterinarian at Seattle's Woodland Park Zoo, and Foster himself was a skilled diver who often brought live fish and octopuses to the Seattle Marine Aquarium. When Don Goldsberry offered him a job with the seal trainers, Foster accepted, and after a quick introduction to his co-workers, he received his first assignment. "There's a bum out back," said one of the trainers. "Go get him out of there." In the aquarium business, this was a serious matter. Over the years, vandals had thrown objects into tanks and even attacked captive animals. The teen made his way to the rear entrance, where he could hear dolphins and seals splashing in nearby pools. There he found a man sleeping beside a dumpster. "He's laid out, he looks pretty big, and I give him my tough voice—my voice that's still cracking," laughed Foster, adding in a high-pitched squeal, "You gotta get out of here!" As he helped the derelict to his feet, Foster was stunned by his size. His hand was "like a baseball mitt," Foster recalled. "He is like six foot six—huge, huge guy." At first, the confused man seemed willing to leave, but as they approached the gate, he decided to take a swing at the teen. Foster managed to duck away, and he never forgot what happened next.

Unknown to Foster, Goldsberry had followed behind to make sure the youngster was safe, and he now grabbed the flailing attacker before he could throw another punch. "The next thing I know this guy is just lifted up and thrown," Foster recounted. "Don picks this guy up and tosses him easily from here to that wall." Yet protection quickly gave way to rage. "He proceeded to kick his teeth in, kicked the shit out of him." Fearing for the man's life,

Foster sprinted to the aquarium office, where he found a woman behind a desk, Goldsberry's wife, Pat. "Don is going to kill a guy!" he yelled. Together, they ran to the back only to find the beaten man staggering away leaving a trail of blood and teeth. "I really thought he was going to kill him," Foster sighed, shaking his head.[1]

The encounter revealed much about Don Goldsberry. On the one hand, he was very much the hard-bitten roughneck he appeared to be. His temper was explosive, and although he stood only five foot ten, his powerful build and menacing manner made him appear much larger. "He was a scary guy," recalled Foster.[2] On the other hand, the ex-fisherman from Tacoma proved a cagey and determined businessman. At times, that meant using his fists to protect his interests. More often, it meant navigating the shifting tides of public opinion and state policy, for which he proved better equipped than his erstwhile partner. Unlike Ted Griffin, Goldsberry was comfortable working with and around government regulation. Although annoyed by the state Game Department, he made peace in order to keep catching whales, and when the US government passed the sweeping Marine Mammal Protection Act in 1972, he joined forces with Sea World. At first glance, it seemed an awkward fit: the gruff Pacific Northwest fisherman and the slick Southern California corporation. But when it came to orca capture, big business offered the perfect answer to big government.

Goldsberry approached the summer of 1972 with high hopes. Eager to leave, Griffin had sold his stake in the aquarium and Namu Inc. for just $10,000—less than half the price of a healthy orca.[3] If Goldsberry could keep the capture business going, he had much to gain, and his willingness to cooperate with the state bore fruit when he received a permit for four whales. In the accompanying letter, dated June 14, 1972, Game Department director Carl Crouse noted that it had been a difficult decision. Not only had the public expressed "emotional" opposition to orca capture, but there were growing calls for a national moratorium on the killing and capture of marine mammals. The Game Commission remained confident that Namu Inc. and the state could continue to work together, but its members had insisted on the conditions that had driven Griffin into retirement. In addition to prohibiting Namu Inc. from captures in Puget Sound, the permit required Goldsberry to deposit $3,000 in escrow—$1,000 each for the three whales he was permitted to sell—and donate the fourth animal to the Game Department as a "research whale."[4] The latter condition must have irked Goldsberry, who suspected

FIGURE 14.1 Don Goldsberry, August 1970. Photo by Wallie V. Funk. Courtesy of Center for Pacific Northwest Studies, Western Washington University.

officials would make the whale available to Terry Newby—the marine biologist who had publicly criticized Namu Inc.[5] But Goldsberry accepted the conditions to get his permit.

As the man who set the nets, Goldsberry had always considered himself the lead whale catcher, but he had relied on Griffin for capture opportunities. Now he had to find the animals himself. Yet, unlike Griffin, Goldsberry wasn't eager to spend days in a runabout following pods and studying their behavior. "Don wasn't an animal person," explained Foster. "He was a fisherman."[6] And that fisherman had lost both his whale spotter and his favorite capture sites. Because the permit barred the company from collecting in Puget Sound, the only viable alternative seemed to be the San Juan Islands.

In hindsight, this might not seem so bad. Today Haro Strait remains one of the best spots in the world to see wild killer whales. But it was hardly ideal for live capture. In addition to deep water and strong currents, the open space made it difficult to drive pods into shallow bays, particularly because older animals had grown wise to their pursuers. "By that time, the whales were familiar with the sound of the *Orca*, with the pitch of the prop," noted Foster. "They would hear us coming, and they would start porpoising across the water."[7] To make matters worse, Namu Inc. faced competition in the area. Bob Wright's Sealand sat on the west side of Haro Strait, just seven miles from San Juan Island. In the summer of 1972, Goldsberry and his team found themselves trying to herd orcas toward their side of the border, while Wright's crew did the same. Sealand curator John Colby repeatedly crossed into US waters in an effort to push the whales westward. "I was driving them to the Canadian side, using every harassment tool at my disposal," he later told me. "Sorry, but that's what I did."[8] By autumn, neither team had made a capture, and the US government was about to change the killer whale business forever.

That summer, the global debate over marine mammals had hit high gear. In the United States, public revulsion at fur sealing, dolphin deaths, and commercial whaling was becoming impossible to ignore. As historian Kurkpatrick Dorsey has noted, "Only the Vietnam War generated more protest letters to the White House in the early 1970s than did whaling."[9] In response, the US government took a more aggressive stance against international whaling, with politicians from the Pacific Northwest playing a key role. In early June 1972, Senator Warren Magnuson of Washington State joined the US delegation to the United Nations Conference on the Human Environment in Stockholm. Along with Alaskan politician and former Interior secretary Walter Hickel, Magnuson championed the proposed ten-year international whaling moratorium. They received a boost from demonstrators, who marched in nearby streets blasting recordings of humpback whales on loudspeakers. Although the conference vote in favor of the moratorium was unanimous, it was largely symbolic. The UN had no power to enforce it, and Japan and the Soviet Union—the two major whaling nations—had boycotted the Stockholm gathering. Yet many observers hoped the vote would spur action from the IWC, set to hold its annual meeting in London later that month.[10]

They were disappointed. At the IWC meeting, the Japanese and Soviet representatives rejected calls for a sweeping moratorium, arguing that no scientific basis existed for protecting nonendangered species of cetaceans. To the chagrin of the US delegates, such arguments swayed their Canadian counterparts. Although Canadian delegates had voted for the moratorium at Stockholm and Department of the Environment minister Jack Davis had issued a press release supporting a whaling ban just days before, Canada abstained from the final vote, along with three other nations. The result was a 6–4 defeat for the US initiative.[11] Ottawa's stance appalled many Canadians, inadvertently boosting the country's antiwhaling movement. For their part, US delegates left London convinced the IWC might never take meaningful action. As Hickel complained, "What the vote shows is the conduct of whaling in the world is still essentially dominated by commercial interests."[12]

The IWC's failure to act proved a final spur for the US Congress. Although the House of Representatives had introduced a marine mammal bill in March 1971, the bill had failed to pass, in part because it contained exceptions for the tuna industry's slaughter of dolphins. Yet public pressure continued to build. As Democratic representative Edward Garmatz of Massachusetts observed that September, "During my 24 years as a member of Congress, I have never before experienced the volume of mail I have been receiving on the subject of ocean mammals."[13] In the spring of 1972, Democrat John Dingell set to work in earnest with his allies in the House, including Republican representative Thomas Pelly of Washington State, and Warren Magnuson lent his support in the Senate. If the international community refused to act, they reasoned, the United States would take unilateral action. Drafted in consultation with Ken Norris, Carleton Ray, and other marine mammal specialists, the final bill moved quickly through both houses of Congress, and on October 21, 1972, President Richard Nixon signed it into law.[14]

By any measure, the Marine Mammal Protection Act (MMPA) was a landmark piece of legislation. In addition to prohibiting the import and export of marine mammal products, it placed a moratorium on the "taking" of marine mammals in US waters—defined as killing, capture, or harassment. It also required the Interior Department to list species that were "threatened," as well as endangered.[15] Perhaps most important, it was the first wildlife protection law in the world to take an ecological approach to species conservation, calling on the US government not only to protect marine mammals but also to safeguard the environments that sustained them. In doing so, it underscored how profoundly public views had changed. As biologist Bob Hofman later observed, the act was the first legislation to state that "marine mammals have values that go beyond subsistence and commercial taking, which includes their functional roles in ecosystems." For his part, Magnuson rightly lauded it as "the most far-reaching animal protection legislation in the world."[16]

The MMPA had profound economic and scientific implications. Most immediately, it threatened to shut down the US government's century-old fur seal harvest, as well as the Pacific tuna industry if it couldn't curb its killing of dolphins. It also placed into question biological research, which included the killing, dissection, and (more recently) live capture of marine mammals. And it posed an existential threat to the display industry. Without the ability to collect marine mammals, how would oceanariums and marine parks acquire their most popular attractions? To account for such interests, the MMPA contained a permitting system, to be administered by the National Marine Fisheries Service (NMFS). In addition to permits for research and public display, the NMFS could issue "economic hardship" exemptions for the capture of protected species. To advise US officials on enforcement, the law also provided for the creation of an independent government agency: the Marine Mammal Commission.

From the beginning, the commission had close ties to the Pacific Northwest. The presence of the Marine Mammal Biological Laboratory, now subsumed by the new National Oceanic and Atmospheric Administration (NOAA), ensured that many leading marine mammalogists resided in Washington State. Governor Dan Evans successfully lobbied the Nixon administration to include scientists from the region on the Marine Mammal Commission.[17] Among its early members were University of Washington statistician Douglas Chapman and Victor Scheffer—the latter of whom would become the first head of the commission. The White House even considered creating a branch in Seattle, though in the end the commission occupied a single office in Washington, DC. Longtime director of research Bob Hofman recalled that the office at 1625 Eye Street, NW, had an eerie feel to it. The previous tenant had been the Committee to Re-elect the President, and rumor had it that G. Gordon Libby and his co-conspirators had planned the Watergate burglary there. Hofman believed it, noting that the entire office was soundproofed.[18]

More than four decades later, I scheduled a meeting at the Marine Mammal Commission with Hofman and its general counsel, Mike Gosliner, to discuss those early years. Traffic on the Beltway was gridlocked, and I arrived late to the office, now located in Bethesda, Maryland. The seventh-floor office in the DC suburbs seemed an odd place to seek answers about Pacific Northwest killer whales. But as I walked to the conference room, I was relieved to see a framed poster of an orca printed by the Whale Museum in Friday Harbor, Washington. Perhaps I wasn't so far from the Salish Sea after all.

"What's the Salish Sea?" began Hofman. "I've never heard of it." I was taken aback, but the question made sense. It was a recently coined term, popular mostly in the Northwest, and Hofman had been retired for several years. After I explained the phrase and its connection to orcas, Hofman had an immediate response: "Have you heard of the Canadian Michael Bigg?"

"Yes," I replied. "I'm writing a whole chapter on him."

"Good, because in my opinion he is the most important figure in the study of killer whales." The big question with orca capture in the 1970s, Hofman noted, was assessing their demographics. "It was Bigg who figured out how to identify individual whales, and that made it possible to determine the management units."

From there, our discussion turned to the MMPA. "It seems to me," I began, "that the law is domestic in focus. After all, it controls US waters."

"I would take issue with that," corrected Gosliner. "There are international mandates in it. It extends to trade, and it has jurisdiction over US citizens abroad as well. It very much had international conditions in mind."[19]

He was right, of course. One of the fundamental reasons for the law's passage had been a desire to influence other nations' policies, and it had worked on some countries. Despite its abstention in the IWC moratorium vote, the Canadian government issued an Order in Council ending whaling on December 20, 1972—one day before the Marine Mammal Protection Act took effect.[20] Four months later, eighty nations signed the Convention on International Trade in Endangered Species, which listed five species of whales. Yet I wasn't entirely wrong: the MMPA had profound domestic implications. Most important of all, it preempted all state legislation pertaining to the management and treatment of marine mammals.

∞

This proved momentous in Washington State, where Olympia had just passed marine mammal legislation in March 1971. Less than two years later, the federal government had rendered it moot. Washington State officials remained hopeful that the Marine Mammal Commission might review state policy and approve its continued operation. But Don Goldsberry faced even greater uncertainty. On the day the MMPA went into effect, his state capture permit was null and void, and he hadn't caught a single whale. He faced mounting debt, a long list of orders, and no guarantee that he would ever be able to collect orcas again. His most eager customer was even more concerned.

In less than a decade, Sea World had come a long way. Restaurateur George Millay had launched the venture with investment from several former UCLA classmates and advice from Ken Norris, former curator of Marineland of the Pacific. When it opened in March 1964, Sea World was a small affair, with fish, seals, and a few dolphins, along with young women dressed as "sea maids" and other exotic attractions. It proved a hit, drawing 460,000 visitors in its first nine months.[21] Yet it was the arrival of the first Shamu in late 1965 that made Sea World a fixture in Southern California's booming tourist industry. In 1970, the company opened a second marine park in Aurora, Ohio, near Cleveland, which enjoyed even higher attendance. Because this midwestern location was open only seasonally, Sea World flew captive orcas and other marine mammals back to San Diego each winter. With the opening of Disney's Magic Kingdom in Orlando in 1971, company officials saw the opportunity for another year-round location, but they needed orcas to make it work. More than any other oceanarium, Sea World had built its identity around the trademarked "Shamu" shows.[22] Since the purchase of its first orca, the company had obtained its signature attraction almost exclusively from Namu Inc. But the MMPA placed that supply in jeopardy. And it wasn't just about stocking the planned Orlando franchise. Captive orcas didn't live forever, and a Sea World without Shamu was no Sea World at all. If the company was to survive and thrive, it needed a steady stream of live killer whales.

~

Sea World staffer Jim Antrim witnessed that concern firsthand. Born in 1947, Antrim had moved to San Diego as a boy and grew up half a mile from the future site of the marine park. "In those days, there were harbor seals and sea lions in Mission Bay," he told me. "We used to see dolphins regularly." One local woman hosted a seal in her backyard, and Antrim went to see gray whales who had washed up on shore. "It sparked an interest and a passion in me for trying to have a career with the ocean," he reflected. As a high school student, Antrim watched Sea World's construction on the site of the old city dump, and some of his friends looked to the marine park for easy fishing opportunities. "What turned out to be the first killer whale pool actually housed yellowtail [jack] at first," he explained. "They had no security, so some of my friends went over there, landed their boats, and just hooked them on rod and reel!"[23]

After graduating from high school in 1964, Antrim enrolled in a local college. But when he withdrew to take a job, he found himself drafted. After a tour in Vietnam, he finished his zoology degree in 1972. That fall, he found himself dressed in a sports jacket and tie, filling out an application at the Sea World office, when a stocky, red-headed man walked in.

"What are you doing here?" he asked Antrim.

"I'm applying for a job."

"Do you know how long it's been since I've seen a clean-cut, nicely dressed young man in this office?" asked the man. "You're hired." It was Sea World president George Millay.[24]

When Antrim began as a veterinary assistant in October 1972, the MMPA hung over the company like a shadow. "There was anxiety. Sea World of Orlando was on the drawing board, set to open in December 1973," Antrim explained, "so they went to the National Marine Fisheries Service and said, 'Look, we've already invested millions in this.'" Yet the company also hoped to gain control over supply, and it approached Goldsberry with an offer to buy the Seattle Marine Aquarium.[25]

∞

Goldsberry was receptive, for obvious reasons. In the past, orca capture had been a local affair, and even under state regulation he had dealt with Washington State Game Department officials. The MMPA had transformed killer whale catching into a federal matter that would be decided in Washington, DC. With national sway and greater resources, Sea World could not only influence the NMFS but also provide extensive funding for capture expeditions. The transaction itself was relatively simple. In return for the aquarium and Namu Inc., Goldsberry received Sea World stock and a contract to run the company's capture operation in Washington State. Henceforth, the Seattle Marine Aquarium would function as a collection arm for Sea World's signature animal. As Vice President Frank Powell later put it, the company had acquired the aquarium and Namu Inc. to ensure a "continued and economical source of killer whales."[26]

In the meantime, both Sea World and its new subsidiary applied separately to the NMFS for permits to capture killer whales. As director of Namu Inc., Goldsberry requested an economic hardship exemption. Noting that the MMPA had voided his permits from Washington State, he requested a federal permit to capture eight orcas: two for display at Pier 56 and six for sale. Sea World submitted its own application to collect eighty-six marine mammals of various species, including four orcas. In doing so, the company emphasized that it had begun construction of its Orlando franchise under the assumption that it could acquire killer whales, and its investment was now threatened by the federal prohibition on the taking of marine mammals. Although the evidence is circumstantial, it seems almost certain that Sea World and Namu Inc. coordinated their applications and that Sea World officials had received assurances from federal officials that their application would be approved. When Bob Wright visited Seattle with Sealand curator John Colby in late

January 1973, Goldsberry told his Canadian rival that "Sea World was lobbying for hardship permits" and that he expected to resume capture expeditions within six weeks.[27] His estimate proved correct, nearly to the day.

When opponents of capture attended the NMFS hearing in Seattle on February 27, they believed the question of the permits remained open. Although the discussion was limited to Namu Inc.'s application, which Goldsberry had reduced to two whales, top Sea World officials spoke on behalf of their new subsidiary. George Millay argued that the acquisition of the orcas was critical to the "economic well-being" of the Seattle Marine Aquarium, but he admitted the facility's main purpose would be to "hold and acclimate the animals for transportation elsewhere."[28] Hoping to win over skeptics, he played footage of killer whale shows at Sea World, but the tactic backfired, convincing many in the audience that the Southern California business was nothing more than a marine circus. The reaction wasn't limited to activists. Having attended the hearing at Bob Wright's request, Colby was appalled. "It was the most embarrassing presentation I ever listened to," he later told me. " 'Shamu Goes to Hollywood,' or some shit like that." Like many Northwesterners in the audience, Colby couldn't help but note the vast divide between local views of orcas and the values of the California company.[29]

On this occasion, however, Sea World prevailed. After the hearing, Goldsberry requested that the record remain open for public comment, thereby stalling the decision on Namu Inc.'s permit. In the meantime, NMFS director Phillip Roedel approved Sea World's application, including permission to capture four orcas. That same day, Goldsberry withdrew Namu Inc.'s application, making it clear that his request had been a red herring all along. Activists and officials in Washington State were stunned. Governor Evans wrote directly to Secretary of Commerce Frederick Dent, pointing out that Sea World's application had received approval without a public hearing, and he asked Dent to countermand Roedel's action. "The people of the State of Washington have developed a great interest and concern for killer whales in Puget Sound," Evans emphasized, and their views needed to be heard "before any unilateral action is taken by members of your office in Washington, D.C."[30]

Public opinion echoed the governor's. Letters poured in from around the state praising his stance, many of them from schoolchildren. One third-grade class in Tracyton, Washington, sent letters of support to Evans. "Thank you for protecting our Killer Whales in Puget Sound," wrote Rowena Goodnight. "I love Killer Whales and I hope you do too." Yet her classmate Merridy Stout inadvertently hinted at the connection between the display industry and public affection. "I think killer whales are pretty," she wrote. "I saw two killer whales in marine land. . . . I hope the killer whale does not become extinct."[31]

Local newspapers also chimed in, with the *Seattle P-I* publishing an editorial entitled "Save the Killer Whale!" Warning that captors might drive the "gentle giants" from local waters, it urged federal officials to undertake an environmental impact study prior to authorizing the permits. Echoing the ecological approach of the MMPA, the newspaper warned: "Destroy the killer whale and there will be a void in the sea, a void in the ecosystem of a planet which may already be teetering on a thin line between harmony and imbalance—and a void in the conscience of man."[32]

For its part, Sea World drew on political connections to safeguard its permits. On March 20, 1973, company vice president Frank Powell appealed directly to California governor Ronald Reagan. "The apparent concern in the Seattle area is not that we have permits for the collection of eighty-two marine mammals," he explained, "but that we have permission to take four killer whales." Powell was particularly worried about the influence of "a few highly vocal conservationists and the elite pressure groups." In addition to noting that Sea World would lose millions if it couldn't catch the whales, he argued that the marine park provided an invaluable educational service to the public. "It would be a great help to us if Gov. Reagan would relate to the Governor of Washington our position and history as a sound and sensible company," he explained. "It is not our intent to disrupt the natural ecology at Puget Sound or anywhere else."[33]

Secretary Dent agreed with Powell. In confirming the NMFS decision on the permits, he noted that prior to the MMPA's passage, Sea World had begun a $17 million investment in its Florida franchise, including public stock offerings and borrowing from private banks. "If the facility could not open on time," he observed, the company stood to lose $10 million, along with "a reduction of 700 people in the planned work force." He also reminded Evans that the Washington State Game Department had issued a capture permit to Namu Inc. for four orcas less than a year earlier, in June 1972. "Since the four killer whales were not taken," Dent noted, "the Letter of Exemption issued to Sea World does not permit the capture of more killer whales than State officials had previously approved."[34]

Hoping to put the controversy to rest, the NMFS then released a short report on orcas. In the process of justifying Sea World's exemption, it offered a revealing summary of the state of killer whale science in 1973. First, it claimed that "morphological studies of the killer whales suggest that the populations freely interbreed." In other words, it was impossible to distinguish "a killer whale taken in the Atlantic Ocean from one taken in the Pacific." The report also estimated that the global population of the species was "in the hundreds of thousands or millions." And while the authors acknowledged Michael Bigg's census, which was now entering its third year, they largely dismissed its results. "In view of the fact that the whale spends so much of its time under the water," the report suggested, "these counts must only reflect a portion of the total whale population, and do not take into account possible annual fluctuations and migrations."[35]

The report then turned to the impact of capture. After estimating orcas' natural mortality at between 7 and 10 percent per year, it observed that over the past decade an average of six killer whales per year had been removed from the Pacific Northwest. As such, "the granting of a permit to Sea World for the capture of four killer whales in Puget Sound is based on the belief that this number does not represent an overutilization of the species and will add only a small increase to the natural mortality." Moreover, it argued, "The available data indicate that the Puget Sound populations interbreed with other populations and that the Puget Sound animals do not form a separate and isolated grouping. This would suggest the possibility that recruitment to the Sound is based in part on animals from the outside, a feature that would provide insurance for the maintenance of the Sound populations."[36] Subsequent research would prove nearly all of these assumptions wrong, but the NMFS document ended discussion of further hearings or an environmental impact statement.

Of course, none of this mattered if Namu Inc. couldn't catch the whales. Although the federal permit reopened portions of southern Puget Sound to orca capture, Goldsberry had no better luck, and the stress was taking a toll. Jeff Foster saw it firsthand in the summer of 1973, when he joined an expedition in Henderson Bay. The team was in hot pursuit of a pod, and Foster rode with Goldsberry on the *Orca*. As the boat bounced along, the seventeen-year-old diver stared at the buoy and net piled at its stern.

"Put it in the hole!" yelled Goldsberry suddenly. Unsure what the command meant, Foster moved aft and looked around. "Move your ass, goddamn it! Put it in the hole!" Perhaps his boss wanted something below deck? As each second passed, Goldsberry grew more furious, pounding the dash with his fists. "No, you fucking dumb shit! Put it in the hole!" At that, the teenager snapped.

"Fuck you, you asshole!" he screamed.

Foster later laughed at the memory. "Right after I said that, I jumped onto the rail ready to dive in, 'cause I knew he was going to kill me. And I was going to swim to shore, I swear to God!" Of course, had Foster jumped in, Goldsberry might have shifted his pursuit from whales to the spirited teenager. Yet the older man didn't make a move. He just stared at Foster.

"No one has ever called me an asshole before," he said finally, nodding his approval.[37]

Goldsberry had settled on his new assistant, but the whales continued to elude him, and Sea World's permit would expire soon, as he confided to Peter Babich.

It was a clear evening in early October 1973, and the *Pacific Maid* was unloading chum salmon on an adjacent pier when the aquarium owner wandered over for a chat. Despite recent tensions between the two men, their history of catching orcas with Ted Griffin remained a bond. The price of killer whales had reached $50,000 apiece, Goldsberry told Babich, and aquariums all over the world wanted them. But the animals were in short supply. Although Goldsberry held the only capture permit in the United States, he couldn't seem to fill it. Babich had tried to help, as he related in a letter to Griffin. On one occasion, he radioed Namu Inc., to report a pod passing by Whidbey Island, but once again Goldsberry failed to make a catch. "I told him that when they do come in and he had such a short time, that he better hire some professional fishermen," Babich wrote.[38]

Goldsberry didn't admit it to Babich, but he was feeling pressure from multiple sides. Sea World needed more whales, and the company had grown so concerned with Goldsberry's failure to deliver that it was exploring new sources, including the Pacific coast of Mexico. Meanwhile, north of the border, a Sealand crew led by John Colby had just caught four orcas in Pedder Bay. And it had turned one of the animals over to Mike Bigg, whose research threatened to end killer whale capture in the Pacific Northwest.

15

The Legend of Mike Bigg

THE CREW WASN'T looking forward to this. It was the chilly morning of October 26, 1973, and a southeastern swell had been rolling into Pedder Bay for hours. But Sealand's twenty-man team was more worried about the task of restraining Taku. The bull killer whale was enormous—at least twenty-two feet long—and although he seemed friendly, no one knew how he would react. Most of the men smoked nervously as they watched the black dorsal fin circle the pen. As the water calmed, Bill Cameron exited the galley of the *Western Spray* and lowered himself onto one of the logs bordering the pen. Bob Wright had hired the Pender Harbour fisherman to help handle the whales. Although a large man, Cameron had a gentle way about him. "You just have to treat them like herring," he instructed. "You can't spook them." At his order, the men aboard three small skiffs slowly began to pull up the net.[1]

Taku didn't like it. As his enclosure shrank, the orca squealed, slapping his pectoral fins on the water. In a nearby pen, another captured whale, Kandy, listened intently. "Dry him up!" yelled Cameron, and the crew pulled harder, drawing the mesh underneath the big orca. As his man-made pond vanished, Taku flopped onto his side, his upturned eye frantic with fear. To the relief of everyone, he didn't lash out, and divers tilted him upright for an explosive breath. "I had a hunch he'd be that easy," said Cameron. "It's the females that are the tough ones." But the hard part was yet to come.[2]

With the big whale secured, federal researcher Michael Bigg stepped away from a group of scientists gathered on deck. In his hands, he held a radio pack, which he had designed using Sealand's captive orca Haida as a model. He planned to mount it on Taku in the first ever attempt to radio-tag a killer whale for release.[3] Bigg handed the device to veterinarian Alan Hoey, who sat astride Taku's back, and Hoey in turn slid it over the five-foot dorsal fin, securing it by threading iron and copper wire through the base of the fin. Bigg planned for it to stay in place for only thirty days, after which electrolysis would weaken the wire and release the spring-loaded pack. But he intended the next step to be permanent. At Bigg's direction, Hoey cut two large notches in the whale's dorsal fin. "It was hard to watch," recalled Sealand curator John Colby. "The blood just came spurting out."[4]

Bigg's two-pronged experiment began the next day, when the crew released Taku. After tracking the animal east toward San Juan Island, Bigg lost the radio signal and never regained it. Weeks later, the pack popped off as designed, but the notches in the whale's fin remained, making Taku—later labeled K1—easily recognizable. And that was the point. By cutting Taku, Bigg hoped to prove that the scars on killer whales' fins were permanent and might be used, along with the patterns of their "saddle" patches, to identify individuals in the wild.[5] It was a pivotal step in the study and conservation of orcas, and it was possible only with live capture.

It is a difficult connection for many critics of captivity to accept. Bigg died in 1990 at the age of fifty-one, and in subsequent years he became rightly recognized as the founding father of killer whale science. It was he who developed the system to distinguish individual orcas, which enabled him to identify the matrilineal ties that structure separate pods. It was he who determined the division between the fish-eating "residents" and mammal-eating "transients"—now called Bigg's killer whales. And he achieved most of these breakthroughs on his own initiative, with virtually no funding or support from the Canadian government. Bigg was a pioneer in the noninvasive study of wild cetaceans, and as such many orca enthusiasts consider his work the antithesis of the oceanarium industry. Yet as the Taku experiment reveals, Bigg's research developed in tandem with killer whale captivity.[6]

Research on live cetaceans represented an astonishing leap forward. Prior to World War II, nearly all zoological science was premised on killing and dissection. This was certainly true of cetacean research, which took place primarily in conjunction with the whaling industry. The only tags used to track animals were discovery marks—cartridges fired into whales with rifles and recovered only if they were killed and processed. Scientific possibilities expanded in the 1950s, as engineers developed new means of tracking live animals, but the technology seemed ill-suited to whales.[7] Not only would early devices not function in water, but tagging cetaceans posed unique challenges. Great whales were nearly impossible to capture, and when scientists attempted to tranquilize smaller cetaceans the animals usually suffocated. As a result, most

FIGURE 15.1 Taku (K1) circles as Sealand of the Pacific staffers pull up the net, October 1973. In author's possession.

researchers considered the live tracking and identification of whales impractical. That was certainly true in British Columbia, where scientists had never even attempted to quantify the orca population. And until the late 1960s, they had little reason to try.

Live capture gave them a reason. Fishermen and other locals had shot at orcas for decades, but such violence had received little media coverage, primarily because locals regarded the species as a pest. With the rise of killer whale display, however, more people regarded the species with affection, and the spate of captures in 1968–1970 raised public concern about the possible depletion of orcas, forcing the federal government to take notice. Yet Canadian officials confronted the same dilemma as their US counterparts. Because killer whales had never been considered a commercial species, few data existed on their life cycle, migration patterns, or demographics. Without such information, no one knew if capture posed a threat to the local population of killer whales—or even if the killer whales *were* local. In fact, most scientists considered *Orcinus orca* to be cosmopolitan. The standard marine mammal textbook at the time, edited by US navy veterinarian Sam Ridgway, described them as "a single species which travels extensively."[8] If this was the case, officials needn't concern themselves with managing a regional stock. To resolve this question, the Department of Fisheries instructed the Pacific Biological Station in Nanaimo to undertake an assessment.

The Pacific Biological Station had been created in 1908 to manage the marine resources of Canada's Pacific coast. For most of its staff, that meant the region's important fisheries. In addition to compiling catch statistics, monitoring salmon streams, and managing hatcheries, its researchers devised programs to eliminate threats to the fishing industry, including the slaughter of basking sharks and sea lions as well as the abortive anti-orca machine gun in 1961. Yet apart from the whaling industry, cetaceans rarely figured prominently at the station. Gordon C. Pike was the only marine mammalogist on staff, and when the Coal Harbour whaling station closed in 1967, he was transferred to Montreal.[9] Yet officials in Nanaimo couldn't help but note the growing popularity of killer whales as the result of display, as well as the rising controversy over their capture. And in contrast to Puget Sound, where the state government initially had authority, the Canadian federal government held jurisdiction over the treatment of sea life. Yet to make any decisions about orcas, officials needed a population count, and that task fell to Pike's replacement: Mike Bigg.

Bigg didn't have deep roots in the region. Born in England in 1939, he had moved to Vancouver with his family at the age of nine. After graduating from college, he entered the master's program in zoology at UBC, where he planned to study water shrews. With the departure of his supervisor, he switched to harbor seals, and his subsequent research involved countless hours shooting and dissecting specimens. But he gave killer whales little thought until the inadvertent capture of Moby Doll in 1964. At the Vancouver Aquarium's request, Bigg brought seal carcasses to the pen at the Jericho Army Base, and although the whale refused his offerings, the young scientist was fascinated. He performed a study of Moby Doll's respiration, and when the whale died, Bigg participated in the necropsy.[10]

Over the following years, Bigg couldn't seem to get away from orcas. In 1970, he accepted a position at the Pacific Biological Station, which at the time seemed a dead end for a marine mammalogist. Whaling in the province had ceased, and some of Bigg's primary duties irked him. He often found himself on seal-shooting expeditions with research technician Ian MacAskie, as Canada was required to examine five hundred fur seals per year under the Northern Pacific Fur Seal Treaty of 1957. Not long after his arrival, however, Bigg was assigned to count killer whales. It was a daunting task, and he realized it required international cooperation. After consulting with Murray Newman, he settled on the idea of the one-day "census."[11]

The first year proved a surprising success. After distributing fifteen thousand questionnaires to fishermen, lightkeepers, and volunteers from Alaska to California, Bigg coordinated the count on Monday, July 26, 1971. The resulting estimate of 549 whales informed the contentious debate over capture permits in Washington State, and in the summer of 1972 Bigg ran a second census, with similar results.[12] The data proved invaluable. In addition to providing officials on both sides of the border with a rough count of the region's orcas, it gave Bigg a sense of where he could focus his field research. But he wasn't the only one making that shift.

Since his firing from the Vancouver Aquarium, Paul Spong had established a small research camp off northern Vancouver Island. Located on Hanson Island, in aptly named Blackfish Sound, the project was an expression of the late 1960s back-to-nature impulse as well as Spong's desire to study orcas in the wild. By this time, he attributed great spiritual meaning to the species, and like many adherents to the counterculture, he looked to indigenous culture for inspiration.[13] Although local loggers and fishermen regarded the hippie researcher with bemusement, Spong fervently hoped to interact with wild whales as he had with Skana. He played his flute, kayaked in local waters, and even brought the Vancouver rock band Fireweed to Alert Bay, hoping their live performance would attract the animals. Yet the orcas passing through Blackfish Sound showed little interest in human contact.[14]

Perhaps for that reason, Spong maintained his ties to the display industry. In 1971–1972, he frequently visited Sealand, whose owner, Bob Wright, hoped Spong would lend scientific credibility to the entertainment-oriented oceanarium. Rather than conduct research, however, Spong spent much of his time trying to connect with Haida and the white whale Chimo, often playing his flute next to the pool. To retain this access, Spong avoided criticizing Sealand, focusing instead on influencing public perceptions of the species. It was in this period that he, more than anyone else, promoted the use of the term "orca." "At a certain point, I felt I understood enough about the nature of this creature that it felt unfair to call it 'killer,' " he later explained. Because the animals didn't attack people, Spong argued, they shouldn't be called "killer whales."[15] It was baldly anthropocentric reasoning, and Mike Bigg and other zoologists would reject the change. But Spong's campaign reflected the profound shift that captivity had brought to the public perception of the species.

It was in the midst of this effort that Spong crossed paths with Farley Mowat. In November 1972, the author came to Vancouver to promote his book *A Whale for the Killing* (1972), based on his failed attempt to save the trapped fin whale in Newfoundland. He met Spong at a downtown hotel, and the two talked through the night, spinning tales of sea wolves and whales. Spong argued that the lifeways of orcas offered solutions to human conflict. "More than any animal that I know of," he asserted, "the whales seem to live without fear and, without aggression"—a statement that would have surprised their terrified prey. For his part, Mowat drew Spong's attention to commercial whaling. What good was the study of a few wild orcas, he asked, if the great whales vanished from the ocean? The next day, the two made a pilgrimage to Sealand. Chimo had died a few weeks earlier, and Haida appeared lonely and ill. The scene appalled Mowat. "If I had the power, I'd put an absolute embargo on putting whales on display before that inflated ape called man," he declared to reporters, and he condemned captivity as "an atrocity."[16]

Spong returned to Vancouver determined to stir opposition to whaling. By December, he and his wife, Linda, were distributing "Save the Whale" pamphlets, and in June 1973 they threw a whale celebration in Stanley Park, just steps from the Vancouver Aquarium. The following month, Spong organized one of the region's first whale-watching cruises. Chartering a sixty-foot vessel, he brought a group of countercultural enthusiasts north to Johnstone Strait. The voyage caught the attention of naturalist Erich Hoyt, there to film wild orcas. "Paddling out in the canoe to greet the mob, we saw some 30 people on deck and another 8 or 10 hanging from the spreaders, all waving, cheering, whistling," he later wrote. "Some were playing flutes; one honked on a saxophone. Dogs on the boat were barking, the children screaming."[17] Little wonder that few whales came to visit.

Yet Spong's efforts in Vancouver gained traction. In the fall of 1973, he convinced Greenpeace to embrace the antiwhaling cause, and at a downtown theater that December he held a "whale show" to raise funds. Convinced his experience with orcas could sway the world to halt commercial whaling, Spong laid plans for a speaking tour of Japan, and he asked permission to announce it at a press conference at the Vancouver Aquarium. Murray

Newman agreed, on the condition that Spong not raise the question of captivity. Held on February 26, 1974, the event proved transformative for Greenpeace leader Bob Hunter. When Spong climbed down to the training platform, Hunter noted, Skana and Hyak "rushed over to him immediately, nuzzling up against him like sleek immense dogs." Spong then invited Hunter to greet Skana. "Before I quite realized what had happened, I was rubbing my forehead against hers, stroking her, feeling nothing but sensuousness," Hunter later wrote. "After several minutes of this kind of delicious contact, I thought there was no fear left in me at all. I trusted her like I would trust my own mother." Suddenly, Skana took Hunter's head in her mouth, holding him gently for several seconds. The moment likely meant little to her. Like other trained killer whales, she had performed the stunt with trainers countless times before. But her impact on Hunter was profound. "I was completely at her mercy. Fear exploded in my chest, yet the feeling of trustful happiness continued in my head," he recounted. "As though satisfied, she let go and sank away—ever so gently." For Hunter, the interaction proved a personal and ecological epiphany.[18] In the following years, he helped lead the antiwhaling crusade that made Greenpeace famous, and as in the case of Spong, his conversion experience came through an encounter with a captive orca.

Spong had less success converting Japan. From March to May 1974, he presented his whale show nineteen times throughout the country. Although conservationists and many children enjoyed the videos and recordings of orcas, he had difficulty bridging the cultural divide. Canadian diplomats in Japan described most local press coverage as "unfavourable," particularly due to the "highly emotional aspects of Spong's presentation."[19] After ten weeks, the disappointed activist left Japan convinced that direct confrontation of whalers on the high seas would be necessary. The tour's impact on Japanese views of cetaceans is difficult to assess. On the one hand, Spong's whale show reached millions and succeeded in stirring debate over commercial whaling. On the other hand, his presentation heightened interest in killer whales, including those at Kamogawa Sea World, and over the following years the Japanese demand for captive orcas continued to rise.

As Spong pushed to save the whales, Bob Wright was angling to corner the market on orcas. He knew the MMPA had hamstrung his US competition. Sea World's permit allowed for the collection of only four killer whales, and even if Don Goldsberry managed to fill it, the animals were earmarked for the company's three franchises. In contrast, Wright held a permit from the

Canadian government for ten orcas, and in October he and top staff members toured European oceanariums to assess future demand. With prices for a healthy killer whale now passing $50,000, the entrepreneurial Wright seemed poised to cash in. But nearly three years had passed since his last capture, and he had just lost his prized white whale. As he explained in a December 1972 memo to his staff, "Due to the recent death of Chimo, the killer whale capture becomes of paramount importance."[20]

Like Goldsberry in Seattle, however, Wright found himself in a catch-22. To retain his permits, he needed to cooperate with research that might ultimately threaten his business. The Department of the Environment (formerly the Department of Fisheries) was already moving toward restriction. In March 1973, after consultation with Mike Bigg, officials revised Wright's permit, allowing him to keep only eight whales and requiring him to make those over the size limit available to researchers. For the inquisitive Bigg, the arrangement made perfect sense. "Mike was a pure researcher and a pure researcher starved for data," explained former Sealand staffer Angus Matthews. "He pretty well would go wherever he had to go to get it."[21] For his part, Wright sought to associate Sealand with Bigg's research. Although he acknowledged that he "would like a mate or two for Haida," Wright maintained publicly that Sealand's main objective was to support the population study.[22] But to accomplish either goal, he needed to catch whales, and in charge of that effort, he placed curator John Colby.

The young American was eager to prove himself. "I was in it for the glory," he later told me, "and my ego was in the wrong place."[23] Throughout the spring of 1973, he and the Sealand staffers scoured the local area, including Haro Strait, with little luck. That summer, they shifted their operation to Pedder Bay, and in the late afternoon of Monday, August 6, after weeks of waiting, reports came of whales approaching from the east. It was K pod—Skana's family—likely returning from summer frolics off San Juan Island. After readying themselves, Colby and his crew watched helplessly as twenty orcas swam by on their way to Race Rocks. Then two of the whales—an older female and a large male—turned abruptly and entered the bay. After waiting for the animals to pass, Colby throttled up the *Western Spray* and made his set. The crew then fastened the cork line to a naval buoy while soldiers from the army ammunition depot helped secure the nets. It seemed a solid catch. The big male appeared too large to keep, but the female was a possible mate for Haida.[24]

Wright was pleased, but he wanted more. Hoping the pod would return for its missing members, he ordered the two whales penned on the west side of Pedder Bay. K pod stayed away, perhaps sensing danger. Two weeks later, L pod approached along the same route, reaching Pedder Bay at 1:00 p.m. on Tuesday, August 21. Once again, the crew tried to use the penned whales as bait. But the two animals, now dubbed Taku and Nootka II, remained silent. At that point, Wright instructed Colby to "prime the pump" by playing recordings made of Taku. Almost immediately, the big male began chirping in response, and his calls carried out to the passing whales.[25]

Had Taku and Nootka II been Bigg's killer whales, the ploy would likely have failed. But they were fellow southern residents, members of K pod, and their extended L pod relatives recognized their calls. "The whales instantly made a ninety-degree turn and headed straight into the bay," recalled Colby. "There were between fifteen and twenty of them, and they were really moving. Some of them were porpoising in." Intent on a reunion, the excited animals showed no signs of fear. "I had the engine running, and half of them were already inside my planned set line," recounted Colby.[26] As before, he began chugging across the bay.

Suddenly, from the depot came a thunderous explosion. Most of L pod panicked and reversed course, fleeing before the *Western Spray* could close off the bay. When it was over, all but two of the whales had escaped. Colby couldn't believe it. He had drilled his crew in the bay all summer and never once heard a detonation. Now an ill-timed drill had robbed him of his catch. "At the time, I was furious. I was really disappointed," he reflected. "In retrospect, I'm relieved. It was probably the best thing that ever happened to me."[27]

Now Wright had four orcas in Pedder Bay, three of them keepers, but he still hoped to fill his quota of eight. After all, he confided to his staff, "I may never get another permit."[28] So he kept the captive orcas in the bay, even as anxiety about their health mounted. Nootka II spent much of her time on the surface issuing soft, whining calls, and she developed a depression behind her blowhole—a clear sign of dehydration. Wright dismissed the crew's concerns, noting that the whales from the 1970 catch had gone more than two months without eating. But some staffers knew what Wright had never admitted publicly—that Scarred-Jaw Cow had died on that occasion—and they continued tossing herring into the pens. By early September, Colby and his crew had all four whales eating.[29]

By that time, Wright had a bevy of interested buyers. He planned to keep Nootka II as a mate for Haida, and because Taku exceeded the size limit, he would be turned over to Bigg. But the two others were healthy young whales: a male dubbed Frankie and a female named Kandy. Initially, Wright hoped to use Sea World's troubles to drive up the price. Aware the company had failed to fill its US government permit, Wright offered to sell the two animals for $240,000 (about $1.3 million in 2018 dollars). Sea World vice president Frank Powell balked, instead offering $100,000 for the pair. In the end, Wright agreed to sell Frankie alone for $60,000, on the condition that Sea World keep the price secret. Powell agreed but then dropped a bombshell: the company's US permit applied to whales it imported as well as captured, and it would expire the following day. If Frankie wasn't across the border by then, the deal was off.[30]

Don Goldsberry arrived the next day, October 20, 1973, to observe what he thought would be a simple operation. Wright's team needed only to get Frankie loaded onto a boat and into US waters by midnight. After a loose log frightened the whale out of his enclosure, however, Sealand staffers spent hours trying to separate him from the other orcas. By 9:00 p.m., the team had again corralled Frankie, but just as they were about to lift him out of the water, Wright misjudged the wind, allowing the *Western Spray* to collide with the pen. As the logs came apart, the net below rose to the surface, threatening to entangle the frightened animal. Furious at Wright, Goldsberry took control, barking orders to Sealand workers, stabilizing the pen, and likely saving Frankie from drowning. But precious time had been lost. After several more hours of work, Goldsberry and his crew loaded the whale onto their boat and made for US waters. By that time, however, it was long past midnight, and Sea World's permit had expired. The company had smuggled Frankie across the border in violation of the MMPA.[31] But that was of little concern to the Sealand staffers, who now focused on their collaboration with Mike Bigg.

⁂

Bigg believed he was on to something, and his enthusiasm was contagious. In addition to coordinating the killer whale census, he and Ian MacAskie had conducted field observations in Johnstone Strait in the previous two summers. In the process, they had noticed an orca with a mostly severed dorsal fin, whom they named Stubbs. The scientists theorized that killer whales might be identified by marks on their dorsal fins and patterns of their saddle patches. If confirmed, this would represent a breakthrough in the study of wild orcas. Although the census had enabled Bigg to make population estimates, it had revealed nothing about the species migration or social dynamics because researchers could not identify and track individual animals. As Bigg later put it, if the markings on killer whales' bodies remained the same over time, they could serve as "natural identification tags."[32] But he needed definitive evidence, and Taku gave him the opportunity to produce it.

After the capture, Bigg made frequent visits to Pedder Bay to observe the whales and plan his experiment.[33] In designing the radio pack, he used Sealand's captive whales, both living and dead. After Chimo's death, Bigg had studied the vascular structure of her dorsal fin to determine the best way to mount the pack, and to design the device itself, he used Haida as a model. Angus Matthews later marveled at the researcher's ingenuity. "Mike Bigg was doing all this on a shoestring. I don't know how he found the money," he recalled. "If it didn't have to do with salmon, the government didn't spend any money on research, and the damn whales were eating too much salmon, so why do we want to study them? But Mike somehow squeezed this stuff out of his budget."[34]

In addition to crafting the radio pack, Bigg weighed his options for marking the whale for release. Initially, he considered freeze branding, but

FIGURE 15.2 Michael Bigg tries out radio pack on Haida at Sealand of the Pacific, October 1973. In author's possession.

conversations with Goldsberry gave him pause.[35] The process had worked poorly on the whales branded in 1970–1971 because their dynamic skin caused the marks to fade and become distorted. As a result, Bigg settled on his plan to make incisions in the animal's fin. Although done for scientific purposes, the procedure was hard for many to watch. It appalled John Holer, owner of Marineland of Canada in Niagara Falls, who had come to purchase a whale from Wright. "As they started to make the cuts," recalled Colby, "Holer was saying 'don't do this, don't do this,' and the water just filled with blood."[36] With Taku's release the following day, one whale remained from Colby's catch.

Kandy was beautiful. Measuring just under eighteen feet, she was about eight years old—nearing sexual maturity. As a healthy breeding female in L pod, she might have given birth to five or six calves over the next thirty years while gaining the age and knowledge to lead her own matriline. Instead, she was lifted onto an army pier, driven by truck to the Victoria airport, and flown to eastern Canada. There she joined Kandu II at Holer's marine park. The two whales seemed to hit it off, and staff expressed hope they might mate in captivity. Yet there were other reasons for their sociability. In all likelihood, both animals hailed from L pod, though they hadn't heard each other since the capture in Penn Cove in August 1971, when Kandu II was removed and sent to Germany. He was later moved to Marineland of Canada, where Kandy joined him in early November 1973.

The reunion didn't last long. Over Colby's protests, Wright had shipped Kandy in a box too small for her body and packed her with insufficient ice. After an eight-hour layover in Vancouver, the young whale arrived at the Toronto airport floating in warm water, her left lung filled with fluid and her skin peeling away in sheets. Wright returned to Victoria soon after, but Colby stayed with her. "For the next ten days, I was in my wet suit with Kandy many hours a day," he recalled. "It was a cold November, and I was frozen every day." Kandy soon succumbed to pneumonia, and the distraught Colby resigned from Sealand. But Wright was more concerned with the wrath of John Holer.[37]

Marineland's owner was not a man to be trifled with. A Slovenian immigrant who had worked in circuses as a boy, he had opened the Marine Wonderland and Animal Farm in 1961, later renaming it Marineland of Canada. Eager to expand, he had bought Kandy for $75,000—more than Sea World had paid for Frankie. Now she was dead, and Wright refused to accept responsibility. Furious, Holer threatened to move into the British Columbia orca-catching business, much as Sea World had in Washington State, and to run the operation, he recruited Colby. By this time, all three whales kept from the August 1973 captures had died, and Colby blamed Wright. Keen to undercut his former boss, he applied for his own permits with Holer's backing.[38] But Sealand's owner would brook no competition. At his prompting, Canadian officials revoked Colby's status as a landed immigrant, forcing him to leave the country just months after his son Jason was born. To appease Holer, Wright promised to replace Kandy when he caught more whales.

Yet ongoing research by the Department of the Environment put future captures into question. Pleased with Bigg's progress, federal officials allocated funds for a more intensive count of killer whales. Stationing boats in seven locations, Bigg coordinated a ten-day survey in early August 1974. Initially, his team struggled to locate orcas around southern Vancouver Island, but they soon organized a network of volunteer observers to notify them when whales appeared. As Bigg later acknowledged, it was a technique "developed by the local killer whale netters."[39]

This synergy between science and capture concerned some observers. In an August 8, 1974, letter to the *Daily Colonist*, Victoria resident A. H. Roberts reminded readers that all three whales taken the previous year had died in captivity, and he singled out Bigg's radio experiment on Taku for special criticism. True, federal researchers were now running an expanded census, he noted, but to what end? "If sending in details of whale sighting is to result in more experiments and deaths from commercial exploitation," Roberts asserted, "the whole census should be boycotted."[40] Despite this skepticism, Bigg's team continued its work, snapping thousands of photos of wild orcas, and one of those holding a camera was twenty-four-year old Graeme Ellis.

Ellis's life seemed to have come full circle. Following the 1970 death of Scarred-Jaw Cow at Pedder Bay, he had quit his job at Sealand and gone sailing off the coast of Mexico. In the summer of 1973, after studying for a year at the University of Victoria, he traveled to Johnstone Strait, where he first crossed paths with Mike Bigg. The researcher hired Ellis to spot orcas and offered him steady work collecting fur seals for the Department of the Environment, but the young man turned it down. "That was really the toughest decision of my life, I think," Ellis later told me. He badly wanted to work with Bigg, but he had quit his job at Sealand over an animal's death, and he couldn't accept another premised on killing.[41]

Yet Ellis was in luck. In August 1974, Bigg hired him as a photographer for the expanded census. The young man took to the job quickly. Having worked closely with killer whales in captivity, he had a subtle understanding of their movements and body language. "That opportunity to work one-on-one with those animals really enabled me when I was working with killer whales in the wild," he reflected. "You really come to know what complex animals they are, and it is because you have dealt with them one-on-one and not just looking at them as a big black-and-white superdolphin." Like many who study wild orcas, Ellis had developed misgivings about captivity, but he emphasized the role it played in his own journey, noting, "The early work helped me a lot to understand the animals that I worked with later." Indeed, it was his time as a trainer that led him to Bigg, a scientist who was creating new ways of seeing killer whales.

"He was a wonderful man," Ellis told me, as he stared out his living room window. "There is not a day that goes by that I don't think about him—still—and he's been dead for twenty-five years."[42]

❧

In the spring of 1975, Bigg had yet to publish his findings, but his research was already having an impact, thanks in part to Don White. Though no longer studying orcas, White followed Bigg's work with great interest, and on April 12, 1975, he summarized it in an op-ed to the *Vancouver Sun*. Researchers had determined that the local killer whale population was probably around three hundred, White noted, not the thousands once assumed, and by beginning the process of identifying individuals, Bigg's team had established that orcas did not wander the ocean. "Instead of a constant stream of new animals passing through the area," White explained, "killer whales tend to remain in preferred home territories." The implication for capture was clear: "If whales are taken from one region for captivity, their capture may directly interfere with the population dynamics of that area." Having spent time at the Vancouver Aquarium and Sealand, White didn't deny that orca captivity had made a positive impact. "Before the capture of Moby Doll, of Namu, and of Skana," he declared, "killer whales as a species were regarded by fishermen as vermin." But the time had come to stop live captures.[43]

Two weeks later, on April 28, 1975, Bigg presented his findings at a "killer whale workshop" sponsored by the US Marine Mammal Commission at the University of Washington in Seattle. The impetus for the meeting, explained Douglas Chapman, was the "transformation in public view of killer whales in the last decade." Once "feared and reviled," he noted, "the killer whale is now highly regarded and sought for public display." The commission wanted scientific consensus on how the species should be managed, and it had invited the few scientists researching orcas as well as representatives of the captivity industry, among them Don Goldsberry and Sea World vice president of research Lanny Cornell. Attending with Ian MacAskie, Bigg summarized his population studies and presented the photo-identification system he had devised. Because the dorsal fin and saddle patch were reliably visible when orcas surfaced, he explained, these features provided the best natural markers, and to simplify matters, the system used only photos of the animals' left sides. Bigg also detailed the alphanumeric system that he had developed for identifying separate pods.[44] Many of his listeners were incredulous. As researcher Alexandra Morton eloquently put it, "It seemed impossible that the young Canadian had discovered a method that allowed more animals to be recognized at lower cost without harming the whales."[45] Indeed, Cornell followed Bigg's presentation by asserting the need for expanded freeze branding to identify individual whales. This resistance from Sea World was understandable: if Bigg's system proved successful, it would decouple much of killer whale science from the captivity industry.

In July 1975, Bigg published an article on killer whale capture in the Pacific Northwest, coauthored by Allen Wolman of the Marine Mammal Biological Laboratory. The first publication to examine orca capture as a cross-border issue, it showed how far international cooperation had come. Bigg and Wolman revealed, for example, that Canadian and US officials now coordinated their issuing of capture permits. The gradual reduction in the number of whales allowed to be removed, the authors observed, "reflects an increasingly cautious policy of the two governments with respect to cropping until research on population abundance can establish harvest limits."[46] The two scientists stopped short of calling for an end to capture, and Bigg himself continued to believe that sustainable capture was possible. But the article underscored the new emphasis placed on killer whale conservation, and that concern was on full display when Sealand netted more orcas the following month.

~

The capture came on the heels of Greenpeace's first antiwhaling expedition. Holding their send-off in late April 1975 at Jericho Beach, where Moby Doll was once held, the activists had unfurled a flag featuring an indigenous image of a killer whale as they departed. Thanks to industrial espionage conducted by Paul Spong in Norway, they located the Soviet fleet off the coast of

California two months later. Walter Cronkite featured footage of the ensuing confrontation on his July 1 newscast, and after being fêted in San Francisco, the activists received a hero's welcome in Vancouver.[47] Their voyage had drawn world attention to whaling and Greenpeace alike, and the group began plans for future campaigns. Members talked of confronting Japanese whalers the following year and of halting the harp seal hunt off Newfoundland. Then, in mid-August, Bob Wright caught killer whales in Greenpeace's backyard.

The Sealand team had been set up all summer in Pedder Bay, but no whales had come. Then, on the morning of August 16, Angus Matthews was sleeping on the bridge of the *Western Spray* when frantic shouts jarred him awake. "Hey, you guys there?" yelled a fisherman from an idling boat. "You still looking for whales?"

"Yeah," responded a groggy Matthews.

"Well, they're *coming*. Look!" Sure enough, a group of orcas was swimming into the bay. Matthews started the engine, sealed off the entrance, and netted six whales—a mammal-eating group that Bigg would label Q pod.[48]

Wright was ecstatic, but his options were more constricted than in the past. Federal officials had issued a permit for only three orcas and required that those above size limits be freed promptly. Wright released the two largest animals, but activists were moving quickly. The BC Wildlife Federation denounced the capture, and within days Greenpeace joined the protest at the urging of Don White. Criticizing Wright as an "irresponsible huckster," Greenpeace vice president Rod Marining warned that the entrepreneur was "passing a death sentence on these beautiful creatures," while the organization's lead ecologist, Patrick Moore, accused the Canadian government of abetting the "obvious commercial exploitation of the whales."[49]

Yet it was Greenpeace's lobbying of leftist premier Dave Barrett that really paid off. On Tuesday, September 9, Hunter and Spong met with Barrett in Victoria and urged him to intervene. Barrett agreed, and three days later his minister of recreation and conservation, Jack Radford, announced a moratorium on killer whale capture in provincial waters. In doing so, he declared that the practice was "neither morally nor biologically justified" and that Wright was "morally obligated" to release the remaining animals. It was a curious announcement. After all, Ottawa, not Victoria, held jurisdiction in the matter, and Radford pointed to the research of Bigg—a federal scientist—to justify his action. But he also drew on English common law to claim that orcas, like all whales, were "Crown property," and hence controlled by the province.[50]

The British Columbia government's stance complicated matters for Wright. By this time, he had released two more whales from the catch and transferred a female to Sealand. This left only a young male orca in Pedder Bay, and he was earmarked for John Holer as Kandy's replacement. Wright had planned to fly the animal out of Vancouver, but Barrett barred BC Ferries from transporting the whale—now dubbed Kanduke—and Greenpeace convinced Air Canada to refuse service as well. Ever resourceful, Wright chartered a private plane from Victoria, and on September 18 his crew loaded Kanduke onto a truck amid a small gathering of protesters, among them Don White.

Five years had passed since Wright had accused White of releasing the two orcas he was paid to guard. Now White was back in Pedder Bay calling for another whale's freedom. With Rod Marining riding shotgun, he tailed Kanduke to the Victoria airport in his pickup truck while local police ran interference for Sealand. "They pulled us over and basically stalled and harassed us," White told me. "It was very deliberate. They knew exactly what they were doing." When the two men finally reached the airport, it was too late. The whale was behind a locked and guarded fence.[51]

In the end, neither White, nor Greenpeace, nor the British Columbia government could save Kanduke from captivity, but Wright saw the writing on the wall. With each catch he made, public opposition had grown, until the most recent spurred provincial officials to challenge federal authority. In the end, however, it was Bigg who posed the greatest threat. Working in part with captive whales, the tenacious researcher had begun to reframe scientific understanding of *Orcinus orca*, even as the display industry's demand for the species grew ever more insatiable.

16

"All Hell Broke Loose"

ON THE MORNING of Sunday, March 7, 1976, Ralph and Karen Munro were still feeling the effects of the previous night's Tartan Ball. Over coffee, the couple chatted briefly about a newspaper editorial: apparently Sea World was trying to catch killer whales in Puget Sound. Although Ralph was special aide to Republican governor Dan Evans, the couple gave the matter little thought as they prepared to go sailing that afternoon with Bill and Pennie Oliver. Leaving from Olympia, the friends enjoyed a leisurely cruise to Cooper Point and were returning south through Budd Inlet at 3:00 p.m. when they spotted orcas off their port side. When Oliver tacked the thirty-three-foot vessel eastward for a look, Karen grew nervous. Would killer whales attack the boat? "I didn't know anything about them at that time," she later noted, "and the name sounded kind of scary."[1] Ralph reassured her. He had followed the story of Ted Griffin and Namu a decade earlier, and like many Northwesterners he now viewed the species with fondness.

But the pleasure boaters quickly realized the whales weren't alone. In pursuit were the seiner *Pacific Maid* and a smaller vessel named *Orca*. The reaction on the sailboat was visceral. "All of a sudden we realized that they were trying to capture these whales," Karen recalled. "They were going to take *our* whales away." Oliver radioed the Coast Guard, but officers responded that they lacked jurisdiction, and when he approached a floating seaplane for help, the pilot said that he was part of the operation.[2] By that time, the animals were cornered in nearby Butler Cove. Determined to intervene, Oliver started his engines and steered into the melee.

"Stay away!" yelled the men aboard the *Pacific Maid*. "We've got a permit!" But Oliver ignored them, and at first the intervention seemed to work. The whales made a break to the north, but the boats cut them off, driving the animals to the east side of Budd Inlet just off Gull Harbor. The two capture vessels set their nets, and then came the seal bombs.[3]

At first, the sailboat's occupants were unsure what they were hearing—a concussive force seemed to be striking the hull. Then they noticed a man aboard the big seiner dropping small charges into the water. "He had a propane torch burning on the stern, and he was lighting those things just as fast as he could," recounted Ralph Munro. "Boom, boom, boom, boom. It sounded like a war." Between detonations, they could hear the whales' cries. The animals behind the nets were calling to others who lingered nearby. "It was sickening," Munro later told me, "like seeing some guy hit his kid."[4]

By 3:45 p.m., the catch was complete, with six of the eight whales behind the nets, and the friends on the sailboat were unsure what to do. Still fuming when he got home, Ralph Munro phoned *Seattle P-I* reporter Mike Layton. "Goddamn it," Layton responded, "what the hell do I care about that?"

"Mike, the whales didn't have a chance," Munro pleaded. "Just go and take a look."

The couple couldn't shake what they had seen. "My wife and I tossed and turned all night long," Munro recalled. "It was just so gruesome, so awful." Still awake when the morning paper arrived, he wandered out to grab the *P-I* and glanced at the headline. "Hey," he hollered upstairs to Karen. "We're winning!"[5]

To Munro's surprise, Layton had followed the tip and somehow filed his story in time for the next day's edition. There it was, big and bold, on the front page: "Five [*sic*] Whales Captured: Hundreds Watch at Olympia Harbor."[6] That morning, Munro arrived at work determined to push the state government into action. What he didn't know was that a Washington State Game Department official had been aboard the *Pacific Maid* monitoring the capture, and that a University of Washington scientist was counting on its success. He also didn't realize that the operation he had witnessed was an act of desperation several years in the making.

Sea World had big plans but too few killer whales. Apart from the Canadian-caught Frankie, now dead, the company had acquired no new whales under its economic hardship exemption, and in October 1973 it had applied for a new extended capture permit. Sea World officials knew of Mike Bigg's three-year census, but they ignored its findings in the application. Rather than the several hundred animals Bigg estimated, the company asserted that the killer whale population in the Pacific Northwest alone ranged "well into the thousands." It also described its proposed capture methods as "allowing the animals to enter a bay or harbor and then closing off the mouth or entrance." To be sure, Namu Inc. had caught whales in this fashion once before—in February 1968 in Vaughn Bay—but it was hardly an accurate description of its usual methods.[7]

Even with such language, the application was no sure thing. Recalling the outrage of the previous year, the new director of the NMFS, Robert Schoning, scheduled a public hearing on the application for February 7, 1974, in Seattle. At the hearing, activists made familiar arguments about conservation and orcas' special meaning for the region, but Washington State officials offered no objection to Sea World's application. "We went into the hearing with the assumption that the issuance of four whale permits would not in any way be detrimental to the population," Game Department director Garry Garrison later explained.[8] For their part, Sea World officials emphasized the role the display industry had played in changing public views of the species. "Before we began exhibiting these animals," asserted Vice President Frank Powell, "killer whales were feared, despised and shot at on sight." It was due to oceanariums, more than anything else, he noted, that the public realized "what basically friendly and good creatures these misnamed mammals are."[9]

As Schoning and his colleagues in Washington, DC, considered the application, newspapers in Puget Sound voiced their opposition. The editors of the *Tacoma News Tribune* called for a ban on capture of the region's killer whales. "Most folks would rather see their black heads bobbing along the waters of the Sound than watch them in captivity," they observed. "Why don't the hunters go somewhere else, say the warmer waters of Mexico? It wouldn't help the whale population any, but the animals could at least have one friendly haven."[10]

In fact, Warren Magnuson was pushing for just such a haven. By 1974, Washington's senior US senator had become a leading voice for environmental reform, and he continued to denounce commercial whaling. As he declared on March 24, 1974, "I do not want to be a member of a generation which visited the moon, while the last great whales, with immense brains, unfathomed and unexplored, died in the world's oceans."[11] Like many Northwesterners, Magnuson now connected the local capture debate to the rising opposition to commercial whaling, and in April he called on Schoning to deny Sea World's application and to designate Puget Sound an orca sanctuary. "The killer whale herd that inhabits Puget Sound is a truly unique natural occurrence," Magnuson asserted. "Nowhere do these magnificent animals have a better habitat within which they can thrive." And now that Sea World owned Namu Inc., Magnuson and others could point to Southern California as the main threat to Northwest orcas. Claiming that federal officials had been "swayed by outside commercial interests seeking to exploit the killer whales for their own personal benefit," he argued that "the feelings of the local populace should no longer be ignored."[12] But Magnuson's proposal to declare Puget Sound a whale sanctuary received no support from Governor Evans, who seemed reluctant to invite more federal regulation into the state.

In his deliberation, Schoning consulted with the Marine Mammal Commission, now headed by Victor Scheffer, who still resided in Bellevue, Washington. Although Scheffer had vehemently criticized earlier orca captures in his home state, he now sided with Sea World. In May 1974, the NMFS issued Sea World a permit to collect four killer whales in the waters of Washington State, Alaska, California, or Mexico—where the company also held a permit. And in recognition of Sea World's ongoing struggles to catch orcas, Schoning extended the permit until December 31, 1976. But in doing so, he imposed several conditions. Sea World could keep only males between eleven and eighteen feet and females between eleven and sixteen, and it was also limited in the areas in which it could operate. In a nod to the Washington State Game Department's earlier decision, the NMFS permit banned collection between Admiralty Inlet to the north and the Tacoma Narrows Bridge to the south. This accounted for the most trafficked recreational areas of Puget Sound, but it left several key sites open, including the southern reaches near the state capital. Finally, as a concession to Magnuson, Schoning announced that the NMFS would "study" the proposal to declare Puget Sound a whale sanctuary.[13]

The decision hardly appeased critics. Magnuson declared himself "bitterly disappointed" and warned that the federal stance would hamper efforts to conserve killer whales. The *Seattle P-I* agreed, questioning the science behind the decision and likening orca capture to commercial whaling. The Japanese and Soviets may be "exterminating" the great whales on the high seas, the editors declared, but that "doesn't grant us a license to share in the greed by decimating the whales on our very own doorstep." Schoning pushed back, disputing the claim that orcas were endangered. "Unlike other whales," he noted, "they have never been hunted commercially with very few exceptions." Claiming that studies indicated that "a minimum of 350 killer whales visit the Sound annually," he concluded that "it doesn't look as though we're running out of them."[14] It was a profound misreading of the data.

∞

Don Goldsberry appreciated the extended NMFS deadline, but it didn't help him catch whales. Despite generous funding from Sea World over the next two years, he continued to come up empty-handed. In the fall of 1974, his team briefly cornered a pod in Henderson Bay, but the whales broke out, racing away at full speed, and when Goldsberry gave chase, he ended up losing a valuable net. "We got in front of them and set the net going thirty miles an hour, and corks just flew everywhere," laughed Jeff Foster. The following spring, Namu Inc. trapped a pod near Bellingham but struggled to control the nets in heavy currents, and Goldsberry nearly drowned in the ensuing scramble. The whale captor was working on the back deck of the *Orca* when he released an anchor, which snagged on his Cowichan sweater, carrying him over the side. "He was terrified," recalled Foster. "He comes up and tries to straighten his glasses, and then he goes back down again with the anchor." By the time Goldsberry's men wrestled him on board, he decided he had had enough and ordered the whales released.[15]

For their part, Sea World officials were growing impatient. When they had acquired Namu Inc., Goldsberry had assured them that he, not Ted Griffin, was the brains behind the operation. For three years, Sea World had funded his expeditions and supplied him with federal permits, but the only whale Goldsberry delivered had been smuggled from Canada in violation of US law. "Sea World [was] getting desperate," explained Foster, "and they were saying, 'Let's re-evaluate this. It has been a couple of years, and you haven't been able to produce.' And that's when your dad came in."[16]

John Colby was desperate, too. In the two years since quitting Sealand, he had repeatedly tried to undercut Bob Wright. After failing to win a permit from the Canadian Department of Fisheries to catch orcas, he proposed construction of a new oceanarium in Victoria, but his former boss blocked the project. With a family to support, Colby approached Sea World in the fall of 1975. Recognizing the former curator as one of the few people who had caught killer whales, company officials leaned on Goldsberry to hire him. Soon after, Colby joined the Namu Inc. team on one capture attempt south of the Tacoma Narrows Bridge. Standing on the back deck of the *Orca*, he watched as Goldsberry threw explosives into the water while seated next to Garry Garrison, a federally deputized state Game Department official who seemed oblivious to the whale hunter's actions. "At least a dozen times, Don lit seal bombs off his cigar when Garrison wasn't looking, and they would explode in the water," recalled Colby. "Garrison would ask, 'what was that?' and Don would say, 'engine backfiring.'"[17] Yet Goldsberry avoided taking Colby on other expeditions. "Don was not happy about your dad coming on because he saw him as a threat," Foster later told me. "That's why he sent him on fliers, like the incident up in Kanaka Bay with those two stupid dolphins."[18]

Their names were Dino and Clicker, and they were part of a hare-brained scheme. By early 1976, Goldsberry had made peace with Peter Babich, skipper of the *Pacific Maid*, and with the fisherman's help he was confident he could make a catch in southern Puget Sound. In the meantime, he instructed Colby to supervise a separate operation in a tiny cove within San Juan Island's Kanaka Bay.[19] There Goldsberry hoped to use bottlenose dolphins to lure killer whales into a trap. Colby was skeptical, but he dutifully prepped the site, and in February the dolphins arrived from Sea World. Although accustomed to clearer, warmer water, the animals adjusted well, and soon they were feeding and frolicking, unaware of their role as blackfish bait. Decades later, Foster hinted at the reasons for choosing them. "They were the two worst animals in the Sea World chain," he noted. "They were notorious for being just miserable, miserable animals."[20] Yet the dolphins proved a welcome diversion, particularly when Colby's toddler son, Jason, visited. Goldsberry, too, came to the cove, and Colby learned the willful ways of his new boss. On one occasion, he watched Goldsberry struggling with a net at the cove's entrance in the face of a southwestern gale. "The wind and rain are just beating down on him," recounted Colby, "and Don is out there screaming at the storm at top of his lungs, 'I was here first!' "[21]

Bizarre though it seemed, the operation at Kanaka Bay had a research component. Following the killer whale workshop in April 1975, the NMFS had allocated funds for the study of orcas in Washington State. This included a grant to University of Washington professor Albert Erickson to freeze-brand and radio-tag up to ten orcas in order to track their migration patterns.[22] Of course, the project required access to the animals that only capture could provide. Working through the university's laboratory on San Juan Island, Erickson established close ties with Sea World, and he used the company's captive killer whales in San Diego as models for his radio packs. He also visited Kanaka Bay, where he planned to hold the orcas for his study. But like Sea World, Erickson could only wait and hope that Goldsberry would make a catch.[23]

Meanwhile, public interest in whales continued to grow. In the wake of Greenpeace's antiwhaling expedition, Indiana University hosted a "national whale symposium" in November 1975. In addition to talks by luminaries such as Ken Norris, the gathering provided an opportunity for Mike Bigg to trade notes with researchers such as Roger Payne, who was developing a similar system for the photo-identification of humpback whales. In hindsight, the symposium helped legitimize methods of noninvasive research, which Paul Spong—another attendee—had long advocated. It also inspired plans for an "international orca symposium," to be held at Evergreen State College in Olympia. The list of speakers for the event, which was scheduled for March 12 and 13, 1976, included not only researchers such as Bigg, Spong, and Terry Newby but also Don Goldsberry.[24] As it turned out, however, Goldsberry would be busy catching orcas just a few miles from campus.

The fateful expedition began more than twenty-four hours before Ralph and Karen Munro and their friends happened upon it. At 9:30 a.m. Saturday, March 6, Goldsberry informed the NMFS that his team had whales in sight and requested an observer. Ninety minutes later, the company's seaplane delivered Dennis Ohlde, a federally deputized Washington State Game Department officer, to the *Pacific Maid*, which was idling a few miles south of the Tacoma Narrows Bridge. During the ensuing chase, the boats lost sight of the animals on several occasions, but the spotter plane stayed with them. Finally, the whales reached Johnson Point and started southward into a narrow inlet.[25] Here was the opportunity Goldsberry had been hoping for. Detonating seal bombs near the entrance, his crew attempted to drive the whales deeper into the inlet, but the orcas seemed to sense a trap and doubled back. Had the captors managed to keep them there, the next day's confrontation wouldn't have happened. But the whales escaped, and darkness forced a pause in the chase.[26]

The entire scenario represented a shocking inversion of the orcas' usual routine. Ordinarily, these eight Bigg's killer whales were the hunters, working in close coordination to tire and devour other marine mammals. Now they were the hunted—pursued by a vessel called *Orca*, no less. The chase resumed the next morning, and after hours of avoiding the boats, the whales fled into nearby Budd Inlet. By the time Munro's party spotted them southbound toward Olympia, the orcas were tiring, and within an hour, the seiners had encircled all but two of them. But that pair included the juvenile whale that Goldsberry had specifically targeted.

That wasn't his only problem. There may be worse places in Puget Sound for a killer whale capture, but it is hard to think of one. In March, Budd Inlet boasts sixteen-foot tides, making for some of the most powerful currents in the area. If the nets weren't anchored properly, the whales could easily tangle and drown. Still worse, the site was within view of Olympia, where state legislators had just been discussing the proposal to make Puget Sound a whale sanctuary. The fact that Goldsberry made the capture so close to the capital revealed just how desperate he was, and he could only hope to wrap up the operation before it drew public attention.

Jeff Foster soon realized that would be impossible. Sent into town the morning after the capture to buy groceries, the twenty-year-old navigated through thick fog and tied up at a local marina. Most stores in Olympia were still closed, but the headline of the *Seattle P-I* caught his eye: "Whales Captured." He was incredulous. How could the story have appeared so quickly? "I read it and thought 'This is really bad,'" Foster recalled. "Instead of getting groceries, I went to the liquor store when it opened up and got all the booze and then went back out there and said, 'We've got a problem here.'"[27] Indeed, they did. By the time Foster returned, Don McGaffin had arrived with a television news crew, and the state's legal gears were already in motion.

⁂

Ralph Munro spent that same morning of March 8 lobbying other state officials, and he talked to every journalist who would listen. The capture was "the most disgusting, rotten thing I have seen," he declared, and he promised that Governor Evans would be "registering a strong protest." What he didn't tell reporters was that he hadn't yet spoken to Evans, who was skiing in Utah. "I was so far out," chuckled Munro. "Way out on a limb!" But he managed to reach Washington State's attorney general, Slade Gorton, who instructed his staff to meet with Munro and explore legal action against Sea World.[28]

Gorton had compelling reasons to take up the cause. In recent years, he had fought a federal lawsuit that aimed to restore the fishing rights of Washington State's Indian tribes, and in February 1974, federal district judge George H. Boldt upheld those rights. It was a stinging defeat for Gorton, who lost an appeal the following year. Denouncing the decision as a federal infringement on state resource management, he refused to enforce Boldt's ruling. His stand won him points with white commercial and sport fishermen, but it alienated the growing number of liberal urbanites in western Washington who sympathized with indigenous grievances.[29] The capture debate offered a chance to recover political capital. Not only could Gorton deliver a symbolic blow to federal overreach, but he could win approval from liberal voters in the process. This was no small matter, as he was considering a run for US Senate against one of the whales' outspoken champions: Democrat Warren Magnuson.

Magnuson was already sounding off. In addition to denouncing the process by which Sea World had gained its permit, Magnuson criticized the Republican-controlled state government. Claiming that Evans had opposed his sanctuary bill, Magnuson called on the governor to "reconsider your previous position and support this effort to protect one of our state's most unique living resources." But Magnuson maintained that there was nothing to be done to stop Goldsberry's present operation, noting that "apparently, this man has a valid permit."[30] If Evans and Gorton could free the "Budd Inlet Six," they could one-up Magnuson.

Yet the cause also spoke to something deeper in a region moving away from its working-class roots. As historian Jeffrey Sanders notes, by the late 1970s urban Puget Sound was becoming "a place oriented to leisure more than production."[31] Many young and middle-class residents in this new Northwest saw sea life primarily as a source of pleasure, and they had little use for the extractive economy and those who worked it. On the afternoon of March 9, more than a hundred protesters appeared around the capture site—in kayaks, in canoes, and on the beach—to demand the whales' release and present Goldsberry with the "First Olympia Bicentennial Bad Citizenship Award." That evening in Seattle, several hundred picketed Pier 56, followed by another rally on the University of Washington campus the next night.[32] Much of the press joined the crusade. "A ban on the capture of these intelligent mammals should have been imposed long ago," asserted the editors of the *Seattle P-I.* Dismissing Sea World's claims that capture contributed to science and conservation, they declared that the company's sole objective was to display orcas at "marine circuses." "Obviously Goldsberry is legally entitled to capture the whales," they conceded, "but enough is enough." Before the orcas were driven from the area, "we must make the Sound a place where killer whales are protected from this sad harvest."[33]

Evans himself had bigger plans. On March 10, he issued a press release calling Sea World's capture permit "another in a long line of federal actions preventing a state from protecting its own unique resources," and that afternoon his administration filed suit in federal court.[34] Within hours, Judge Morell Sharp of the Western District Court issued a temporary restraining order preventing Sea World from moving the whales, and he assigned Ralph Munro and his colleague Mel Murphy to serve it. Munro knew Goldsberry was a rough customer and that the task might be dangerous. "We asked for the biggest game agent we could find, and they sent us this guy about the size of that credenza over there," Munro gestured to me. It was 1:00 a.m. when the three men tied their skiff to the *Pacific Maid*. Murphy was the first to climb aboard. "The whales were right there, and one of the whales surfaced and blew," recalled Munro. "Scared the shit out of Mel—I thought he was going to jump overboard!" In the vessel's galley, Goldsberry took the restraining order and stared silently at the trio of trespassers. "He was just pissed off," noted Munro, "but then he said, 'You boys want a cup of coffee?' So we actually all sat in the cabin and had a cup of coffee."

"Tense?" I asked.

"Tense."[35]

Two days later, on March 12, Judge Sharp held a hearing on the case. By this time, public interest had skyrocketed, and activists lined the streets around the courthouse. For the Republican Munro, the cause brought odd alliances. "We were walking down the sidewalk, and all these creepy-looking people were picketing," he explained. "My wife said, 'Who *are* these people?' and I said, 'Honey, I think they're on our side.'"[36] Within the courtroom, the primary issue was the capture methods used by Sea World. Because the company had a federal permit, the state's lawyers had to prove the operation had violated its terms. Initially, they claimed the captors had dropped explosives from their spotter plane, which was prohibited by federal law.[37] When no evidence for that emerged, they zeroed in on two other lines of argument. First, they emphasized that Sea World had used different methods than those described in its application. Rather than "allowing" the whales to swim into a bay or inlet, the capture boats had forced them in. Second, they argued that the use of seal bombs represented an "inhumane" treatment of the animals. Both were questions of interpretation, and Officer Ohlde of the Game Department testified that at no time did he perceive a violation of the permit. But he also admitted that, in his mind, "inhumane" treatment was limited to physical harm of the animals.[38]

Judge Sharp disagreed. After criticizing the use of a state Game Department official to enforce federal regulations, he ordered the whales released. His justification, however, was not that Sea World had violated its permit but rather that conditions at the capture site posed a danger to the animals. It was a curious position. Two days earlier, he had signed an injunction preventing the company from moving the whales; he now ordered them released because the location was unsafe. Sea World immediately filed an appeal. Within hours, Judge Eugene Wright of the US Ninth Circuit Court of Appeals issued a

FIGURE 16.1 Protest sign, March 1976. Courtesy of Washington State Archives.

stay on Sharp's order, granting Sea World permission to move the whales to a "safer location."[39]

Amid the protests and legal wrangling, the capture team sought to keep the whales safe, a task that fell primarily to John Colby. On the evening of Monday, March 8, the twenty-eight-year-old had been headed to Victoria to visit his family when he learned that Goldsberry needed him in Budd Inlet. "All of a sudden, I'm on my way to Olympia," he recounted, "and the next day, all hell broke loose."[40] In the early afternoon, as Colby and the crew of the *Pacific Maid* worked to stabilize the makeshift pen, the protesters in canoes and kayaks appeared and began berating them. Amid the confusion, one whale escaped under the net, to the cheers of onlookers.[41] Three days later, on March 12, Colby received two separate visits from federal marshals sent to enforce the contradictory rulings of Judges Sharp and Wright. "The first marshal came and ordered me to open the nets immediately," Colby explained. "I refused, telling him the currents would tangle the nets and drown the whales. I had just gotten rid of him when another marshal comes down and tells me, 'Absolutely, do not release these whales.' It was chaos."[42]

Granted a reprieve by Judge Wright, Sea World staffers attempted to isolate individual orcas for measurement. Vice President of Research Lanny Cornell spent most of Saturday, March 13, trying to snare the orcas with a hoop net, and in the confusion, two more escaped. Most of the media welcomed the news. "What protesters and a court battle couldn't do, two of the five captured killer whales did for themselves yesterday," exulted *Seattle Times* reporter Erik Lacitis. "They ripped their netting and swam to freedom." But Lacitis noted that NOAA scientist Mark Keyes had a different view of the whale captors' critics. "When they see an animal performing in a show or in a zoo, they don't realize what's involved . . . that these animals have to be captured in the wild," he observed. "I think that's what's happening here. Here's an actual capture that's been taking place in full public view."[43]

In later years, many would claim that the public outcry and lawsuit forced Sea World to free the Budd Inlet Six. In truth, the whales themselves determined the outcome. Three had escaped, and a fourth measured by Cornell exceeded the federal size restrictions and was released. On March 14, the Sea World crew managed to lift one of the two remaining animals to the deck. A male, he measured eighteen and a half feet—six inches too long. That same day, protesters in scuba gear circled the pen while US marshals and local police watched for attempts to free the whales.[44] One marshal suggested to Foster that he station divers with spearguns around the nets to shoot anyone who approached. "I told him I thought he was insane," recalled Foster, shaking his head, "but that's how crazy it was getting."[45]

That evening, with the court's permission, Sea World moved the too-large male to the Seattle Marine Aquarium in preparation for his transfer to Al Erickson's research project on San Juan Island. The following day the team measured the last whale in Budd Inlet. At seventeen feet, eleven and a half inches, it was just within the size limits for a male, but it was a female, and her arrival to Pier 56 didn't go well. As staffers prepared to lower her, the cable holding her sling gave way, and she plunged twelve feet into the drained pool. The ball holding the cable then gave way. "It weighed at least 350 pounds," Foster noted, "and it came down and just pounded her on the back."[46]

Soon after, Foster entered the pool to check on the animal. "She was screaming," he recalled, "so we tried to raise the pool level and make sure she was all right." But the moment she was afloat, the angry female came after Foster. "I hear Lanny yell, 'Watch your ass!' and she came across the pool," Foster recounted. "You could see her eyes—they get this thick eye, this really red eye when they are really pissed." The young diver jumped clear of the pool just as the orca lunged for him. "She hit the side of the pool, and I thought she was going to go right through to the seal tank."[47] Thankfully for the seals, she didn't.

Meanwhile, Sea World tried to defend itself from the barrage of criticism. "Ten years ago you wouldn't have had this protest," declared public relations director Polly Rash. "But we taught people about killer whales, that they're friendly. . . . Now, that's turned against us."[48] In a sense, she was right. Captivity had indeed transformed the public perception of orcas. Yet as many listeners knew, it was Northwesterners, not Sea World, who had initiated that change, and the company's chances of collecting killer whales in Washington State waters continued to shrink. On March 16, the US Senate Commerce Committee unanimously approved a bill designating Puget Sound as an orca sanctuary. "The time to act is now to see that no more killer whales are molested in Puget Sound," declared Warren Magnuson.[49] Three days later, the Ninth Circuit Court of Appeals in San Francisco remanded Sea World's case to Sharp's district court.

By this time, rhetoric against the company had grown more vehement. In a letter to the *Seattle P-I*, one Port Townsend resident wrote, "I am petitioning the Great Whale Master in the sky that when Don Goldsberry dies he be permitted to return to earth as a killer whale, so that he may be hunted, captured, and placed in an aquarium where the public can feed him screws, razor blades and jagged glass bottle tops." Others likened orca capture to social injustice. "We have tried to enslave our black brothers, kill our red brothers and corrupt our yellow brothers," observed one Seattle resident. "Now we seek to encage our cousins and make them perform as fools in a Disneyland of the Sea for our own amusement." Bellevue resident G. D. Graham was especially appalled to learn that government officials had been observing the capture. "What a comfort!" he wrote. "Perhaps we should expect to hear next that police officials are supervising mugging and murder to see that they are done properly."[50]

Yet the debate was hardly one-sided. Many locals questioned the need to protect orcas. Seattleite Virginia Bencel complained that the protection of certain animal species was "tak[ing] precedence over the needs of mankind" and called for Puget Sound to be "famous for more productive sea life than the killer whale." Renton resident Maxine Banker agreed, asserting that killer whales served "no earthly purpose."[51] Others denounced the activists themselves. "Don Goldsberry is getting a raw deal from the 'glory hunters,'" asserted Seattle resident Ryoji Mihara. "He and a few others took all the risks, made all the efforts and dispelled all the myths about killer whales. Now the 'environmentalists' are taking all the bows. Perhaps it's time scientists captured a few of them for study."[52]

Others voiced the extractive values that had once dominated the region. One retired fisherman rejected the new warm-and-fuzzy image of killer whales. Noting that *Orcinus orca* had the same linguistic root as "orc," he reminded readers that the species was "a very fierce dolphin" that hunts in packs and kills for sport. "Outside of man, they are the most destructive creatures on this earth," he asserted. "I suppose they have a place in the life cycle of sea creatures, but just what it is, I'm not sure. Maybe they are the official executioners of the marine world."[53] Sport fisherman A. G. Schille used the debate as an opportunity to denounce the recent court victory for indigenous fishing rights. "If Senator Magnuson has his way and Puget Sound is made a haven for killer whales, it's another nail in the coffin of the salmon," Schille declared. "I think Judge Boldt should now rule that the killer whales are entitled to their share of the fish. Perhaps they should get their 50 percent first. No doubt, they were here before the Indians."[54]

The following Monday, March 22, Judge Sharp held another hearing. Federal officials testified in support of the NMFS permitting process, while Sea World maintained that western Washington was the only viable collection area. But Attorney General Slade Gorton was adamant that all captures cease in state waters.[55] So was Ted Griffin.

∞

This time, Griffin stood not in the spotlight but in the shadows. Since his departure in 1972, he had moved his family to eastern Washington, trying his hand as a gas station attendant and then a real estate agent. By 1974, he was divorced, alone, and back in Seattle. Bearded and gaunt, he bore little resemblance to the bold adventurer of the 1960s, and even some friends failed to recognize him. He went commercial fishing in Alaska under the alias Irving Irving, lived among hippies near the University of Washington, and closely followed the efforts of his erstwhile partner, Don Goldsberry. Listening to a marine band radio as he drove an old, beater car, Griffin followed the chatter of Namu Inc. hunting for whales with its Sea World permits. Although he couldn't bring himself to visit Pier 56, he accepted an invitation to go on one capture operation, during which he intentionally foiled a catch. "I was furious at the whole situation," he later told me. "That Sea World's power, money, and influence allowed them to do things I couldn't do." But he was skeptical that Goldsberry could succeed. "I thought that Don wouldn't catch the whales and that Sea World would give up."[56]

When Goldsberry made his catch in Budd Inlet, Griffin felt a compulsion to do something. He attended Greenpeace meetings near the university, followed the twists and turns of the Budd Inlet Six, and listened to his ex-partner defend the operation in the news. "I'm not ashamed," Goldsberry told reporters. "We have done more for killer whales and their welfare than all the environmentalists put together." Griffin couldn't stand it. "Don was reaching for some justification, some way to solidify his position," he asserted. "In no way did he have the relationship with the whales that I had." As the hearings resumed in Sharp's courtroom, Griffin decided to act, using intermediaries to pass information to Gorton's office about Goldsberry's methods and the circumstances of the 1970 Penn Cove capture. "What happened next," he told me, "felt like justice."[57]

It is unclear what effect Griffin's intervention had on the outcome, but without a doubt, information on earlier orca captures concerned Sea World. Although no killer whales had ever died during its own capture operations the company worried that renewed scrutiny of the Penn Cove tragedy would decrease their chances for future permits. So when Gorton threatened to discuss past captures in court, Sea World offered a settlement. In return for the state dropping its lawsuit, the company promised never again to catch orcas in Washington. "Our desires for killer whales are the same as those of the people of the state," declared Sea World executive George Becker after striking the bargain. "We hope some day the people will understand this." For his part, Judge Sharp marveled that he had "never presided over a case with more public interest."[58]

In retrospect, the Budd Inlet controversy was a defining moment for the role of killer whales in the environmental politics and regional identity of the new Northwest. Like the controversy over Sealand's catch seven months earlier in British Columbia, local officials and activists had concluded that orcas were too important to be removed for display and that distant federal officials could not be trusted to protect them. In the case of the Budd Inlet Six, Washington State had forced change not only on Sea World but on the US government. In late March, the Marine Mammal Commission reversed its position of two years earlier, declaring its opposition to capture in Puget Sound. Like the Washington State Game Department before it, however, it framed the decision in terms of public relations. As commission chairman Victor Scheffer put it, "The capture of killer whales for public display and research would be legal and appropriate in places other than highly populated regions." When done in public, he noted, "it bothers people."[59]

❧

On the evening of Sunday, March 28, Ralph and Karen Munro hosted a victory party at their home—three weeks to the day after they witnessed the capture. It was a celebration of the Pacific Northwest's new environmental values that still made both smile four decades later. Although the couple was now divorced, their role in the Budd Inlet debate remained a bond between them. For Karen—now Karen Ellick—the event had been transformative. Prior to that moment, she had worked as a Republican staffer—first for the Reagan administration in California, then for Republicans in Washington, DC. Although she had environmentalist leanings, she had never been to an oceanarium or thought much about killer whales. Then came Budd Inlet. "It changed my life," she told me. "I was an activist immediately when this happened."[60]

For Ralph, it became an even bigger source of identity. After our interview at his home near Olympia, he took me on a tour of his converted barn, which he calls the "Ralph Munro Mausoleum." Inside were dozens of photos of family and friends, as well as artifacts from the Budd Inlet capture and a range of orca art and bric-a-brac. Once a boy on Bainbridge Island who shot at killer whales, Munro was now fiercely proud of his role in their protection, and he told his story with wit and moral certainty: the capture of orcas was a brutal practice that had to be stopped.

"In historical perspective," I asked, "do Ted Griffin and others who caught them deserve any credit for changing people's views?"

"Oh, Ted does—all these people do," he replied. "Every little kid who comes to Sea World—they don't want anything to happen to those whales. They want to protect them. You know, I even give Sea World quite a bit of credit, much as I hate the bastards."[61]

❧

One week after the Munros threw their party, the remaining female orca joined the male in Kanaka Bay. At first, she seemed happy to see him. But when she saw him accept herring from trainers, she chased him aggressively around the cove. Although observers at the time were puzzled, recent research provides clues about her behavior. Like the others caught in Budd Inlet, the two animals were Bigg's killer whales, and scientists now believe that the pair, later named Flores (T13) and Pender (T14), were likely mother and son. In this light, her parental scolding may have been the equivalent of a vegetarian mother reminding her son that their family doesn't eat meat.[62]

But Al Erickson wasn't interested in the whales' food preferences—he wanted to get his radio packs on them. And unlike Mike Bigg, he planned to make the devices permanent additions to their bodies by mounting them with stainless steel bolts. After considerable effort and several beers, John Colby persuaded him to use corrodible nuts, which would allow the packs to fall off. He also shared slides that Bigg had made of Chimo's dorsal fin, warning Erickson that the whales could lose too much blood or suffer an infection if the devices weren't attached carefully.[63] In subsequent press interviews, Erickson depicted his radio-tagging project as cutting-edge science. In retrospect, it represented an approach to killer whale research that Bigg's photo-identification system was making obsolete. Rather than reading the animals' natural markings, Erickson had decided to alter the whales physically to facilitate tracking. On Sunday, April 25, the biologist branded the whales and mounted the packs.

The following morning, as television reporters milled about, Erickson carefully checked the radio packs and positioned his research boat, SS *Propeller*, to follow the whales after their release. As researcher Kenneth Balcomb wrote in his log, "Dramatically, at 13:54, John Colby cut the last rope holding the net,

FIGURE 16.2 Pender (T14), branded and mounted with radio pack in Kanaka Bay, April 1976. In author's possession.

and people on shore pulled it away from the north side of the bay." After lingering for a while, the whales started west toward the afternoon sun. Worried the animals would outpace his slow-moving vessel, the eccentric Erickson gave Sea World staffers one last unintended laugh. "They're heading toward the sun," he yelled frantically. "Quick! What direction is that?"[64]

∽

By July 1976, Erickson's experiment had fizzled, and Sea World had decamped from Kanaka Bay. But Dan Evans was looking forward to a getaway at his summer home on nearby Dinner Island. One afternoon, as he swam in a little cove, a young harbor seal came for a visit, circling several times and peeking at Evans through his facemask. Delighted, the fifty-year-old governor played with the animal, and he was thrilled to see the little seal waiting for him the next morning. Pulling on his wetsuit, Evans hurried down the path, but just as he arrived, a killer whale entered the cove. "The baby seal is sitting on a rock," explained Ralph Munro, "and the whale goes over and slaps its tail. Wham! The seal jumps in the water and gulp! Whale ate it."[65]

The intruder may not have been one of the Budd Inlet Six, but it gave Evans new perspective on the predators he had helped save. The following Monday, the dignified governor walked quietly by Munro's desk and disappeared into the next room. "Munro!" he bellowed. "You and your GODDAMN whales!"[66]

17

New Frontiers

SKANA LOOKED SICK. On September 18, 1980, she failed to finish her show, and the next day she remained sluggish. Murray Newman and his staff were concerned. Along with Hyak II (formerly Tung-Jen), she was the Vancouver Aquarium's biggest draw. In the thirteen years since Ted Griffin had captured her, Skana had been the star of Stanley Park, giving millions their first close-up view of a killer whale. And through her impact on Paul Spong and Greenpeace, she had helped reframe the international whaling debate. She may well have been the most influential cetacean in history, but she grew weaker each day, and despite heavy doses of antibiotics, she succumbed on Sunday, October 5.[1] The necropsy revealed a fungal infection in her reproductive tract. Although aquarium officials were correct in noting that she had lived longer in captivity than any other killer whale, she was still young—no more than twenty. She might have lived fifty more years in the wild.

Skana's death left Hyak alone. He had come from Pender Harbour in 1968 as a small, frightened calf, and now he was a sexually mature male in need of a mate. Yet the acquisition of killer whales was no simple matter. The Department of Fisheries had stated that it would allow wild capture to replace orcas who died in captivity, but the Vancouver Aquarium hadn't caught a killer whale since Moby Doll in 1964, and if it tried now, activists would surely oppose it. "I knew it would be unpopular for us to try to capture a live killer whale locally and felt a little frustrated about it," Newman admitted. "To my mind, the entire awareness of the killer whales' right to live was brought about by aquariums exhibiting these animals."[2] With nearby waters out of play, he looked to Iceland, which had become the primary source of captive orcas in recent years.

After receiving the Canadian government's permission to import whales, Newman boarded a plane for Iceland, arriving at Keflavik International

Airport in the early morning of December 13, 1980. He immediately visited the nearby holding facility where the US-based International Animal Exchange (IAE) had five young orcas for sale. Newman and his team selected two females for purchase and agreed to transport two more for marine parks in California and Japan. When they learned of the deal, Greenpeace and other organizations tried to block it, but on December 20, the four orcas arrived at the Vancouver Aquarium.[3]

Jeff Foster was waiting. No longer working for Sea World, the veteran orca catcher had volunteered to come to Vancouver to help handle the animals. As he watched the aquarium's new whales, however, he noticed something amiss. "They had them in the pool," he recalled, "and I'm thinking, 'That doesn't look like a female—that's not a female.'" He shared his concerns with the facility's veterinarians.

"No, you can't be right," they replied. "It's a female." Soon after, the aquarium introduced the pair as Finna and Bjossa, two young females from Iceland, and thousands came to see them. But one day in January, a visitor caught a glimpse of Finna's penis. Having worked with the whale for a month, the staff was forced to admit that she was a he.[4]

The aquarium's struggle to sex orcas may have seemed old hat, but its mission to Iceland wasn't. Just a few years earlier, buyers around the world had considered the Pacific Northwest the only source of killer whales. Yet so profoundly had views of the species changed that capture in the region had become almost unthinkable. As a result, Sea World and other oceanariums sought new sources even as Northwesterners tightened their embrace of local orcas.

Prior to the controversial 1976 capture, Sea World hadn't planned to leave Puget Sound. In fact, it had considered expansion. In late 1975, company officials explored the possibility of building a new franchise in Tacoma's Point Defiance Park and even hosted members of the city's park commission in San Diego. But the Budd Inlet fiasco scotched those plans and brought the closure of the Seattle Marine Aquarium. Acquired as a platform to capture local killer whales, the facility now held little value for Sea World, particularly with the publicly funded Seattle Aquarium set to open a few piers away.[5] In the summer of 1976, as Spong and Greenpeace carried out their second confrontation with the Soviet fleet off the California coast, the aquarium staff prepared to close shop. Out of the country seeking other sources of wild orcas, Don Goldsberry assigned John Colby to oversee care of the remaining animals.

It was a sad time on Pier 56. Conjured from Ted Griffin's imagination fifteen years earlier, the Seattle Marine Aquarium had played a pivotal role in altering popular views of killer whales, particularly during the magical year of Namu. But its capture expeditions had made it a political lightning rod, and many in Puget Sound considered the small, undercapitalized aquarium an embarrassment. As the staff sifted through old papers, the swiftness of that decline was palpable. Photographs of the aquarium's heyday seemed drawn from a different era. Most went in the trash, but Colby kept one of Griffin cradling a young harbor seal as a memento. As a teenager, Colby had watched Namu and Griffin frolic in Rich Cove. Now he wondered what had become of the aquarium's creator. That fall, he, Foster, and other staffers loaded the remaining animals for their plane rides to San Diego. The ensuing convoy of flatbed trucks, with its menagerie of belugas and other sea creatures, drew stares from nearby motorists as it headed south to Sea-Tac Airport. It was a fitting end to Sea World's time in Puget Sound.[6]

Yet leaving Washington State hardly resolved the company's killer whale shortage. Some Sea World officials believed captive breeding held the key, but others were skeptical. Up to that time, no live orca calf had been born in captivity, let alone survived. Whales held by other marine parks offered a limited source, and in October 1976 the company purchased Winston from the Flamingo Park Zoo in Great Britain. Netted in Penn Cove six years earlier, the animal had grown too large for his North Yorkshire pool and was on his way to San Diego. Sea World officials hoped he might impregnate one of the marine park's captive females, but they weren't counting on it, and they had already sent Goldsberry on a mission that must have felt strangely familiar.[7]

⁂

Like many Northwesterners, the people of Iceland had long drawn their living from the sea. When Vikings first settled the island in the ninth century, they brought with them traditions of fishing and sealing, as well as driving pilot whales into fjords for slaughter. Over the centuries, Icelanders hunted larger species as well, until the ruling Danish government outlawed whaling in 1915. Achieving independence three years later, Iceland initially kept the law, but it reversed itself in 1948, and over the following decades whalers operated out of the southwestern port of Hvalfjord (Whale Bay). Like Coal Harbour on Vancouver Island, it was a small operation focused mostly on the Japanese meat market, and it rarely targeted killer whales.[8] But orcas did draw the attention of local fishing vessels.

Icelandic fishermen had pulled fish from the frigid North Atlantic for centuries, and by the 1950s they were feeling squeezed. Herring stocks were falling, while the number of European trawlers and marine predators in local waters seemed to grow each year.[9] They were especially concerned with killer whales, which they claimed threatened fish and gear alike. It was this anxiety that prompted the Icelandic government to ask the US naval airbase for help in September 1954. One *Time* magazine reporter framed the request as a chance to help a NATO ally. In recent weeks, he explained, "the largest packs of killer whales in living memory" had "terrorized" the Icelandic coast. Stealing fish and destroying nets, the predators had driven dozens of fishermen out of work. In response, "79 bored G.I.s" climbed into boats and launched a series of raids against the "savage sea cannibals." The bloodiest action came near Grindavik, where they herded orcas into a tight group before machine-gunning them. By the end, "the sea was red with blood," recounted another journalist, who estimated that more than a hundred orcas died. "It was all very tough on the whales," he quipped, "but very good for American-Icelandic relations."[10]

Twenty years later, Icelanders and Americans found a new use for the ocean's greatest predator, and it all stemmed from the herring fishery. Every October, large schools of herring gathered along the southern coast of Iceland for their winter fattening, and with them came the killer whales. Herring fishermen still considered orcas pests and didn't hesitate to shoot them, but the whales grew more brazen, and in late 1974 one found itself trapped by a herring net. In previous years, the fishermen would have killed it, but news of the species' value to the display industry had reached Iceland, and they hoped to profit from their windfall. In this sense, it was similar to the capture of Namu a decade earlier. The fishermen contacted Jón Gunnarsson, director of a small marine zoo in Hafnarfjörður. Although Gunnarsson and his team concluded that the whale was too large for captivity and recommended its release, they noted the potential to catch more orcas. Just as the Pacific Northwest's live killer whale fishery was ending, Iceland's had begun.[11]

Unlike their US and Canadian counterparts, Icelandic officials immediately imposed a permitting system, but early collection methods remained primitive. In 1975, Roger de la Grandière, a French collector for Marineland of Antibes near Nice, netted a female orca off the coast of Iceland. Without the experience and equipment to handle her, however, his crew lifted the animal by the tail and broke her back.[12] The following year, de la Grandière succeeded in catching and transporting another female whale to France, but soon after, the Icelandic government announced that it would issue permits only to its own citizens. The first to receive one was Gunnarsson. Working closely with Dudok van Heel of the Harderwijk Dolphinarium in the Netherlands, Gunnarsson's team netted three small orcas in October 1976. Uncertain of his staff's ability to handle the animals, he decided to release them, and that was when Goldsberry came calling. That same month, the veteran orca catcher, now Sea World's director of collection, participated in his first Icelandic capture, netting two young females: one for van Heel's aquarium in Harderwijk and one for San Diego. Sea World, it seemed, had found a new source of killer whales.[13]

In the meantime, another challenge had arisen. The runaway success of the summer blockbuster *Jaws* (1975) had left Universal Pictures' parent company, MCA Inc., flush with cash, and that fall it initiated a hostile takeover of Sea World. In response, company executives turned to publishing giant Harcourt Brace Jovanovich (HBJ), which purchased the marine park chain for $46 million.[14] Henceforth, Sea World would be owned by larger conglomerates. Eager to develop the company's new supply of orcas, HBJ chairman William Jovanovich approved an expedition to Iceland for the fall of 1977. Sea World vice president Lanny Cornell then came to an agreement with Gunnarsson, who offered the use of his permits for $25,000 per whale. That October, Goldsberry arrived in Iceland with a hand-picked team, including Jeff Foster.

⁂

Foster told me all about it at a cafe in Bellevue, Washington, near Whalers Cove, where whaling ships had once passed their winters. Truth be told, it was more reunion than interview. In 1979, Foster had lived on my dad's abalone boat in Alaska. I was a five-year-old with endless questions, and he was a young diver willing to answer them. Like most who met Foster, I had marveled at his skills in the water. "He has gills," my dad often quipped, and I sometimes looked for them. I hadn't seen Foster since, and he was now nearing sixty. But as we talked, tales of Iceland came pouring out of him.

By necessity, capture there had been a cooperative affair. Gunnarsson provided the permits, Sea World the capital and whale-handling expertise, and van Heel the holding facilities in the Netherlands. Yet as in Puget Sound, local fishermen would prove critical, and Goldsberry contracted Jón Gíslason, skipper of the sixty-seven-foot herring seiner *Gudrun*. The seasoned fisherman not only had extensive contacts but also knew local waters and understood how to apply fishing methods to orca capture. "It was a totally different technique than we used in the sound," explained Foster. "In the sound we would drive them into shallow water and then the [net] would sit on the bottom until we could seine it up." Capture in Iceland meant bigger boats, deeper waters, and larger nets.[15]

Herring fishing itself created the opportunities. As seiners made their sets, killer whales often gathered nearby hoping for an easy meal. "We would get a call from another boat, 'OK, we have whales,' and we would find the whales just sitting there," noted Foster. For the next hour or so, orcas and captors would wait patiently as the vessel pumped fish into its hold. "When they got down to their last couple of tons, we'd say, 'We'll buy your herring now if you dump the net,'" Foster explained. "They would dump the fish, and the whales would get into kind of a feeding frenzy. And you could just set the net around them."

"How do you separate the animals you want?" I asked.

"Well, you get a line, and as they swim around in the nets, you grab them by the tail."

"You had a diver in there?"

"That was me. I was the only one in the water [in the 1976 expedition]."

"So your job was to go into the water in the set?"

"Yeah."

"And hope these whales, like the ones in the Northwest, don't eat people?"

"Right."

"Will they relax for you sometimes?"

"Absolutely. And sometimes they don't, and they kick the shit out of you," Foster laughed. "I got grabbed and pulled underwater. I broke a lot of bones. I got caught in the net one time and came close to really drowning. I had to be resuscitated."[16]

For Foster, catching whales in the late fall in Iceland didn't conjure fond memories: "It was horrible. It was dark, the food was shit, nobody spoke English. It was not like it is now." Still, the work paid well, and Sea World faced none of the criticism it had in the Pacific Northwest, as locals still viewed orcas through the lens of the fishing industry. "Iceland was somewhat behind the times, and killers were still considered a nuisance," explained Jim Antrim, who took part in Sea World's 1977 and 1978 expeditions. "The captain and crew on our capture vessel looked upon this venture as an alternative way to make money, and the few other Icelandic fishermen I talked to thought we were crazy to jump into the water with these animals."[17]

Hollywood wasn't helping. That summer, Paramount Pictures released *Orca* (1977), a thriller in which a vengeful killer whale stalks a fisherman played by Richard Harris. Harris's character, Captain Nolan, is a Goldsberry-like figure who uses his fishing boat to collect marine life for aquariums. But when he inadvertently kills a female orca and her unborn calf, he unleashes the wrath of the whale's "mate," dooming his crew and ultimately himself. Filmed mostly in Newfoundland, the movie used captive orcas from marine parks in California for several scenes, and its plot partly reflected shifting public attitudes toward the species. In contrast to the shark in *Jaws*, which devours hapless victims at random, the male killer whale seeks righteous revenge for his slain family. Shortly before his demise, Captain Nolan himself comes to regret his abuse of this sentient being. Yet the film's depiction of orcas attacking people harkened back to earlier fears of killer whales. It may have fit the views of many Icelanders, but it didn't jibe with what researchers in the Salish Sea were learning.

By the time *Orca* hit theaters in the Pacific Northwest, Mike Bigg had submitted his report on killer whales, which recommended that future captures be limited to replacing animals who died in captivity. The Department of Fisheries accepted his conclusions and cut his funding, ordering him to focus on seals. Yet Bigg had been bitten by the killer whale bug. Over the following years, he would continue to study the species on his own time and dime, much to the annoyance of his superiors. In addition to working with Graeme Ellis to refine his photo-identification system, Bigg provided critical support to others researching killer whales in the region.[18] Among them was Kenneth C. Balcomb.

⁂

Balcomb started a long way from the Salish Sea. Born in Clovis, New Mexico, in 1940, he grew up near Sacramento and graduated from the University of California, Davis, with a bachelor's degree in zoology in 1963. He had a keen interest in cetaceans, but unlike many of his peers who took jobs at oceanariums, Balcomb gravitated to the whaling industry. He began by sifting through the carcasses on the whaling docks of Richmond, California. "Some of the first whales I saw were humpback and blue," he recalled. "They were bringing them in, and I was seeing some pretty neat stuff." Soon he had a job with the US Fish and Wildlife Service firing discovery marks into whales with a rifle off the Pacific coast. In the process, he came to know Dale Rice of the Marine Mammal Biological Laboratory in Seattle, and he participated in the killing of orcas for Rice's research.[19]

The first was a female, harpooned off San Miguel Island in January 1964. Along with Rice and Japanese scientist Masaharu Nishiwaki, Balcomb helped dissect the animal on board a whaling boat, finding mostly the remains of blue sharks. "Whale research was done postmortem" in those days, Balcomb reminded me. "It was a specimen, you know?" The following year, he served as the lab's lead collector, and on March 22, 1965, his chartered whaling boat *Sioux City* shot and killed a twenty-four-foot male orca near Point Conception, California. In the animal's stomach, he found the blubber of a minke whale. The vessel stayed on the hunt, and two months later it came upon Bigg's killer whales attacking a sea lion southwest of the Farallon Islands, but the orcas eluded the whalers' harpoons. As Balcomb wrote in his report, "We could not get a shot at the killers, so we went on our way."[20]

That summer, the Namu craze hit Seattle, and Balcomb couldn't help but take notice. In 1966, he visited Sea World in San Diego to see Shamu, and he eagerly read Ted Griffin's *National Geographic* article, "Making Friends with a Killer Whale." "There was an immense amount of science that was learned from animals in captivity," he reflected.

"Is it fair to say that that period, the initial six or seven years of killer whale captivity, really did transform the image of this species?" I asked.

"Hell, yeah," responded Balcomb.[21]

At the time, however, he had more pressing concerns, having received his draft notice. After managing to defer service for a year by working on a federal project tagging seabirds in Hawaii, he enlisted in the US Navy, and by 1969 he was working as a technician for the navy's SOSUS array on the coast of Washington State. The whale calls he heard there made a powerful impression, and after leaving the navy Balcomb enrolled at the University of California, Santa Cruz, to study under renowned marine mammalogist Ken Norris. But graduate school didn't suit him, and he took a job at the Williamsport Whaling Station in Newfoundland. Amid the shifting popular views of cetaceans, however, he now found the slaughter hard to watch. "Inside I was almost sobbing when a whale was killed," he told me, but "it wasn't manly to start crying on deck when you got all these guys who make a living shooting whales." The whaling station burned down in 1972, and Balcomb returned for another stint in the navy, this time in Japan. But in 1975, Dale Rice informed him of a research opportunity: the US government had allocated funds to assess the Pacific Northwest killer whale population. With little experience studying live whales, Balcomb proposed to utilize Bigg's photo-identification system. He won the contract and enlisted help from his friend Rick Chandler, with whom he had tagged birds in Hawaii.[22]

Now living in Washington State, Chandler was aware of the capture debate but had no strong feelings on the issue. Like Balcomb, he was trained in the shoot-and-dissect school of biology, from which captivity seemed a progressive breakthrough. "People learned a lot from captured whales," Chandler later told me. Like many interested in orcas, he and Balcomb had been preparing to attend the symposium at Evergreen State College in March 1976 when Sea World made its catch in Budd Inlet. "Nobody could believe the immediacy of it," Chandler marveled. "It was like a Hollywood movie." The two biologists distanced themselves from the protests. "It seemed to me like a hippie demonstration," explained Balcomb. "I was trying to stay away from this controversial crap . . .. I didn't have sufficient empathy for the animals' point of view."[23] Yet time with the whales quickly changed that.

Balcomb and Chandler began their research on April Fool's Day, 1976. Over the following seven months, along with Balcomb's wife, Camille Goebel, they spent long hours in their Boston Whaler following up reports of sightings and taking photos of whales. For Chandler, it proved a profound experience. "The first year we spent so much time with J pod. We knew J pod, and J pod knew the sound of our engines," he explained. "It got to be really personal." At the end of October, their federal contract ran out, and Balcomb and Chandler wrote up their report. By that time, they had learned that Sea World had shifted its capture efforts to Iceland. "My reaction to that was 'OK, that's better. Open ocean whales—not our whales,'" Chandler reflected. "I had developed a possessiveness about our whales."[24]

In hopes of continuing their project, which they now called Orca Survey, Chandler and Balcomb published a summary of their findings in the May 1977 issue of *Pacific Search* magazine, along with an advertisement for "chartered killer whale excursions." The notice raised concerns at the NMFS that the researchers intended to make commercial use of the whales, and the agency moved to shut them down.[25] Soon after, Chandler moved on, but Balcomb remained on San Juan Island, continuing Orca Survey and founding the Whale Museum in Friday Harbor in 1979. Initially intended to display the bones and baleen he had collected over the years, the museum soon became an exemplar of the new Northwest's values, catering to middle-class locals and tourists fascinated with killer whales.[26]

Pro-orca sentiment developed more slowly in Iceland. With the country's small whaling industry still taking hundreds of great whales per year, few Icelanders concerned themselves with live killer whale capture. Nor did Greenpeace initially seize on the issue. In 1976–1977, the organization continued to harass the Soviet whaling fleet while Paul Spong doggedly lobbied the IWC for a moratorium on commercial whaling. When Greenpeace did turn its attention to Iceland in spring 1978, the organization focused its effort on commercial whaling, not orca capture. Departing from London, the activist ship *Rainbow Warrior* reached Icelandic waters in June and spent several weeks trying to confront the much faster whaling ships. Although the protest annoyed the Icelandic government, it made little dent in its policies, and Greenpeace was long gone when Goldsberry's team arrived for another expedition.[27]

That fall, Gunnarsson had permits for ten whales, and the *Gudrun* filled them easily. But to select desirable animals, Sea World needed to determine sex as well as size, and this required Jim Antrim, now lead diver, to get intimate with the whales. "As soon as we got an animal on the deck, I'd have to give it a cursory physical examination, and then I'd have to sex it," he explained to me, "and the only way to do that was by physical examination of the genitalia with your hands."[28] By the end of the month, the team had captured ten orcas. Sea World brought five back to the United States, while Gunnarsson kept the remaining five in his new holding facility in Hafnarfjörður to await shipment by the IAE to Japan. But the IAE failed to arrange transport before the Icelandic winter arrived. By February 1979, all five animals had developed severe frostbite.[29] After two died of pneumonia, Gunnarsson "released" the remaining three. "More than likely they died, too," observed Foster. "I was told by one of the people there that they were close to death when they were let go."[30] As rumors of the debacle leaked out, Greenpeace called for investigations into Sea World's involvement, and Icelandic officials reeled from the international criticism.

Soon after, Sea World received another surprise. As the company geared up for the 1979 capture season, Gunnarsson informed it that he was terminating their arrangement. He had instead struck a deal with the IAE. Previously known for trading in African wildlife, the Michigan-based enterprise was pushing to get into the marine mammal business and it had courted Gunnarsson aggressively. "Sea World was shocked," declared Foster.[31] Having established Iceland as a source of orcas, the California company was now cut out of the trade. Over the following two years, the IAE sold whales to marine parks in Mexico, Japan, Canada, and the United States. But international scrutiny continued to mount, and Greenpeace ramped up its criticism. In July 1980, as the IWC opened its annual meeting in London, US senator Warren Magnuson called for a worldwide ban on the capture of killer whales.[32]

∽

Yet the IWC meeting revealed something far more troubling than live capture. Soviet delegates reported that their whaling fleet had harvested 916 orcas off Antarctica in the previous season. It was a stunning figure. Over the previous decade, the Soviets had killed only about 24 killer whales per year, which they usually sold to Japanese processing ships. In the 1979–1980 season, however, they had targeted orcas as never before, and the antiwhaling campaign may have been partly responsible. At the previous year's meeting, in the face of pressure from Greenpeace and other groups, the IWC had sharply reduced sperm whale quotas. Soviet whalers had responded by hunting killer whales, whose oil resembled that of sperm whales and whose commercial harvest was not regulated. And the total of 916 didn't tell the whole story. Of the 349 dead females whom Soviet researchers examined, 141 were pregnant, and many of those not pregnant left behind nursing calves. For researchers in the Pacific Northwest, who had painstakingly identified individual whales and their family structures, the reported carnage was unimaginable. "It just shocked us, and it is still not clear to me how they managed to kill that many," observed Graeme Ellis. "You don't just drive up and harpoon one after the other. They're not like a bunch of sheep." The Soviet report offered some answers to this mystery. "If a killer whale was harpooned, the group did not

FIGURE 17.1 Orca carcasses on deck of Soviet whaling ship, ca. 1979. Courtesy of Robert L. Pitman, National Oceanic and Atmospheric Administration.

leave the wounded animal for a long time," noted researcher M. V. Ivashin. Instead, they stayed to help injured podmates and were in turn struck down.[33]

The Soviet hunt was a sobering reminder that not all the world shared the Pacific Northwest's growing reverence for orcas. Yet it also created an opportunity for scientists in the region to influence international policy and research. "That harvest led to an urgent workshop on killer whales at the IWC Scientific Committee meeting in Cambridge, England, in 1981," explained Canadian researcher John Ford. Because studies of wild orcas were still confined almost entirely to the Pacific Northwest, most of the experts invited were from British Columbia and Washington State, among them Ford, Balcomb, and Mike Bigg, who presented his research on the population dynamics and photo-identification of killer whales. Paul Spong also attended as an observer.[34] As a result of the workshop, the IWC placed a freeze on the harvest of killer whales during its general session in 1981—a policy that became permanent the following year when it voted to enact the long-sought moratorium on commercial whaling. For Greenpeace—now a sprawling international organization based in Amsterdam—the 1982 moratorium vote represented the culmination of a long international struggle. But it owed much to Spong's encounter with Skana and the Pacific Northwest's growing embrace of killer whales.

Murray Newman believed he saw similar changes taking place in Iceland when he came shopping for whales in 1980. Still stinging from Greenpeace's criticism, the Icelandic government had reduced quotas for orca captures and mandated improvements in holding facilities. In Newman's mind, this indicated a new attitude toward the species. Killer whales had "always been looked at as enemies" with no commercial value, he reflected; now "fishermen were refraining from shooting the orcas because of the aquariums' interest in collecting them."[35] It was a self-serving argument, to be sure, but Newman was largely correct. As reports of killer whale captures circulated, fishermen grew less likely to shoot them, and the public debate over permits and holding conditions only strengthened the trend. Although the captures continued, Iceland also made initial efforts to assess the wild population. Government scientists considered utilizing Bigg's photo-identification system, and in October 1982 researchers and fishermen cooperated in a three-week killer whale census.[36] That same month, permit holders captured a young female orca. Purchased by Sealand of the Pacific in British Columbia, she would become known as Haida II, named after a southern resident killer whale who had helped shape these changes.

18

Haida's Song

HAIDA DIDN'T KNOW it, but Bob Wright was thinking of setting him free. A longtime attraction at Wright's Sealand of the Pacific, the male orca had reached twenty-three feet, and despite being paired with several females, he had failed to impregnate any of them. In June 1982, the oceanarium's director, Angus Matthews, proposed an exchange to the Canadian government. In return for a permit to capture two young killer whales in local waters, Sealand would release Haida to the wild.[1] It was a bold plan, made possible by recent scientific breakthroughs. Using Haida's own calls, researchers had deduced that he was a member of L pod, one of the three southern resident pods identified and named by Mike Bigg. But the orca's training for release would begin only after Sealand acquired new whales. Following a successful capture, Matthews explained, the oceanarium would move Haida to a pen in Pedder Bay, where the long-captive orca could learn to catch live fish and make acoustic contact with his family. But Matthews cautioned that success would ultimately depend on the whale himself. "Haida will be given his own choice," he emphasized, "of joining his old pod and becoming a born-again whale, or returning to his friends at Sealand." In late August, the Department of Fisheries and Oceans (DFO, formerly the Department of the Environment) approved the project and assigned Bigg to supervise it. The respected scientist cautioned the public that there was no guarantee Haida would survive, but he argued that the release "needs to be tried."[2]

Critics disagreed. Some accused Sealand of plotting to abandon Haida now that he had served his purpose. Likening the plan to "throwing out the family pet when it is no longer young and amusing," one local woman warned that Haida's "trust in humans will probably result in a bullet from a gun-happy fisherman." The fiercest opposition came from Greenpeace, which denounced the entire proposal. Declaring rehabilitation "unlikely,"

Greenpeace Canada president Patrick Moore argued that to move the imprisoned whale to a "halfway house" in Pedder Bay would be "to condemn him to death—alone."[3]

Yet even as handlers, officials, and activists debated Haida's fate, none took the full measure of his life. He had been born free at a time when locals still shot blackfish. Captured in Puget Sound in 1968, he barely missed induction into the US Navy's Marine Mammal Program and resided briefly on the Seattle waterfront. He then came to Sealand, where he gave many visitors their first intimate view of killer whales. In his thirteen years at the Oak Bay marina, he challenged trainers, drenched tourists, and met figures such as Bigg, Graeme Ellis, and Paul Spong, who were transforming scientific and public views of orcas. He also welcomed a procession of four female "mates," few of whom he could ever have known in the wild. Yet Haida sired no offspring, and he heard all four of the females die in Sealand's pens. All told, it had probably been a sad life. But through his long captivity, Haida had touched the lives of millions, and the debate over his freedom revealed just how much the image of orcas had changed in his lifetime.

❧

We can only guess at Haida's early years. In 1982, observers estimated his age at twenty-two, but he was almost certainly younger. Judging by his size at capture, he was likely born around 1963—a tense time for southern resident killer whales. Fishermen and locals routinely shot at them, and Ted Griffin had begun his pursuit in Puget Sound. Yet Haida likely experienced little of this. In contrast to J pod, K and L pods spend much of their time off the west coast of Vancouver Island, where the young whale would have encountered less human harassment. Like other orcas, he nursed for his first eighteen months while learning to eat the salmon shared by older relatives. Although he usually swam beside his mother, orca calves—like toddlers—can be exhausting, and moms need breaks. So his older siblings or perhaps his grandmother sometimes looked after him.

By all accounts, the most exciting time for southern resident orcas is when their pods meet in the Salish Sea. After performing the "greeting ceremony," they frolic, socialize, and copulate with multiple partners. Too young for sex play, Haida would still have shared in the joy when L pod encountered its relatives. J and L pods were likely engaged in such a gathering in early October 1968, when Namu Inc. corralled the little whale and two dozen others off Manchester, Washington. Although several of the K pod whales had experienced capture before, this was probably the first time that L pod had been behind a net. Unsure what to do, the little whale would have stuck close to his relatives as interested buyers came calling.

Among them was Sam Ridgway. Head veterinarian of the US Navy Marine Mammal Program, Ridgway had been attending a conference at Stanford University when he heard of Griffin's catch, and he caught the next flight for Seattle. By this time, the navy program had performed cutting-edge research on cetacean acoustics and physiology, and it was exploring the use of marine mammals in the deep-water recovery of military ordnance. Program officials had tried sea lions and bottlenose dolphins for the task, but these species had limitations. Sea lions cannot echolocate, and bottlenose dolphins rarely descend below one thousand feet. Hoping larger animals would dive deeper, navy officials leapt at Griffin's offer to donate two orcas to the program. Ridgway and his colleagues may have considered young Haida, but they opted for two older males, whom they named Ahab and Ishmael. A week later, as Haida learned to eat herring at Pier 56, the navy's new cetacean conscripts—possibly his podmates—were on their way to Point Mugu, California.[4]

From the beginning, Ahab and Ishmael puzzled their navy handlers. To help the animals adjust to their new schedule and feeding procedures, trainers placed each in a pen with a bottlenose dolphin. They believed this carried some risk: orcas were still regarded as a fearsome predator—the only whale known to eat other marine mammals. Instead, the dolphins bullied and harassed the two killer whales, forcing trainers to separate them. "It very much surprised us," Ridgway told me decades later, that these "great ocean predators" would let dolphins pick on them. One reason for this may have been orcas' little-understood social structure. Ahab and Ishmael were still young, and they hailed from matrilineal pods in which males spend their lives following their mothers. As such, they were accustomed to being on the bottom rung of the social ladder. Yet Ridgway suggested another factor likely in play. "Later on," he laughed, "we found out that we had the fish-eating killer whales, not the dolphin-eating killer whales."[5]

After the whales' basic training, the navy flew Ahab and Ishmael to the program's new facility on Oahu. There they learned to eat dead mackerel and bonito and became the first killer whales to be mounted with radio harnesses. Yet, as participants in the ordnance recovery program, Project Deep Ops, they proved disappointing. In contrast to the program's pilot whales, who dutifully followed their trainers' torpedo boat in open water and dove deeper than two thousand feet, the orcas seemed ill-suited to life in the navy. It may have been that the water was much warmer than they were used to, or perhaps that their background as salmon hunters rarely required deep diving. Whatever the reason, they grew difficult to control, and in February 1971 Ishmael simply swam away during a training session. Far from his home waters, without his mother to guide him, he almost certainly died. Four months later, Ahab, too, nearly went AWOL. Refusing to heed the recall signal, he led his handlers on a slow fifteen-hour chase around the northern end of Oahu, at one point forcing them to wait while he took a three-hour nap. Soon after, trainers discontinued Ahab's swims in open water. Despite these setbacks, Ridgway remained hopeful of the capacity of orcas and other marine mammals to "work under human control in the open sea," and in his widely read text *Mammals of the Sea* (1972) he suggested that in the future, "killer whales might be used as 'sheep dogs' to herd the larger baleen whales." But Ahab wouldn't be part of such plans. He died of pneumonia in 1974—thousands of miles from his home waters.[6]

Haida's life took a different path. Initially dubbed Junior by Griffin, the young whale spent seven months at the Seattle Marine Aquarium before arriving at Sealand in April 1969. His first trainer, Graeme Ellis, marveled at the young whale's responsiveness. "I think we had him trained up in about a week to do a show," Ellis recalled. "He was a great animal." Sealand's unique setting allowed Haida's colorful personality to emerge. In contrast to the enclosed pool surrounded by terraced bleachers at Sea World, for example, Sealand's open-water pen enabled visitors to stand directly over the water, ensuring that its whale shows were the wettest in the world. Whatever he thought of his confinement, Haida loved to drench the audience. "He wanted to interact with people all the time," Ellis noted. "He'd pick people out in a crowd if someone looked a bit different, and he would target them to get them wet."[7] Like other captive orcas, Haida breached on command, producing enormous splashes, but his favorite trick was to shoot seawater from his mouth at high speed. "He would nail you when you weren't expecting it, and it felt like a fire hose," recounted John Colby. "Once he hit me so hard that it blew my contacts to the back of my eyes."[8]

Haida initially used the trick to scold his trainers. "He tended to do it if he didn't think we were rewarding him fast enough," explained Mark Perry, who had come to Sealand from the Vancouver Aquarium to work with Ellis and Haida. The two trainers sought to incorporate the trick into the show, but it backfired on the first attempt. In the middle of one performance, Perry pointed to the far end of the pen, and on cue Haida sank out of sight. Seconds later, he surfaced, spy-hopping directly in front of two nuns, dressed in full habit. Before the women could react, the whale doused them with gallons of seawater. "I'll never forget the sight of Haida, in his black and white outfit, right in front of the nuns, in their black and white outfits," recalled Perry. "I half expected a lightning bolt to blast me into oblivion."[9]

In contrast to many marine parks, Sealand trainers didn't perform with the whales in the water, and Ellis and Perry built shows around what they considered natural behaviors. That meant giving Haida creative leeway. "I thought it would be interesting to include some 'free time' in the show," noted Perry, "so I decided *not* to reward Haida when he approached the training platform." The results took the young man by surprise. "Haida waited for a few moments, treading water, then went off and tried one of his regular 'tricks'—no reward," recounted Perry:

> Finally, he became so frustrated that he sped off as fast as I'd ever seen him move, circling the pool near the surface and creating a wave so big that the entire floating structure around the pool began moving up and down and creaking loudly. The staff in the gift shop all came to the windows to see what was happening. It was the one time working with orcas that I felt *fear*. "Maybe he [is] trying to wash me into the pool!" He ultimately slowed down and came back to the platform with his mouth wide open. I was so impressed that I gave him the entire remaining bucket of herring, about ten pounds in one shot.[10]

A year after Haida arrived at Sealand, he received a pen-mate. In early March 1970, Wright caught the white whale, Chimo, and four other Bigg's killer whales at Pedder Bay. Three weeks later, he transferred her and a young female named Nootka to Sealand. Fearing that Haida would attack the newcomers, the staff initially placed them in a partitioned section of the pen, but he did just the opposite. After watching the Sealand team lower Nootka into the water, he decided to welcome her. "Within moments of her being free and swimming in the pool," recorded veterinarian Alan Hoey, "Haida took a herring from his trainer and swam to the divider net, subsequently pushing it through the divider net towards Nootka who then . . . swam to the divider net, stopped, and took the herring from Haida's mouth." Haida did the same when Chimo arrived. By the end of the day, the young whale had convinced both females to eat ten pounds of herring. "This is a very remarkable piece of animal behavior," marveled Hoey. "It appeared that he showed or somehow told the other whales that they had nothing to fear from their trainers and that readily accepting food from man was not harmful."[11]

Hoey didn't know the half of it. In the wild, fish-eating orcas such as Haida rarely, if ever, interact with mammal eaters such as Nootka and Chimo. But in this captive setting, Haida drew on the common orca practice of sharing food to convince them to eat fish—probably for the first time in their lives. In doing so, he broke through a cultural and evolutionary divide that researchers believe is 250,000 years old. Reflecting on the scene after decades spent researching wild killer whales, Ellis could only shake his head and marvel at Haida's behavior. "Knowing what we know about them now," he chuckled, "what the hell was that all about?"[12]

Eventually, the females began performing at Sealand, and Haida seemed to enjoy the company. But the good times didn't last. Nootka's instincts as a dominant female emerged, and she grew aggressive, particularly toward Chimo, who suffered from a variety of ailments. Wright had a choice to make. On the one hand, he hoped to breed orcas in captivity, and Nootka was the healthier female to pair with Haida. On the other hand, he was loath to part with his prized white whale. In April 1971, he sold Nootka to the Japanese Village and Deer Park in California. Yet he would come to regret his decision. Young and vigorous, Nootka lived twenty more years in captivity, whereas Chimo's health continued to decline.

Hoping Haida and Chimo would mate, Wright invested heavily in the white whale's care. By 1972, researchers had determined that she suffered from Chediak-Higashi syndrome—a rare cellular disorder found among several species, including humans, cattle, and killer whales. With heavy doses of vitamins and antibiotics, Chimo grew stronger, and the staff hoped to announce a pregnancy soon. Wright knew that other marine parks had attempted artificial insemination, but he was resistant. "I'd like to put my faith in Haida right now," he explained, "and only if Haida fails would we consider that possibility."[13]

It wasn't that simple. During interpod gatherings, male southern resident killer whales rely on their mothers to find their female sexual partners, who can range from recently matured females to menopausal grandmothers. But Haida's only potential partner was Chimo, and she was a poor match. Although the Sealand staff couldn't have known it, Chimo was only four or five years old in the fall of 1972. In the wild, she would not have participated in sexual activity for several more years. At Sealand she was penned with an older male who was becoming sexually mature—and sexually frustrated. In captive settings, female orcas normally establish dominance over males, but Chimo wasn't normal. In addition to having a weakened immune system, she had impaired echolocation abilities and was sensitive to sunlight. Prior to capture, she had likely relied heavily on her pod for guidance and survival. Now she followed Haida's lead.

For the Sealand staff, this was most obvious during the shows, as the big male prompted Chimo to follow trainers' commands, but he also initiated her in sexual activity. It began in the spring of 1972 and picked up that summer, when Haida began performing what Colby and the staff dubbed the "morning ballet." "During this courtship dance, Haida would emit a wailing, squeaking, rather mournful cry, not unlike that of a prowling Tom cat," Colby wrote, and Chimo would respond by rubbing his genital slit. At that, Haida would break his ballet, and the two would clasp their pectoral fins around one another, swimming in circles as they copulated.[14]

Colby and Wright were on their tour of European oceanariums in late October when the whales' sexual activity became more violent. On October 28, Colby's brother Peter, a Sealand diver, noted that Chimo was "cut up with a bleeding nose," and he expressed concerns that "Haida is abusing her." Trainer Jill Stratton agreed. On October 30, she observed that Haida was preventing his pen-mate from approaching the training platform for food. "Saw a lot of sexual and vigorous activity," she noted in the trainers' log. "His bunting of her was very violent." Two days later, matters grew worse. "He seemed to be getting more and more frustrated," staffers observed, adding that Chimo's skin was "badly mangled in a grid network of deep scratches." In London, Wright and Colby attended a performance of *Othello*, in which the dark-skinned title character kills his wife Desdemona in a blind rage. That same night, five thousand miles away, the same scene may have been playing out at Sealand.[15]

When staffers arrived the next morning, something didn't seem right. Haida was breaching loudly in the pen, and Chimo was nowhere to be seen. Manager Rob Waters sent Peter Colby and another diver into the water. After a brief search, the men found Chimo upside down against the containing net. She was dead, and Haida's blood was up. "When the divers went in," recorded Waters, "Haida was nudging her and had an erection." And he wasn't backing down. "Haida would not let the divers near her," Waters wrote, adding that the agitated orca "bit Peter on the hand." An hour later, staffers managed to distract Haida long enough to raise Chimo to the surface. What they saw troubled them. Trainers identified "several new deep scratches near her genital area," as well as "an orangish-brown fluid extrud[ing] from her genital-anal slit."[16]

What had happened overnight? Had Chimo's weakened body simply given out? Had she fallen victim to a bacterial infection, as Wright told the press? Or had Haida's sexual frustration driven him to kill her? Only Haida knew for sure.

Ordinarily, oceanariums move dead whales some distance from their penmates before performing a necropsy, but Wright ordered the staff to dispose of the carcass immediately. Perhaps he couldn't bear to see his prized white whale dead. Perhaps he didn't want observers to note the injuries Haida had inflicted on her. Whatever the reason, the Sealand staff dissected her body right next to the pen she had shared with Haida. "When they cut inside, it was boiling like a cup of tea, boiling like water," recalled fisherman Tor Miller. "And talk about rich! You could smell it all the way to Glenlyon School."[17] Although the staff erected a plywood barrier to block Haida's view, the orca surely understood what was happening. He had watched staffers pull Chimo's body from the water, and now her blood was pouring into his pen. Afterward, he wasn't the same.

When Wright and Colby returned from Europe, they learned that Haida wasn't eating. Some staffers thought he was sick; others believed he was mourning. "For nine days he was completely off food," Colby told a reporter. "He just lay still in the middle of the pool." Concerned he was about to lose a second whale, Wright asked Don Goldsberry for help, and on the Seattle whale catcher's advice, Sealand mixed up a unique "eggnog" for the ailing orca. In addition to five dozen eggs and large doses of vitamins, the beverage included thirty-six bottles of Guinness Stout. After pumping it into Haida's stomach, staffers released the tipsy whale into his pen. Colby took watch that evening, perhaps becoming the first person in history to pass a night with a drunken whale.[18]

Days later, jazz flautist Paul Horn paid Haida a visit, on the suggestion of Paul Spong. Years earlier at the Vancouver Aquarium, Spong had used music to connect with Tung-Jen (Hyak II), then a lonely calf. Throughout 1971 and 1972, he had visited Sealand on many occasions, sometimes inviting the musician to come along. "Haida really responded to my playing and seemed to favor the alto flute," recalled Horn. "He would lie on the top of the water near me all the time I was playing and occasionally answer back with a great variety of sounds."[19] The musician was so moved by the experience that he included recordings of Haida, paired with his flute, on his new album, *Inside II* (1972). Along with the eggnog, Horn's music seemed to pull Haida from

FIGURE 18.1 Staffers at Sealand of the Pacific prepare to cut apart Chimo, November 1972. In author's possession.

his doldrums. "Time and a skillful flute player heal all wounds," quipped one reporter, "including a killer whale's heartbreak."[20]

But more heartbreak was in store. In August 1973, Colby's Sealand team made its captures of four whales in Pedder Bay, including another female to pair with Haida. The two had likely crossed paths before. Named Nootka II, she was probably from K pod and as such would have recognized some of Haida's calls. But she was much older than he—in her thirties or even forties—and like most adult orcas, she didn't do well in captivity. Despite asserting dominance over Haida, she struggled to adapt to life in a pen and died of a ruptured aorta in May 1974. The following year, in August 1975,

FIGURE 18.2 Paul Horn plays flute for Haida at Sealand of the Pacific, 1972. Courtesy of Mark Perry.

Angus Matthews led Sealand's third catch at Pedder Bay, which netted six Bigg's killer whales and brought resistance from the provincial government. It was Kanduke's sale to Marineland of Canada that stirred the controversy, but Wright kept a second whale for Haida. Named Nootka III, she died of an infection the following year. Then came Miracle.

⁂

On August 2, 1977, a diver looking for crabs in Menzies Bay, just north of Campbell River, nearly bumped into a small orca just beneath the surface. The encounter intrigued Bill Davis, a local mill worker and sport fisherman, who jumped into his boat for a look. Davis offered the young whale a handful of herring, which it gobbled down eagerly. He returned each day for a week, soon venturing to pat and stroke the animal. But Davis worried about its health, as well as the attention it was drawing from locals. On August 8, he called the Vancouver Aquarium, which in turn contacted Mike Bigg. The researcher himself had spotted the animal off Nanaimo a month earlier—the first orca calf he had ever seen swimming on its own. He phoned Bob Wright, and soon after Bigg, Wright, and Matthews boarded a seaplane to take a look.[21]

When the trio met Davis in Menzies Bay, the whale was nowhere to be found. But when Davis started his engine, the whale surfaced next to his boat seconds later. Bigg was stunned by the behavior, and after examining the whale he agreed that it would die without immediate treatment. Fearing public criticism, however, his superiors in the Department of Fisheries were reluctant to issue a capture permit. "We don't have time to wait for them," declared Wright. "We'll take care of the whale and worry about the permit later."

"Okay," agreed Bigg, "but just don't tell me about it."[22]

To Bigg's relief, Ottawa issued the permit the following day, and with the help of local loggers and a crane barge owned by timber giant MacMillan Bloedel, the Sealand team netted the whale and loaded it onto a truck. "If you had set out to design a barge to be used specifically to lift a killer whale out of the water, you couldn't have designed a better one," marveled Angus Matthews.[23] It was a remarkable scene. Sixteen years earlier, before any orcas had been displayed in captivity, locals in the area had pushed the Department of Fisheries to install a machine gun to kill blackfish. Now sport fishermen and loggers cooperated to rescue a young orca in distress.

Driving through the night to avoid the summer heat, the team arrived in Victoria in the early morning of Wednesday, August 10, 1977, and lowered the whale into an outdoor saltwater swimming pool at the Oak Bay Beach Hotel, located just south of Sealand. Soon after, the calf went into shock, sinking to the bottom of the pool. As staffers struggled to hold the limp animal at the surface, veterinarian Alan Hoey declared that her heart had stopped. With massages and shots of adrenaline, the plucky calf sprang back to life, and over the following hours staffers began to realize the extent of her injuries. In addition to being emaciated and covered in fungus and parasites, she had propeller slashes on her back and rifle wounds in her side. Even as he injected her with antibiotics, Hoey warned that the animal would likely die. Yet each day she grew stronger, and it soon became clear that the team had managed to save the little whale. "It was our chance to apply everything we knew in a really righteous way," reflected Matthews, "so it was a real turning point for a lot of us, and I think everything about that became a very positive thing." Celebrating the animal's recovery, the local press dubbed her Miracle.[24]

It sounded like a Hollywood script—and, in truth, Hollywood may have been partly responsible for her injuries. That summer, the movie *Orca* had played in local theaters, and scientists and activists had warned that the film's violent depiction of killer whales might provoke attacks on the species. The shooting of this calf seemed to justify such fears.[25] Yet the effort to rescue her came to symbolize a new relationship with cetaceans. Noting that "the last of her larger relatives are being slaughtered in the Pacific by the Japanese and Russian whaling fleets," Greenpeace Seattle spokeswoman Julie MacDonald declared that "Miracle is fortunate to be the recipient of human kindness."[26]

FIGURE 18.3 Angus Matthews "walks" Miracle at the Oak Bay Beach Hotel, August 1977. Photo by Jim Ryan. Courtesy of Angus Matthews.

It was a curious turn for Sealand. Just two years earlier, Greenpeace had denounced the oceanarium for its capture at Pedder Bay; now it praised Sealand's efforts as the antithesis of commercial whaling. In fact, Miracle had already become a local celebrity. Over the next three months, more than three thousand people a day came to see the little whale, among them Paul Horn and Bill Davis—her friend from Menzies Bay. In the process, many locals grew attached. A thirteen-year-old local boy named Tony Akehurst worried the calf would be shot again if released and pleaded for her to remain in the oceanarium's care. "I love Miracle dearly," he wrote, "and I would like to see her grow up in Sealand and safe in Victoria!"[27]

Most experts agreed. Although some animal rights activists called for Miracle's release, Bigg regarded that as a death sentence. "It would be like putting a 6-year-old child out in the wilds and saying, 'Look after yourself,'" he warned. Moreover, he and other researchers noted that science had much to learn from such a young calf. The fact that Miracle was teething, for example, revealed that killer whale calves were born without front teeth, likely to facilitate nursing. "Scientists had always thought that young killer whales are born with a full set of teeth," noted one observer. "They're rewriting the textbooks."[28] By February 1978, Miracle had nearly doubled in weight, and with winter storms threatening the animal's outdoor pool, Sealand airlifted her to its main facility. Wright may have hoped that she would eventually mate with Haida, but he had learned from his mistakes, and he kept the two animals separate.[29] As Haida had done before, he accepted the newcomer, and the two began calling to one another. Over the following years, Miracle settled in to life at Sealand. But the little whale seemed troubled, at least to Alexandra Morton.

Like most orca researchers, Morton got her start in the captivity business. Raised in Connecticut, in 1976 she had moved to California, where she worked with John Lilly—the controversial theorist who had experimented on captive dolphins in the 1950s and 1960s. Morton herself researched dolphins at Marineland of the Pacific, and it was there that she saw her first killer whales—Corky II and Orky II, both originally caught in Pender Harbour. In 1980, Morton moved to British Columbia, hoping to observe orcas in the wild. Yet she maintained close ties with oceanariums, and in 1981 she received permission from Sealand to undertake acoustic research on Miracle. Like Spong before her, Morton enjoyed the close contact with the species that was not possible in the wild. "At night I stroked her as she lay beside my tent," she later wrote. "This lone baby whale craved too much attention for me to maintain a passive scientific observer relationship." Ultimately, Morton concluded that both of Sealand's orcas suffered from acoustic deprivation. Like Miracle, Haida had been separated from his pod as a young whale and retained few of its calls, and Miracle seemed to suffer from having lost her family so young in the wild. "She was an adorable, affectionate little whale who exhibited extremely strange behavior," noted Morton. "After every show she rushed over to a particular spot and banged her head against the underside of the dock."[30]

By the beginning of 1982, the Sealand staff had other concerns. Miracle's playful habit of pulling on strands of her outer chain net had cut a hole in the inner net, threatening to entangle her. After a seal who shared her pen drowned in the net, Matthews warned Wright of the danger. "I went up to see Bob and said, 'This is a disaster waiting to happen,' and Bob dragged his feet and dragged his feet," recalled Matthews. Then, on the morning of January 12, trainers found the little orca dead, her body wedged in a hole in her net. Publicly, Matthews maintained that Haida didn't know Miracle had died, but that seemed far-fetched. Despite being held in separate pens, the two whales regularly called to one another. Haida would have heard Miracle's distress and certainly noticed her silence.[31]

In the aftermath, Wright claimed that activists had cut the net, and it certainly seemed plausible. Animal rights and environmental groups had called for Miracle's release on several occasions, and a year before her death, Sealand staffers had chased a diver away from the perimeter of the pen.[32] But the story ultimately revealed more about Wright than about the accident. "He was a harsh entrepreneur—no question about that," observed Matthews, "but he was so anthropomorphic towards those animals. It was amazing—he couldn't even watch us draw blood." The rescue of Miracle had been the proudest moment of Wright's career, and he couldn't bring himself to accept responsibility for her death. "He really had trouble processing that. For all his rough exterior, he was heartbroken—absolutely heartbroken," reflected Matthews. "The long and short of it is that he really, really wanted to believe that somebody had done it, and I actually think he did believe that."[33]

The little whale's death saddened many. "Miracle was special, like the baby in the family," wrote the editors of the *Victoria Times-Colonist*, noting that the entire community had embraced her. "There are those who believe no wild creatures should be penned," they noted, "but the killer whales at Sealand and elsewhere have awakened many people to the richness of life in the sea and have spurred a respect for magnificent creatures like killer whales."[34] That argument would be tested when Sealand proposed trading in Haida for two new whales.

⁂

The idea had originated with Greenpeace. In October 1974, in consultation with Paul Spong, Bob Hunter had written a column in the *Vancouver Sun* suggesting that oceanariums move toward a rotating system of captivity, with the release of killer whales after a period of several years and their replacement with new whales.[35] A few months after Miracle's death, Matthews took up the idea. Haida had spent his youth opening the eyes of millions to his species, the director explained to reporters, and now the mature whale deserved a chance to swim free. But in order to release him, Sealand needed new orcas to display. Some viewed it as a cynical maneuver to exchange an older whale for younger animals who would be cheaper to maintain. Yet many, including Bigg, believed Haida was a good candidate for release. He was healthy, he had never left his home waters, and scientists had identified him as a member of L pod—making reintroduction to his family possible.

⁂

Thirty-five years after making the proposal, Matthews met me in Oak Bay, less than a mile from where Sealand once stood. Like most who worked with Haida, he had fond memories of the oceanarium's star orca. "Haida was an incredibly confident, curious, engaged animal, and the working relationship was a game he really liked to play," Matthews told me. "We think we are so smart and in such great control of our environment, but when you encounter an animal that is incredibly highly developed in its environment, you learn a whole new level about your smallness." By the early 1980s, however, he was concerned that the big whale's time was running out. "Bob wanted to apply for a permit to get more whales," explained Matthews, "and it started as how could we come up with PR—a justification that the government could swallow." For Matthews, however, it became something more: he genuinely wanted Haida to have the opportunity to rejoin his family, even if Wright hoped the whale would choose to return to Sealand. "I really saw [captive whales] as ambassadors, and the notion of an ambassador as a fixed-term posting," Matthews explained. "It was a pure and well-intentioned concept."[36]

It was the proposal to catch more whales that really galled Sealand's critics. From across Haro Strait on San Juan Island, Ken Balcomb likened the plan to "capturing two lions in the African veldt and putting them in cages." Rich Osborne, the curator of Balcomb's Whale Museum in Friday Harbor, agreed, declaring, "We basically look at southern B.C. and upper Washington State as one giant aquarium." The best way to see whales, he argued, was from tour boats in their natural environment. In fact, the museum itself now offered whale-watching cruises in cooperation with Greenpeace. Pitched as an alternative to the profit-driven oceanarium industry, it charged twenty-eight dollars per person—four times the admission price of Sealand. For his part, Matthews ignored this irony, instead cautioning against the environmental impact of large-scale whale watching. In the previous year, he noted, seven hundred thousand people had visited Sealand; if Osborne "thinks you can take 700,000 people out to see orcas in the wild without totally desecrating their environment, he's got to be kidding."[37]

The most effective argument against Sealand's plan was that the southern resident killer whales belonged to both nations. "The Canadian government is trying to make a unilateral decision about animals which don't belong to either country," asserted Greenpeace Seattle spokeswoman Julie McCullough.[38] Ralph Munro agreed. In 1980, four years after leading the charge against Sea World in Puget Sound, Munro had won election as Washington State's secretary of state. But he remained an outspoken advocate for orca conservation, and he was appalled that Bigg and his DFO superiors had approved Sealand's proposal. "It's a sad day for the Pacific Northwest," Munro told reporters. "They might as well have issued a permit to shoot the whales."[39]

Yet Canadian officials refused to give in. Although they prohibited Sealand from taking whales from J pod, which had been culled most extensively in the past, they maintained that the limited capture posed no threat to the southern residents. Despite public criticism, Bigg himself continued to support the project, arguing that the pods that lost members would compensate by living longer and having more calves. "The reason is simple," he theorized. "There's more food available when there are fewer animals so they reproduce faster and die less."[40]

Greenpeace stepped up its opposition in late August. Having celebrated the IWC's approval of a whaling moratorium a month earlier, members in the Northwest now turned their attention to the capture debate. In early September, activists set up camp near Pedder Bay and launched skiffs to interfere with Sealand's efforts. On at least one occasion, they claimed to have frightened orcas away from the capture area.[41]

Yet Sealand had unexpected allies, among them Paul Watson. A former Greenpeace activist, Watson now led the rival Sea Shepherd organization, and he had recently won fame for ramming and disabling the infamous pirate whaler *Sierra* in a Portuguese port. For years, Watson had called for protection of the world's whales, but he dismissed the protest of Sealand's plan as misplaced. His position was partly informed by his friendship with Matthews, who had participated in Greenpeace's antinuclear demonstrations years before becoming a target of the organization.[42] Reminding everyone that Greenpeace itself had first suggested temporary captivity, Watson praised the project as "a revolutionary move on the part of an aquarium." He argued that, in addition to setting a precedent for other releases, the exchange would "break the back of the captive orca industry in Iceland, where whales are killed in an ongoing capture program." Watson expressed hope that aquariums would "one day phase out the practice of holding captive orcas," but he also noted that "it cannot be denied that the few dozen orcas in captivity have been excellent ambassadors for all whalekind." And he reminded readers of recent history:

> Remember that only a decade ago, the killer whale was feared and hated, shot at and killed by ignorant and irate fishermen. The whaling

> industry slew tens of thousands of whales each year. The fact that orcas are now totally protected by law and the slaughter of other whales has decreased is to a large extent because the public was given the opportunity to meet, know and love whales.[43]

In doing so, Watson pointed to a larger conservationist ethic that joined the two sides of the debate. However bitter it may have seemed, the argument was over means, not ends. Some saw oceanariums such as Sealand as obsolete; others viewed them as essential to public education. Yet all agreed that orcas—the once dreaded blackfish—were worth protecting.

In the end, the debate made little difference to Haida. His release permit expired on October 1, 1982, and that same day his behavior took an odd turn. The big whale seemed listless, and the following day he roamed his pen in a distracted manner, eating only six herring. The next morning, Sealand staffers were preparing to dry him up for an examination when he swam to the far corner, rolled, submerged, and failed to resurface. "When he went down, he went slowly down and bubbles came up," recalled trainer Cees Shrage. "So many bubbles."[44]

My family heard the news that afternoon, as we drove to Tacoma. We were going to visit the Point Defiance Zoo, and I was excited. As I listened to music on the radio, I tried to imagine the animals I would be seeing. We were just a few miles from where Haida had been caught fourteen years earlier. State Route 16 took us southeast past Gig Harbor and onto the Tacoma Narrows Bridge. The song ended, and a reporter broke in with a news flash: "The killer whale at Sealand of the Pacific in Victoria died early this morning." "Haida!" my mom cried, so loudly it startled me. My dad remained silent. After crossing the bridge, he pulled off at the first exit and stopped the car. Then the tears came.

That evening, he brought me to the living room, thumbed through his records, and pulled out an LP with a dark cover. He started the turntable, lowered the needle, and told me to listen. The song was ethereal, haunting. I would later understand that it was Paul Horn's flute, accompanied by the voice of the whale who had died that morning. It was Haida's song.

Mark Perry had listened to the record many times. It had been a decade since he'd been Haida's trainer and nearly two since he sneaked onto Jericho Beach to see Moby Doll. Now working in television news, he viewed his time with orcas as part of a distant past. Perry had been back to see Haida only once, and the visit saddened him. "I'd been able to move on with my life," he later told me, "but Haida, still trapped in a small enclosure and haunted by the loss of so many of his pool mates, had no chance of that." The news of the whale's death hit him hard. "I felt like I'd lost an old friend."[45]

Others felt the same. "Haida has played a special role in my childhood," wrote one Victoria teenager, who explained further:

> I was four years old when he came to us and I have little recollection of life without him. I didn't go to Sealand very often, usually once or twice a year, but I often sat on the rocks outside his pool. It was always comforting to be able to hear him blow and splash when coming up for air. He was so big and powerful, that it was like magic that he could live with us at all. I always wanted to show him off to visiting friends and they always came away thinking as much of him as I did.[46]

A woman from Courtenay, 140 miles up island, agreed in a letter to the *Times-Colonist*. "The whales of Sealand opened my eyes (and those of my family) to the wonder and grace of those huge animals," she wrote. "I was totally ignorant of them before but after Sealand I loved and respected them." The capture of wild killer whales entailed sacrifice, she acknowledged, especially for the animals, but the benefits that a whale like Haida brought outweighed that cost. "How else," she asked, "can the average person see a whale close up?"[47]

19

The Legacy of Capture

IT WAS A quiet day in the summer of 1983 when the shots rang out in Robson Bight. Just minutes earlier, researcher Dave Briggs had been watching A4 pod visiting the rubbing beaches. The orcas had then headed in the direction of nearby purse seine vessels, and something had clearly gone wrong. Rushing down to the water, Briggs signaled a nearby whale-watching boat, which picked him up and motored out to investigate. Soon after, two orcas approached the vessel. It was the pod's matriarch, A10, and her young calf, A47, both of whom had been shot. The horrified passengers watched as the injured mother pushed her child toward the tour boat. "We could see the wound oozing blood," Briggs recounted. "It really seemed that she was showing us: Look what you humans have done."[1] Jim Borrowman learned of the shooting within minutes. An environmental activist and whale-watching entrepreneur based in nearby Telegraph Cove, he jumped into his Zodiac—a small, inflatable boat—and raced to the area. "I saw A10 with a bullet hole in the side of her face," he recalled. "I just couldn't believe it." Over the years, Borrowman had seen many gunshot wounds on orcas, but this seemed a senseless act of violence reminiscent of an earlier era.[2]

Appearances aside, the incident underscored how far the human relationship with the species had come. No longer the indistinguishable black masses of the past, each orca in the Pacific Northwest now had an alphanumeric label, a family tree, and even affectionate nicknames. Once considered menacing pests, killer whales had become symbols of the region's new environmental values and prime attractions for its tourist industry. Writing in the early 1980s, naturalist Erich Hoyt had little doubt that the display of killer whales at oceanariums had caused this shift. "The most important result of the captive-orca era has been the almost overnight change in public opinion," he observed. "People today no longer fear and hate the species; they have fallen in love with them."[3] In the following years, that link between captivity and public affection was gradually forgotten, and as whale watching grew in importance, Pacific Northwesterners came to regard killer whale capture as the antithesis rather than the origin of their fondness for orcas. Yet the knowledge and interest generated by captivity became ever more critical to the struggle to protect killer whales.

Jim Borrowman himself embodied that connection. Born in 1949 and raised in Victoria, he had cared little about orcas until the Vancouver Aquarium acquired Skana in 1967. Over the following years, he went to see the captive whale perform every chance he got, and he also visited Haida at Sealand. "I was like the [other] people in the stands. I was just amazed by this animal," he later told me. "Even in Victoria, no one knew anything about whales. We barely ever saw them. These whales were made *accessible* to people." In later years, Borrowman trained as a commercial diver, working on docks and oil rigs before taking a job at a sawmill in tiny Telegraph Cove. It was there he met Erich Hoyt, who asked if the diver could photograph killer whales underwater, and the two began regular expeditions in the area. But in 1978, the provincial government announced that it was opening the Tsitika Valley to logging by timber giant MacMillan Bloedel. The plan threatened the last pristine watershed on the east coast of Vancouver Island, and it called for a log boom site in Robson Bight, virtually on top of the rubbing beaches frequented by northern resident orcas.[4]

Borrowman leapt into action. Along with fellow mill worker and diver Bill Mackay, he launched a campaign to save Robson Bight. Drawing on research provided by Mike Bigg and Graeme Ellis, the pair crisscrossed British Columbia giving presentations on the importance of the rubbing beaches for northern resident killer whales. But Borrowman also learned the perils of the rising public affection for orcas. In 1979, he and Hoyt coauthored an article in *Diver* magazine entitled "Diving with Orcas," about their experiences in Robson Bight.[5] In response, swarms of divers rushed to the area, eager to have their own encounters with killer whales. "It drove us nuts," Borrowman recalled. "They were coming up on the dive boats, they were leaping off the bow of the boats in front of the whales, and it was just a gong show." The following year, he and Mackay founded a company called Stubbs Island Charters, initially intended to cater to visiting divers. But the partners also found themselves taking journalists, government officials, and visitors to see the whales. In 1982, their efforts paid off, as the British Columbia government declared Robson Bight an ecological reserve.[6] It was a remarkable sign of shifting values: a province built on logging had prioritized an orca cultural site over the timber industry—at the urging of two former logging employees.

By that time, Borrowman and Mackay had realized the commercial potential of whale watching. Their business grew slowly but steadily, in large part due to the allure of Robson Bight.[7] By the mid-1980s, other companies began to appear, and although their tours enabled tens of thousands to see orcas in the wild for the first time, some observers had misgivings about the incipient whale-watching industry. Having followed the same pods around in her Zodiac for years, researcher Alexandra Morton now worried about the growing traffic. "Whenever a whale jumped," she claimed, "a dozen tour boats home in like flies to a wound."[8] Borrowman, too, noticed this troubling pattern. "Ninety-nine percent of people were fascinated by the whales," he reflected. "The closer you got, the more fascinated they became. The closer you got, the closer they wanted to get. These are the downsides of whale watching—the pressure to get closer is horrendous, and you are trying to satisfy people."[9]

Ralph Munro observed similar trends south of the border. Through the 1980s, public desire to see killer whales in Washington State continued to grow, and San Juan Island was becoming a popular destination. In 1983, as Washington's secretary of state, Munro led the push to establish Lime Kiln Point State Park—a prime whale-watching site on the west side of San Juan Island. The following year, the Whale Museum announced its new Adopt an Orca program, which boosted donations by giving individual animals catchy names such as Granny and Oreo to go along with their alphanumeric designations. Yet as in British Columbia, the species' iconic status had a dark side. People came from far and wide to see whales in Haro Strait, and not all the sightseers proved considerate. Munro had a close-up view of this problem when he joined Ken Balcomb on a research cruise in the summer of 1985. "Frankly, I was shocked and disgusted at the innumerable speed boats, power boats, sailboats, and tour boats that were constantly harassing the whales," he wrote in a letter to the *Seattle Times*, and he was especially concerned for the safety of a newborn in K pod. It would be "an absolute tragedy," he asserted, "to lose [the calf] to the propeller of a speed boat trying to get 'one more picture.'"[10] Rather than fishermen or marine parks, it seemed public affection now posed the greatest threat to Northwest orcas. But Munro wasn't done fighting Sea World.

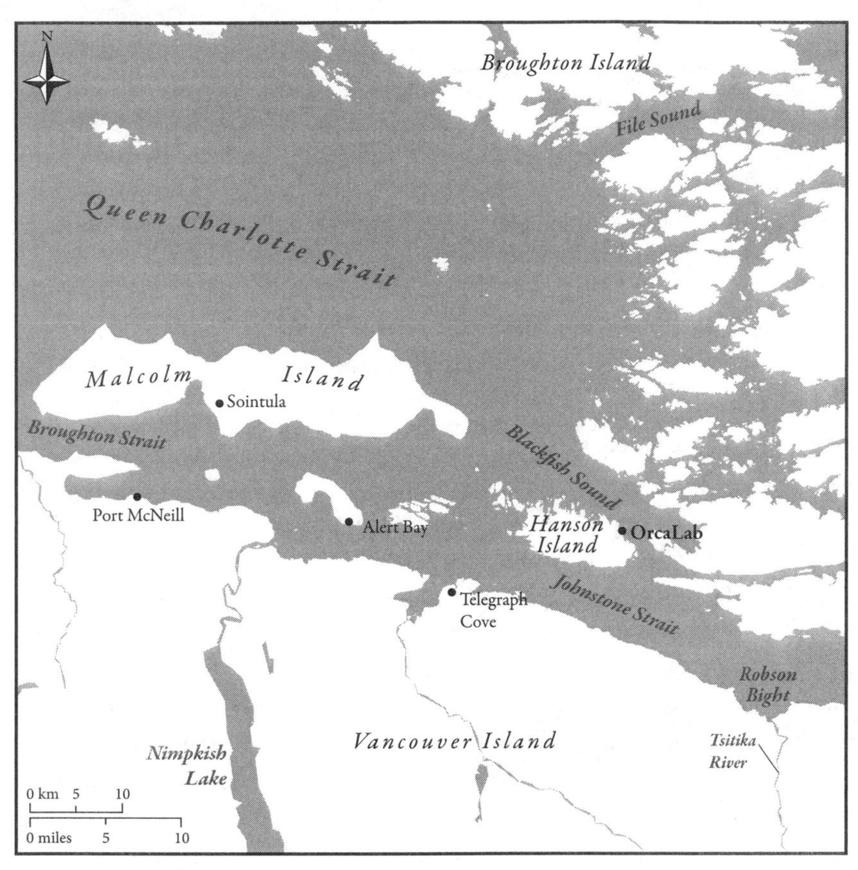

FIGURE 19.1 Johnstone Strait.

By the early 1980s, the company again faced a limited supply of its signature animal. Shut out of the Icelandic trade, it applied for a federal permit to collect killer whales in Alaska. Senior Vice President Lanny Cornell framed the project as a research operation, explaining that Sea World would capture and run medical tests on a hundred orcas, thereby enhancing scientific knowledge of the Alaskan population. But his plan also called for the company to keep ten of the animals for display and breeding at its three locations. After all, Cornell noted, "Shamu is our Mickey Mouse." Activists in Washington State and British Columbia immediately mobilized in opposition, and they had a key ally in Munro. "We're sick and tired of these Southern California amusement parks taking our wildlife down there to die," he declared to reporters. "They'd be laughed out of town if they tried to do business here."[11] Once again, Munro enlisted the help of longtime friend Slade Gorton. Now the junior US senator from Washington State, Gorton hadn't forgotten his clash with Sea World, and he pressured NMFS officials into holding public hearings in Seattle. "They were furious," laughed Munro. "I think they wanted to issue the permits and be done with it."[12]

Sea World was equally appalled. Company officials had carefully cultivated Alaskan support, including flying leaders of the indigenous Tlingit nation's Killer Whale Clan to San Diego.[13] Now they had to defend their proposal in Seattle—a hotbed of anticaptivity activism. The crowded hearing in August 1983 began with a lineup of prominent Washington State politicians—including Gorton, Munro, and Governor John Spellman—criticizing the proposal, and it ended with other witnesses denouncing Sea World. "We beat the shit out of them," Munro proudly recalled. Yet the company had its supporters, among them University of Washington scientist Al Erickson, and it initially seemed to prevail. In November, the NMFS approved the capture permit, but it also required Sea World to fund a study of Alaska's orca population.[14]

Coordinated by Sea World's Hubbs Institute and modeled on Bigg's photo-identification system, the project produced the first intensive survey of killer whales in Alaskan waters. To run the study in Southeast Alaska, the company hired Balcomb, and for Prince William Sound, it contracted fisherman and naturalist Craig Matkin. From April to September 1984, Matkin and his team took thousands of photographs of local whales, which they sent to Graeme Ellis in Nanaimo, British Columbia. Working largely pro bono, Ellis identified ten resident pods and a total of some 180 animals in Prince William Sound. As information about the state's orca population emerged, so did local opposition to capture. "It was probably inevitable," Ellis reflected. "Once the whales became individuals with labels and pods, people there began thinking about them as 'our whales.' "[15]

Yet activists accelerated the process. Even as the study proceeded, Paul Spong and others reached out to Alaska's small environmentalist community, which formed a group called Organized Resistance to Captures in Alaska (ORCA). For his part, Munro testified to the Alaska legislature, describing Sea World's 1976 operation in Budd Inlet and playing up regional resentment. "If you want these marine circuses to come up here and steal your whales and take them to Southern California and charge your kids nine dollars to get in and see them," he declared, "then go right ahead."[16] The strategy worked. Alaska governor Bill Sheffield announce his opposition to a "California bunch" removing his state's whales, and his administration followed Washington State's precedent, filing suit in federal court. In May 1984, a district court judge issued an injunction, and the following January he voided the permit, citing the NMFS failure to conduct an environmental impact study.[17]

Yet the marine park chain hardly represented the greatest threat to Alaska's killer whales. In March 1989, the tanker *Exxon Valdez* struck a reef in Prince William Sound, spilling nearly eleven million gallons of crude oil. Sea World's photo-identification study now enabled researchers to track the spill's catastrophic impact on local orcas, once again with the help of Graeme Ellis. Fourteen of the thirty-six members of one salmon-eating pod died over the next two years, while a unique population of mammal hunters, known as the Chugach killer whales, lost all their reproductive females, setting them on the path to extinction.[18]

~

Meanwhile, Sea World still had an orca supply problem. Captive breeding offered a possible solution. In September 1985, the Icelandic-caught Katina gave birth to a calf at Sea World Orlando, and five months later another whale gave birth in San Diego.[19] Yet the program faltered in April 1986 with the death of Winston, the company's only breeding male. With construction underway on its new San Antonio franchise, Sea World went on a shopping spree. First, it purchased Kanduke and Nootka from Marineland of Canada—two animals originally caught by Sealand. It then approached Marineland of the Pacific with an offer to buy Corky II and Orky II. Despite being from the northern resident A5 pod and hence closely related, the pair had bred in captivity. Without the assistance provided by female relatives in the wild, however, their calves had died quickly. Still, the pregnancies proved that Orky was virile, and when Marineland refused to part with the whales, Sea World simply bought its longtime rival. At first, company officials promised to keep the newly acquired park open, but in early 1987 they moved Corky and Orky to San Diego and shuttered Marineland.[20] Meanwhile, Sea World vice president Lanny Cornell was angling to resume captures in Iceland, and he approached Bob Wright for help.

Since Haida's death, Sealand's owner had been busy. In 1983, he had acquired three orcas from Iceland—a young male and two females. The male died soon after, and the oceanarium replaced him with another Icelandic whale, to be named Tilikum. In 1986, Wright hosted a conference on the captive breeding of cetaceans, and soon after he struck a deal with Cornell to act as liaison to Helgi Jónasson, an Icelander who had permits to capture four orcas. Working through a shadow subsidiary called BoLan (for "Bob and Lanny"), staffers would supervise captures without implicating Sea World directly. To avoid conflict with the NMFS, which was restricting the importation of wild whales, Cornell also made an arrangement with Kamogawa Sea

World in Japan, which agreed to hold the four whales temporarily in return for being able to keep one of them. Cornell planned to import the remaining three animals at a later date, thereby classifying them as already captive.[21]

It seemed an elegant plan. But when Don Goldsberry and Jim Antrim arrived in eastern Iceland in October 1987 to help capture the four whales, they hit a snag. Someone at the San Diego franchise had tipped off Sea Shepherd and the Humane Society of the United States (HSUS), as well as the new Whale Friends of Iceland. When activists appeared in Seyðisfjörður, Goldsberry and Antrim decided to pull up stakes. Almost immediately after they left, the vessel they had hired captured four whales.[22]

Jeff Foster received the call in late October. It was Goldsberry, his father-in-law. "We need you in San Diego right away," he said. Upon arrival, Foster was ushered to Sea World's back entrance and told he would be flying to Chile immediately to explore capture possibilities. But the next morning, Goldsberry surprised him at the airport. "I'm standing in the line for South American airlines," Foster recounted, "and Don comes in and says, 'Oh, wait a minute, you've got the wrong ticket—you're going to Iceland.' "

"I don't want to go to fucking Iceland," Foster protested. "I have no interest at all in going to Iceland."

"You're going to Iceland," repeated Goldsberry. "We'll buy you anything you need."[23]

Carrying $5,000 in cash and traveling under the name Jim Jeffries, Foster landed at Keflavik Airport and made his way to Seyðisfjörður. There he found himself caring for the four young orcas. Goldsberry had told him to sit tight: someone would be in touch in a few days. But the call never came, and the bills started to add up. "I had to pay for all this!" Foster told me. "Had to rent a crane to get them in and out, had to make vitamins for the whales, had to buy the fish, had to pay off all these people." Within a week, his cash was gone. "Pretty soon, I'm having to put stuff on my credit card, and I am not hearing from anybody," he explained. "I try to get ahold of Don, and I can't get ahold of Don. I try to get ahold of Lanny, and I can't get ahold of Lanny." Finally, Foster managed to reach a Sea World secretary, who informed him that Cornell and Goldsberry no longer worked for the company.[24]

The reasons for Cornell's firing remain murky. Although the Iceland scheme may have played a role, the primary factor was a series of injuries sustained by trainers at the San Diego franchise—the worst of which occurred on November 26, 1987, when Orky landed on top of a trainer riding Nootka.

Such incidents belied the gentle image of orcas that had become an industry trademark, and some insiders questioned the wisdom of people performing with such powerful predators. William Jovanovich, chairman of Sea World's parent corporation, HBJ, spoke out about the incidents. "In my opinion," he asserted, "human beings should never again enter the [killer whale] pool."[25] Whatever the reason, by December 1987, Cornell was gone, Goldsberry had quit, and Foster had four unclaimed whales and mounting credit card debt.

Finally, he received a call from Ed Asper at the Orlando franchise. "So, Jeff," he asked, "what do you have over there?"

"I'm sitting on four black and whites, and I'm out of money," the desperate Foster replied. "What am I supposed to do?"[26]

Asper wired funds, and Foster moved the whales to Hafnarfjörður. Soon after, Sea World officials met with Wright in Victoria to untangle the mess. The two sides reached an agreement that transferred control of the whales to Wright, who sold all four to Kamogawa Sea World. In March 1988, Foster helped load the animals onto planes, and he left Iceland convinced that his whale-catching days were over. The following year, however, Wright hired him for another round of captures, selling four more orcas to parks in France and Japan. "Bob never came to Iceland," laughed Foster, "but he made a fortune."[27]

Yet Wright's high times in the killer whale business came to an abrupt and tragic end. For years, there had been signs that Sealand's three Icelandic whales posed a danger to trainers, with one of the large females proving particularly rough. During a routine show on February 20, 1991, part-time trainer Keltie Byrne—a twenty-year-old marine biology student at the University of Victoria—slipped into the pen, and over the following minutes Tilikum and his pen-mates drowned her in front of screaming onlookers. It was the first time an orca had killed a human being in captivity, and it ended killer whale display in Victoria. Facing protests and government pressure, Wright closed Sealand and put its whales up for sale.[28]

Sea World leapt at the opportunity. The deaths of Orky II in 1988 and Kanduke in 1990 left the company's four franchises with thirteen killer whales between them, only one of whom was male. If the breeding program was to succeed, it needed more males, and when Sea World veterinarians examined Sealand's animals, they were pleasantly surprised: both females were pregnant. It was a bitter irony for Wright. After twenty years of trying to breed orcas, he had succeeded just as he was closing shop. But Sea World was thrilled. The pregnancies proved not only that the females were fertile but also that Tilikum might serve as a breeding stud. "Sea World has only one other male," explained *Seattle Times* reporter Eric Nalder, "so Tilikum is essential to the company's breeding program, especially if his prowess in Victoria is any sign."[29] In January 1992, the company received an emergency NMFS permit to import Tilikum on the grounds that he might attack the females or his offspring. Having paid Bob Wright millions for Sealand's whales, Sea World dismissed activists' calls to return the big male to Iceland and instead flew him to Orlando.[30] Little did the company know that the push to free another orca was about to turn their industry upside down.

His name was Keiko. Captured in Iceland in 1979—four years before Tilikum—he was first displayed at Marineland of Canada before being purchased in 1985 by Reino Aventura, an amusement park in Mexico City. Keiko drew millions of visitors over the years, but he remained isolated, the only orca at the park. Kept in a small pool, much warmer than his native waters, he lost weight and developed a skin disease.[31] He was in this condition when filmmakers Jennie Lew Tugend and Lauren Schuler-Donner first saw him in 1992. The pair was searching for a whale to use as the animal lead in *Free Willy*—a family film about a boy who befriends a lonely killer whale at a Pacific Northwest marine park. Yet they struggled to find an orca to fill the part. Rejected by oceanariums in the United States and Canada, the two women arranged to use Keiko at Reino Aventura. Schuler-Donner later described her first impression of the sickly whale. "It broke my heart," she told an interviewer. "It was life imitating art."[32] Yet the movie she and Tugend produced unintentionally highlighted the close ties between the display industry and pro-orca sentiment.

Released in July 1993, *Free Willy* seems at first glance an anticaptivity parable. It opens with callous fishermen capturing Willy, who is acquired by a penny-pinching aquarium owner. But the story centers on Jesse, a troubled orphan who initially fears but later bonds with the whale. He plays his harmonica for the lonely animal and even takes part in his training. Most moviegoers remember *Free Willy* for the orca's triumphant leap to freedom at the film's end. Yet the personal transformation of Jesse, on which the story hinges, is the direct result of his time with a captive whale. Indeed, the film's most touching scene, in which Willy saves the unconscious Jesse from drowning, was inspired by a real event at Reino Aventura. And throughout filming, the director relied on veteran orca trainers to elicit the desired behaviors from

Keiko. In sum, the movie's plot as well as its production depended on the very industry it critiqued.

Free Willy was a surprise hit, earning more than $150 million, and Warner Brothers approved plans for a sequel. In the meantime, public concern for the movie's animal star grew, and with it came competing efforts to help the whale. Encouraged by the producers, Ken Balcomb and his half-brother, Howard Garrett, approached Reino Aventura with a proposal to return Keiko to the wild. Garrett went so far as to claim they could identify the whale's original pod, though no photo-identification studies had been conducted in Icelandic waters. Sea World then entered the discussion, offering to improve Keiko's conditions at the Mexican park and pointing out that neither Balcomb nor Garrett had the experience or training to care for a killer whale.[33] Amid the debate, entertainer Michael Jackson, who sang the closing song for *Free Willy*, offered to fund Keiko's rescue, on the condition that Keiko be moved to his Neverland Ranch. In November 1994, the Earth Island Institute created the Free Willy-Keiko Foundation, which declared as its mission the "care, treatment, and potential future release" of the celebrity whale. Drawing on donations from Warner Brothers, the HSUS, and thousands of Keiko's adoring young fans, the foundation arranged to build a $7 million rehabilitation facility at the Oregon Coast Aquarium and convinced Reino Aventura to donate the whale to the project.[34]

Keiko arrived in Oregon in early 1996 amid much fanfare. It was supposed to be a short stay, but the famous orca proved such a lucrative attraction that the aquarium grew reluctant to part with him. And, some wondered, would it be so bad to keep Keiko there? He seemed happy and healthy, and children everywhere wanted to meet the *real* Willy. That would hardly be possible if he were returned to Iceland. Eventually, the aquarium relented, pending approval by the US and Icelandic governments. Initially, Icelandic officials rejected the project, but they soon changed their minds. It was a striking reversal. Four decades earlier, Iceland had asked US troops to help slaughter orcas in its waters. Now it threatened to sue the US government if it didn't allow Keiko's return.[35]

The whale's arrival in Iceland in September 1998 thrilled supporters of the project, many of whom believed it would pave the way for the freedom of other captive orcas. Yet experts warned that Keiko was a poor candidate for release. Captive for nearly twenty years, he had never caught wild fish, and his chances of integrating with local pods, none of which had been identified by scientists, were nil. "Keiko's been a big teddy whale all his life," quipped

Balcomb. "He may be a winner in people's hearts, but to other whales he's a loser with a skin condition."[36]

Funded generously by cell phone entrepreneur Craig McGraw, the Free Willy-Keiko foundation installed the whale in a state-of-the-art facility in Iceland's remote Westman Islands. Yet, from the beginning, competing priorities hampered the effort. Because no one with the HSUS or Earth Island Institute had any experience caring for killer whales, the foundation was forced to rely on the same industry that it and the film had villainized. Indeed, the team hired to reintroduce the whale to the wild often resembled a Sea World reunion, featuring project leader Jeff Foster, former company vice president Lanny Cornell, and ex–Sea World trainers Robin Friday and Mark Simmons. Upon joining the project, Friday and Simmons initiated essential steps in Keiko's progression toward freedom, including open-water swims with accompanying vessels, but the whale failed to catch fish or socialize with wild orcas. Frustrated with rising pressure to release Keiko, despite evidence that he wouldn't survive, the two trainers quit in 2000, followed soon after by Foster.[37]

Leadership then fell to Jim Horton, a veteran of Sea World's rescue and rehabilitation program in Florida. Like the others, Horton found conditions in Iceland challenging. "Sometimes it's blowing a hundred miles per hour, and you're crawling on your hands and knees," Horton later told me. "Keiko would get sick two or three times a year, and you had to catch him to the point of jumping on his back and stabbing a needle into him to get antibiotics into him as he is swimming twenty miles per hour." Nor did many local Icelanders embrace the venture. "It was usually some drunk old whaler coming up to you in a bar," Horton laughed. " 'You fucking Keiko bastards. I hate you. I love Keiko, but I hate him. I kill him.' " Yet the biggest challenge was the tension between the expectations of the project's sponsors and the reality that the long-captive whale showed little desire for freedom. "I think everybody thought Keiko would see killer whales, swim right over, and bam," Horton noted, "but it didn't work that way."[38]

By 2002, the project had consumed more than $20 million, and pressure was mounting to bring it to a close. HSUS official Naomi Rose remained adamant that Keiko's story end like *Free Willy*: with a whale triumphantly swimming free. When Keiko became separated from his escort boats during a storm that August, she celebrated it as a successful release, claiming he had swum off with wild whales. But twenty-two days later, Keiko appeared alone off the coast of Norway. Local children rushed to swim with the famous orca, and Keiko welcomed the attention. Hoping to avoid a spectacle, his handlers moved him to a nearby bay, but Rose and the HSUS insisted that they refrain from feeding him, and over the following months he steadily weakened. As it became clear he wouldn't survive, the Miami Seaquarium offered to return him to captivity as a companion for its lone orca, Lolita, and the US government supported the proposal. But HSUS resisted the plan, with the backing of Norway. As in the case of Iceland, it was quite a reversal. In recent decades, Norwegian vessels had regularly slaughtered whales, including orcas. Now Norwegians embraced Keiko, and they mourned with the rest of his fans when the whale died of pneumonia in December 2003.[39] By that time, however, experts in the Pacific Northwest had turned their attention to other lost whales.

I met John Ford in Nanaimo on a sunny morning in March 2016. As he answered a call from a fishing boat skipper, I perused his office at the Pacific Biological Station. Books and file folders lined the walls, along with reels of recordings. On the table next to me rested a jar of black-eyed peas sporting the label "Instant Killer Whales—Just Add Water." (That might have come in handy for Sea World, I mused.) On the bookshelf sat a copy of Ford's *Marine Mammals of British Columbia* (2014)—an opus dedicated to his mentor, Mike Bigg. Although a renowned expert on killer whale acoustics, Ford was modest and soft-spoken, and he seemed reluctant to acknowledge his role in the story I was writing.

Born in Victoria in 1955, Ford fished with his family often as a boy and on at least one occasion had a frightening encounter with blackfish. "I have the vivid memory of a group of killer whales swimming all around the boat," he told me. "We were all terrified, especially me." In 1963, his family moved to Vancouver, where the following year he joined thousands at Burrard Dry Dock to catch a glimpse of Moby Doll. The little orca intrigued Ford, and when Skana arrived three years later, he gravitated to the Vancouver Aquarium. Once, as a teenager, he lingered beside the pool between shows, hoping to connect with the young whale. "Nobody was around, so I stuck my hand down and patted her on the head," he recounted. "I thought, 'Wow, that's what they feel like.'" In his first year at the University of Victoria, Ford met fellow student Graeme Ellis, who told him tales of catching and training killer whales, and in the summer of 1974, he himself worked as a trainer at the Vancouver Aquarium. That December, he attended Paul Spong's whale show in Vancouver.[40]

The following summer, Ford joined an aquarium expedition to the Arctic, where he recorded narwhal calls, and by 1978 he had begun his doctoral research on killer whales, working closely with Ellis under the mentorship of Mike Bigg. He hoped to establish that killer whale pods have distinctive calls, but Bigg was skeptical. Not only was there no known example of a mammalian species possessing separate dialects, but various enthusiasts had approached Bigg with claims that they could speak or communicate telepathically with orcas. "Mike had this folder in his filing cabinet entitled 'strange people,'" chuckled Ford, "and we used to joke about me being in there initially." Yet Ford's research eventually swayed Bigg. Although the region's pods lacked language in the way an anthropologist would define it, Ford established that each had a unique set of calls passed between generations. It was an astonishing discovery, and it offered insight into the natural history of *Orcinus orca*. By analyzing the degree of acoustic similarity between matrilines, for example, Ford could chart their degree of relation.[41]

Ford's research played a key role in policy debates. In 1980, he drew the maps that assisted Jim Borrowman and Bill Mackay's campaign to save Robson Bight, and the following year he participated in the IWC workshop that stopped the commercial harvest of killer whales. It was his acoustic findings that made it possible to identify Haida's pod, and his research also offered revelations about long-dead orcas. Listening to recordings of Namu, for example,

FIGURE 19.2 Pioneers of killer whale science (left to right) Graeme Ellis, John Ford, and Mike Bigg in Johnstone Strait, 1987. Courtesy of Linda Nichol.

Ford determined that Ted Griffin's famous whale had hailed from the northern resident C1 pod, and he discovered that Namu had cycled through northern resident dialects in an effort to make contact with passing orcas. "They know each other's calls; they can make them," explained Ford, "but there is a great inhibition against doing so, unless you are sitting there trying every frequency, every wavelength, so you can get a response from somebody."[42]

In 1988, Murray Newman hired Ford as curator of marine mammals at the Vancouver Aquarium. The position gave the young scientist an intimate view of captive orcas and their impact on the public. As he told one writer in the early 1990s, "To see them, hear them, smell them, there's a major difference and I think for young kids that is the hook. Seeing the real thing, they develop an emotional bond."[43] The aquarium also offered unique research opportunities. Ford conducted systematic studies on the hearing of the captive whales, and he looked forward to observing the interaction between mother and newborn. In November 1988, Icelandic-born Bjossa gave birth to a calf—the first ever at a Canadian facility. Although the youngster died soon after, Newman remained determined to establish "a breeding colony of killer whales."[44]

At the time, Ford saw nothing wrong with that goal. Sea World's capture proposal in Alaska had troubled him, and like many scientists he hoped captive breeding would end the push for wild capture. He also found himself eager to observe the transmission of acoustic culture to a young orca. But the death of Bjossa's second calf in January 1992 dashed those hopes and shook Ford's comfort with captivity. "It was a really challenging time for the aquarium," he recalled. "Bjossa had two calves, both died, and everybody hated us." Public misgivings toward the display industry only grew with the release of *Free Willy* the following year. "The producers of the movie approached us to film it at the Vancouver Aquarium," Ford told me. "I got a copy of the treatment and thought, 'What a stupid movie this is. It'll never go anywhere if they do manage to make it.' That shows what I know about that business!" Indeed, the film gave an enormous boost to anticaptivity activism in Vancouver, particularly after Bjossa lost a third calf in 1995.[45]

By that time, Ford had turned his focus back to wild killer whales, and his aquarium office became a clearinghouse for acoustic and genetic research on the killer whales of the Pacific coast. In the process, he found new ways to fund his research. Following the Whale Museum's example, he initiated the Wild Killer Whale Adoption program, which involved giving the northern residents names in addition to their alphanumeric designations. "I didn't want to use the cute, anthropomorphic names that were given to the southern residents," explained Ford. "Instead, I looked for First Nations and geographic terms." Together with Ellis, he continued to catalog the northern residents, and each year the two kept a lookout for newborns. In June 2000, they spotted a small calf traveling with A4 pod, which Ford named Springer, for Springer Point.[46] The following year, after the aquarium sold its last killer whale to Sea World, he accepted a position as the Pacific Biological Station's marine mammal scientist, vacant since Bigg's death in 1990. He had barely settled in when reports appeared of a lone orca in Puget Sound.

The little whale surfaced off West Seattle in mid-January 2002 and soon became a frequent sight. It didn't seem to be hurt, but it was clearly lonely—approaching pleasure boats and regularly greeting the ferry *Evergreen State* when it pulled into its berth on Vashon Island. At first, Ford assumed it was Luna—a young male from L pod who had turned up alone in Nootka Sound months earlier. But Luna was still there, and the whale in Puget Sound didn't match any photographs of southern residents. Using recordings made by Seattle-based researcher David Bain, Ford and Ellis identified the animal as Springer. The two-year-old calf had vanished with its mother months earlier and was presumed dead. What was the youngster doing three hundred miles from its home range? And what, if anything, should be done to help it?[47]

If the first question was theoretical, the second was pressing. Experts on both sides of the border agreed the whale would die without intervention. Veterinarians Jim McBain of Sea World and David Huff of the Vancouver Aquarium concluded that Springer was suffering from malnutrition and skin ailments, and others noted that she seemed heedless of the danger posed by boats. With locals eager to catch a glimpse of the young whale, an accident seemed inevitable. Yet many Seattleites argued that the animal should be left alone, and the NMFS weighed its options carefully. Nearly thirty years had passed since the US government had authorized a capture in Puget Sound, and officials recognized the potential for backlash, particularly if the animal died in a rescue attempt. Further complicating matters was the fact that Springer was a northern resident orca, whose "repatriation" would require Canadian support.[48]

Ford and other DFO officials worked closely with their US counterparts to reach a solution. Finally, after months of discussion, the two governments announced a rescue plan and hired Keiko veterans Jeff Foster and Jim Horton to enact it. After three days of close observation (during which they determined Springer was female), Foster and his team captured the youngster and placed her in a sea pen. As Vancouver Aquarium specialists administered medication and screened her for pathogens, Seattle resident Ted Griffin could only marvel at how far human relations with orcas had come. "It's fabulous that we're now looking at this whale from a totally different point of view," he told a local reporter.[49]

Springer recovered quickly. Wormed and fed live fish, she gained two hundred pounds in a month. On July 13, Foster's team loaded her onto a high-speed catamaran and delivered her to Johnstone Strait, near Paul Spong's OrcaLab. As if on cue, wild killer whales appeared, and Spong's hydrophones revealed them to be members of A4 pod. The little orca called to her relatives, and the whales returned to investigate the next day. On Ford's signal, the divers pushed Springer under the beam and into open water. The reunion wasn't immediate, and Springer's rescuers fretted as the whale visited with boaters who came to see the famous orca. But within weeks she had rejoined her pod, and a decade later Ford and Ellis spotted Springer, once lost and near death, swimming with a calf of her own.[50]

By any measure, it was a heartwarming story. In the summer of 1983, the same pod—A4—had lost its matriarch and her newborn calf to gunfire in Robson Bight. Now US and Canadian scientists, officials, and activists had joined together to help a young orphaned orca rejoin her family. In the process, many observers marveled at how far the region had come since the dark days of killer whale capture. Yet few acknowledged that it was the legacy of capture that made the operation possible. Not only had live display transformed orcas into regional icons, but the main actors in the drama—Bain, Ellis, Ford, Foster, Horton, and Spong—had gotten their start in the captivity industry. Now their knowledge and commitment gave Springer a second chance.

In July 2017, Jim Borrowman invited me to Telegraph Cove for the fifteenth anniversary celebration of Springer's rescue. If it was figures such as Griffin and Spong who launched public interest in killer whales, it was enthusiasts like Borrowman who had carried it forward. In addition to leading the effort to save Robson Bight, he and his former partner Bill Mackay had taken hundreds of thousands of paying customers to see orcas on their whale-watching vessels, and a great number of other visitors for free.[51] Borrowman and his wife, Mary, sold their company, Stubbs Island Whale Watching, in 2011 and had since devoted themselves to their new Whale Interpretive Center (WIC)—a breathtaking exhibit of whale bones and history that pays tribute to the legacy of Mike Bigg. But Borrowman insisted on taking my family to see live whales. Forty-five minutes out of Telegraph Cove, we lucked upon two groups of

Bigg's killer whales swimming and frolicking off the north side of Malcolm Island. Despite having led thousands of similar tours, Borrowman seemed just as excited as my young sons. "What a show," he grinned and shook his head. "I wish everyone in the world could see these whales." And over the radio came reports of inbound orcas, perhaps A4 pod—Springer's family. This was big news. Just a few weeks earlier, Spong's OrcaLab had confirmed that Springer was traveling with a newborn calf—her second.[52]

Days later, the WIC held its Springer celebration. The audience listened in rapt attention as participants in the rescue told their stories. Among the speakers were Ford, Spong, Lance Barrett-Lennard, and NOAA scientist Brad Hanson, who had gotten his start with the 1976 radio-tagging project in Kanaka Bay. The presentation was powerful revealing just how deeply this single whale had affected participants and observers alike. But notably absent were the Sea World veterans who had proved critical to the operation, among them Jeff Foster and Jim Horton—the men who had captured and handled Springer. As such, it was easy to forget that the rescue wouldn't have been possible without the knowledge and skills gained through captivity. Toward the end of the presentation, Spong rose to share some final thoughts. Now in his late seventies, he spoke in a soft voice, his New Zealand accent barely noticeable. "The rescue of Springer was a wonderful accomplishment," he reflected, "but we must also remember those orcas whom we failed to help."[53]

Most of those present knew Spong was referring to Luna. In the wake of Springer's release, many called for a similar effort to help the young male orca, but Ford and his colleagues were hesitant. Noting that Luna was thriving in Nootka Sound and may have been cast out from his pod, they resisted calls to return him. Enthusiasts flocked to see the attraction. Many of them, such as journalist Michael Parfit, expressed a longing to connect that echoed Griffin's sentiments four decades earlier.[54] Yet over time, Luna's presence stirred anxieties reminiscent of the old Northwest. Sport fishermen complained he was eating up steelhead at the mouth of the Gold River, and some reported dangerous behavior. "Luna began seeking out boats for social contact," Ford later told me. Likely out of loneliness, the young male began pushing small vessels and even seaplanes around, often with occupants inside.[55] Finally, in 2004, officials announced a rescue plan, but indigenous locals protested. Claiming Luna was a reincarnated chief, they blocked plans to capture him, with Spong's support. In the face of this protest, the Canadian government relented, and two years later Luna collided with the propeller of a tugboat and died.[56]

His demise was part of a worrisome trend. Since the mid-1970s, the population of the southern resident killer whales had enjoyed slow but steady growth. By 1995, it had reached ninety-eight—probably nearing its precapture numbers. But it then began a two-decade decline.[57] Meanwhile, the *Free Willy* movies and high-profile rescues of Keiko and Springer had raised public interest in orcas to new heights. In summer, tourists crowded onto whale-watching boats, and more than two hundred thousand people visited San Juan Island's Lime Kiln Point State Park each year.

Yet the whales they came to see were vanishing. And while guides told incredulous tourists of the days when men used to catch orcas, it was becoming difficult to blame marine parks for the decline. Live capture in the Pacific Northwest had ended decades earlier, and Sea World was breeding its own whales. Moreover, the northern resident population was increasing rapidly. From 132 whales in 1975, it exceeded 200 by the mid-1990s, and by 2017 it had passed 300. Although there were many reasons for the recovery, the end of capture wasn't one of them, as few northern residents had been taken into captivity. Rather, the central factor seemed to be the decline of fishermen and other locals shooting them, as a direct result of growing public affection. John Ford had little doubt about this explanation. "Their numbers have doubled since the 1970s," he told me bluntly. "They had to be recovering from something."[58]

To be sure, capture brought far greater losses to the southern residents, and they felt its effects for many years. But they, too, benefited from reduced human violence. The difference was that the southern residents spent much of their time in the increasingly urban Salish Sea, and their staple food was becoming scarce. In the late 1990s, local chinook salmon runs began appearing on the US government's endangered list, and in 2005 the southern residents joined them.[59] In 2008, the NMFS presented a $50 million recovery plan for the imperiled population. Written in the Northwest Regional Office, home of the old Marine Mammal Biological Laboratory, the report emphasized how closely the species had become tied to the region's identity. "Killer whales are an icon of the Pacific Northwest," it observed, adding that "many people feel a kinship or connection to these family-oriented mammals."[60] Not so very long before, locals had considered blackfish a menace, and scientists at the lab had thought nothing of harpooning them for research. Now few Northwesterners could imagine the region without them, and the southern resident killer whales had grown so accustomed to whale-watching vessels that some observers considered the animals partly tamed.

But had Northwesterners embraced their orcas only to lose them?

Epilogue

"You become responsible, forever, for what you have tamed."
Antoine de Saint-Exupéry, *The Little Prince* (1943)

IT WAS A warm Sunday in January 2016, and Lolita swam slow circles as a crowd gathered and snapped photos. She seemed healthy—I couldn't argue with that. Although her pool was small, as critics claimed, it looked clean. The performance itself was familiar to anyone who has seen a killer whale show, but Lolita was unique—the only southern resident killer whale still alive in captivity. Caught in Penn Cove in 1970, she had resided at the Miami Seaquarium ever since, and many in the Pacific Northwest wanted her back.

Ken Balcomb launched the campaign in 1994. After withdrawing from the Keiko project, he called for Lolita's "retirement" to her home waters, and his brother Howard Garrett took up the cause. Garrett and his partner, Susan Berta, insisted on calling her Tokitae, which they argued was her original name, and they enjoyed support from Ralph Munro and other Washington State politicians. Garrett and Berta worked tirelessly to publicize the campaign, even convincing the state government to name a new ferry after the captive whale. Launched in 2014, the *Tokitae* ran between Mukilteo and the Whidbey Island port of Clinton, crossing the same waters through which Lolita and other orcas fled in the minutes before their capture. And momentum had continued to build. In September 2015, the Seattle magazine the *Stranger* published a long article entitled "It's Time to Free Lolita."[1]

The project had its limitations. Lolita was indeed a member of the endangered southern residents, as NOAA recognized, but her return wouldn't help the population—she was long past her reproductive years. Her retirement might offer a feel-good story, but it could well end like the Keiko project:

a media circus with millions spent and a celebrity whale's death. Yet as I watched her, I couldn't help but ponder her life. She had precious few years in the wild before she was taken from her family, severing what we now know are lifelong connections, and she hadn't touched or heard another killer whale since Hugo's death in 1980. What does that do to a complex social animal hardwired for acoustic stimulation? Do young orcas fear silence the way children fear the dark? Had Lolita been living a nightmare all these years?

And did she have any recollection of the old man standing beside her pool?

No one there recognized Ted Griffin, who passed for just another visitor to the park. As Lolita swam by, he reached his hand over the glass like a trainer, and an usher asked him to step back. After the show, Griffin tried to strike up a conversation with the young staffers. "I caught this orca in 1970," he told them, "and I haven't seen her since." But they grew nervous and formed a circle around him. The Miami Seaquarium had seen its share of protests lately, and they seemed to assume this was one. Soon a sheriff's deputy arrived, escorting me, Griffin, and his wife from the stadium. It was a reminder of how charged the issue of killer whale captivity had become, and as I walked out with Griffin, I wondered how we should remember this man.

Mark Funk had long wrestled with that question. It was his father Wallie's photos, more than anything else, that shaped the memory of the Penn Cove

FIGURE E.1 Staffers at the Miami Seaquarium surround Ted Griffin, January 2016. Photo by author.

Roundup. In the 1980s, Mark, then a journalist himself, attended an arts camp in Oregon, where he wrote a poem about zucchini, a poem about relatives in North Dakota, and a poem about Ted Griffin. "I couldn't find it," he later told me, "but the first line was something like, 'You're gonna go to hell, Ted Griffin, for what you done to the whales.' " Yet Funk's views had softened since then. "Ted did help reshape Northwest politics and the environmental movement. I think today there is no doubt about that," he reflected. "When people argue that individuals don't impact history, Ted *did*."[2]

Many of Funk's fellow Pacific Northwesterners still have a settled opinion: Ted Griffin was a callous profiteer who stole their beloved orcas. Yet the story of how they became beloved can't be told without him. When he set out to befriend a killer whale in the early 1960s, his quest had seemed foolhardy. Blackfish were considered vermin and dangerous ones at that. The debate at the time wasn't between whale watching and whale catching; it was between whale catching and whale killing. There was no Marine Mammal Protection Act, no NOAA, no Greenpeace, and virtually no research on the species. Fishermen routinely shot orcas, and scientists thought nothing of killing them to examine their stomach contents. Capture and display had changed all that, as even some of Griffin's outspoken critics later admitted. "Seeing them in aquariums individualized these creatures," observed Victor Scheffer in 1994. "They weren't just whales in the abstract."[3] When pressed, most activists acknowledge that the captivity of Lolita and other southern residents transformed popular views of killer whales, but few are willing to forgive the man who caught them.

"How can you even talk to Ted Griffin?" one young woman demanded of me at the Langley Whale Center on Whidbey Island. "He almost wiped out our orcas."

No, but he certainly could have.

I had begun my journey planning to write about the role of business and science in the shifting human treatment of killer whales. Yet as I watched Griffin ushered out of the stadium at the Miami Seaquarium, I couldn't help but reflect on the importance of personal character. After Namu Inc. made its first catch in Washington State in 1965, the demand for captive orcas was ravenous, and for six years no law prevented Griffin from selling as many as he could—his Canadian counterparts had done just that. Yet he didn't. Even in the infamous Penn Cove Roundup, which netted Lolita, he insisted on releasing nearly all of the whales caught. Had Griffin made different choices, that event alone could have annihilated the southern residents. In the end, the man who did the most damage to the population also prevented it from being extinguished—long before figures such as Mike Bigg, Paul Spong, or Ken Balcomb made their marks. But unlike many who caught orcas, Griffin had refused to renounce his past, and many orca enthusiasts found that unforgivable.

"I've told my story many times," Griffin told me after seeing Lolita. "People just want me to tell a different story."[4]

The next day, I visited a man who told the story most wanted to hear. Like Griffin, Ric O'Barry (formerly Richard O'Feldman) started out catching and training cetaceans in the 1960s—in his case, bottlenose dolphins at the Miami Seaquarium. But he spent the following decades denouncing captivity and had recently won fame for *The Cove* (2009), an award-winning documentary on the dolphin slaughter in Taiji, Japan. He also organized regular protests at the Seaquarium calling for the release of Lolita—likely the reason for the staff's panicked response to Griffin. O'Barry spoke with passion, criticizing conditions at the marine park and recounting his time with Hugo—Miami's first orca. "I know they are not going to free Lolita," he told me. "If you break it down to one word, it's all about 'greed.' "[5]

After the interview, O'Barry showed me his office, where two photographs of Hugo hung prominently on the wall. In the first, O'Barry strums his guitar next to the pool; in the second, he plays the flute while standing on Hugo's back. It was a striking reminder of the intimate link between captivity and those who denounce it. Like Spong, O'Barry had made his greatest impact protesting the killing of wild cetaceans. Yet it seems unlikely that either would have taken up that cause without their encounters with captive animals.

I didn't raise the question. O'Barry was determined to voice his critique of captivity. Privately, I questioned the simplicity of his narrative, but I couldn't doubt his courage. Two days after the interview, he was arrested in Japan for protesting the ongoing dolphin hunts.

For my part, I returned to the Salish Sea, whose economy and identity seemed ever more bound to the fate of its killer whales. On both sides of the border, cities that once boasted immense canneries and fishing fleets now relied on orcas to attract billions of tourist dollars each year. This was the new Northwest, where member businesses of the Pacific Whale Watch Association offered tours from nineteen different ports in British Columbia and Washington State.[6] On San Juan Island alone, with only 6,400 residents, nearly twenty whale-watching and kayaking ventures operated, and in nearby

FIGURE E.2 Ric O'Barry in his office in Coconut Grove, Florida, January 2016. Photo by author.

Victoria, the tour boats of Prince of Whales and other companies motored steadily out of the inner harbor. The trend even reached Campbell River, which had long claimed the title of "Salmon Fishing Capital of the World." In 1960, Painter's Lodge fishing resort led the push for the anti-orca machine gun on Seymour Narrows; now it offered whale-watching cruises and sold "Namu the Orca" stuffed animals in its gift shop. It was all part of what the manager of one San Juan Island kayak shop declared "an orca-based economy."[7]

Yet there was no guarantee it would continue. In global terms, the species was doing fine, even spreading to new ranges in the Arctic thanks to melting ice cover. But the southern resident killer whales, who taught us to love their species and now drove the Salish Sea's tourist industry, had declined to their lowest numbers since the mid-1970s.[8] Experts warned that the population had reached a tipping point, but in January 2017 the British Columbia government joined Ottawa in approving a massive expansion of the Kinder Morgan pipeline.[9] Carrying oil from Alberta to Vancouver, it would bring a sevenfold increase in tanker traffic. In addition to heightened marine noise, the decision raised the likelihood of an *Exxon Valdez*–like spill, which could poison the approaches to the Fraser River—the southern residents' last reliable food source.

It drove home a hard reality. The demand for Lolita's return may sound good, but the main threat to Pacific Northwest orcas wasn't the Miami Seaquarium or Sea World; it was us—the growing region that loves them. In its preindustrial state, this coast was ideal for specialized predators who fed on fat, abundant chinook salmon. But the great Sacramento and Columbia runs were now shadows of their former selves, and the Salish Sea was becoming an urban, saltwater lake—increasingly loud, empty, and polluted. And because they continued to rely on chinook salmon, which spend much of their lives in inland waters, the southern resident killer whales were now one of the most toxic marine mammal populations on earth.[10] In 2017, the state of Oregon joined Washington State and British Columbia in declaring June Orca Awareness Month. The embrace of southern resident killer whales had become a part of our identity as Pacific Northwesterners. But did we care enough to save them?

Historical perspective has its rewards. In his superb study of wolves and people in North America, Jon Coleman observes, "Alone among the planet's organisms, humans impose both categories and values on their fellow living creatures."[11] In the case of killer whales, the shift in values was so rapid and profound that most have forgotten that it happened at all. Northwesterners assume that we have always loved orcas and that those who once caught them ought to have known better. Yet hindsight can blind us to the process of change. Regardless of the different roles they played, fishermen, captors, divers, researchers, activists, and orca lovers are all part of the shared story of how we came to love these remarkable creatures. And the human encounter with captive killer whales—Paul Spong's with Skana, above all—helps explain why we came to care so much about the great whales as well. As historian of Greenpeace Frank Zelko has eloquently put it, "Anyone concerned about the fate of the world's whales can't help but wish that a few Japanese whalers and American admirals could have a close encounter with someone like Skana."[12] And she was, after all, a captive whale.

In the end, of course, none of this negates the damage done, nor does it make it easier to live with regrets.

I was nine years old in the summer of 1983 when my father took our family on a trip to Victoria. He brought us to Sealand, where I saw my first killer whale show—performed by two orcas recently imported from Iceland. We then drove to Pedder Bay and rented a boat. Exactly ten years earlier, he had netted four orcas there for Bob Wright—two from K pod and two from L pod. Unlike Griffin, Wright sold all he could, and it might have been many more if not for the timely explosion at the army depot, which frightened away the other L pod whales. My dad had just finished telling the story when a dozen orcas entered the bay and approached the boat. For the next thirty minutes, they rolled and frolicked around us, apparently just sharing a moment. One big male with a towering dorsal fin swam so close I could have touched him. My father was dumbstruck. "They never did this back then," he said, "How can they be here now?" And blissful tears rolled down his face.

If you spend enough time talking with "orcaholics" in the Pacific Northwest, you'll hear some mighty strange ideas. Some argue that killer whales are smarter than people; others claim the animals can peer directly into your soul. Ralph Munro insists that orcas have a mystical knack for appearing at significant moments. I tend to dismiss such notions, but as I think back to that encounter in Pedder Bay, I understand their appeal. What were the odds of those whales intersecting with my father at that time and place? A decade earlier, at that very spot, he had taken three orcas from the Salish Sea. As it turned out, they were the last three southern resident killer whales ever removed for the display industry, and their deaths haunted him long before he turned against captivity. But on that day, in his mind at least, the whales had come back. "We forgive you," they seemed to tell him, "and it will all be OK." And maybe it will be.

Acknowledgments

MUCH LIKE MYSELF, this book is the product of patience and generosity on both sides of the US-Canadian border. My parents, Jan and John Colby, filled my young life with animals and adventure. As a child, I waddled with emperor penguins, lived on fishing boats, worked on a mussel farm, and helped raise dippers—little river birds who bounce to a rhythm all their own. Although employed as a fisherman, my father was a trained biologist who understood animals better than he did people. A veteran of zoos and aquariums, he never turned away the sick and injured birds delivered to his door, and our house was often filled with convalescing fowl. One year, our Christmas tree hosted a saw-whet owl with a broken wing. Another year, we raised mandarin and wood ducks in a homemade aviary. Such a childhood taught me and my sister, Allison, many lessons, above all about our connections and obligations to fellow creatures.

Although I grew up on animal tales, I never considered writing one of my own until I came to the University of Victoria. My return to the Salish Sea reminded me of where I belonged and the story I needed to tell. Yet it took the support of colleagues, friends, and a range of institutions to make it happen. It was over coffee in 2011 that the prolific Rachel Cleves informed me, in no uncertain terms, that I *had* to write this book, and other members of the department echoed her enthusiasm. The research itself was made possible by a five-year Insight Grant from the Social Sciences and Humanities Research Council of Canada, as well as internal research funding from the University of Victoria. I was greatly assisted by many librarians, archivists, and curators. For their extraordinary help chasing down leads and images, I especially thank Ben Helle of the Washington State Archives, Gavin Hanke of the Royal BC Museum, Scott Daniels and Scott Rook of the Oregon Historical Society, and David Middlecamp of the *San Luis Obispo Tribune*.

Far more than anything else I have written, this book was born of relationships. I was initially a stranger to most of those I interviewed, and I am grateful that they entrusted me with their memories. Time after time, I left interviews with photos and documents, as well as new insights into a story that grew ever more complex. It saddens me that several informants passed away before the book appeared in print, and I hope my writing does them justice. Of those interviewees who went the extra mile to help me locate material and make other contacts, I want especially to thank Ted Griffin, Mark Perry, Terry McLeod, Ralph Munro, Sam Ridgway, John Ford, Jeff Foster, Jim Antrim, Mark Funk, and Jim and Mary Borrowman.

Like science, history is built on the work of previous scholars, and this book is no exception. I was only eight the first time I thumbed through Erich Hoyt's book *Orca: The Whale Called Killer* (1981), and I was thrilled to meet the author-activist on Saturna Island more than thirty years later. Hoyt wrote the first great book about people and orcas, and he remains a compelling voice for marine conservation. More recently, historians such as Jeff Bolster, Kurk Dorsey, Ryan Jones, and D. Graham Burnett provided inspiring models of marine environmental history, while Mark Leiren-Young and Mark Werner—two scholars of British Columbia orca history—were welcome sources of documents, knowledge, and camaraderie.

I also benefited from wonderful professional support. Transcriber extraordinaire Tara Neufeld somehow converted my long, rambling interviews into clear texts and in the process became an integral part of the project. At the University of Victoria, I received help from two enthusiastic and tenacious research assistants, Isobel Griffin and Blake Butler. As the manuscript took shape, it received thoughtful readings from friends and colleagues. First in line was *mi mejor amigo*, Josh Gartner, whose early encouragement meant more than he could know. In the following months, Blair Woodard, James Martin, Rachel Cleves, Freda Zimmerman, John Lutz, and Jill Walshaw all offered superb suggestions. For Oxford University Press, three anonymous readers gave useful feedback, and Susan Ferber proved the ideal editor. Susan and I have discussed doing a book with Oxford for a dozen years, and I'm glad it turned out to be this one. I can't imagine a more understanding editor for a story so personal.

Finally, my family. My wife, Kelly, and sons, Ben and Nate, experienced the journey of this second book far more directly than the first, in ways both good and bad. As always, Kelly proved an extraordinarily supportive partner. Of all the memories of this project, my favorite will be sitting down each Friday evening to read her my writing of the week. For their part, Ben and Nate remain the most challenging yet magical part of my life. My research brought many opportunities for them to travel and meet fascinating people, including the luminaries of killer whale science and the founder of whale watching in the Northwest. But it also drew their father away too often—both physically and emotionally. At times, I'm sure they wondered if I would ever finish "the whale book." But now I have, and I can finally write the most eagerly anticipated dialogue of all:

"Can you come play banjo for us in the raspberry patch?" asked Nate.

"Yes," replied Daddy. "I'll be right there."

Notes

INTRODUCTION

1. Although now known as SeaWorld, the company's official name was Sea World in the period covered by this book.
2. Carmel Finley, *All the Fish in the Sea: Maximum Sustainable Yield and the Failure of Fisheries Management* (Chicago: University of Chicago Press, 2011); Callum Roberts, *The Unnatural History of the Sea* (Washington, DC: Island Press, 2007), esp. 174–183.
3. "Iceland: Killing the Killers," *Time*, October 4, 1954.
4. Russell Owen, "Byrd Locates His Flying Base for Polar Trips," *Seattle Times*, December 31, 1928.
5. Whale-watching figures drawn from conversation with Brian Goodremont, owner of San Juan Safaris, July 13, 2017.
6. According to historian John Lutz, the 1960s were perhaps the most economically desperate times in Coast Salish history. See Lutz, *Makúk: A New History of Aboriginal-White Relations* (Vancouver: University of British Columbia Press, 2008), 279–280. On indigenous fishing rights in Washington State, see Fay G. Cohen, *Treaties on Trial: The Continuing Controversy over Northwest Indian Fishing Rights* (Seattle: University of Washington Press, 1986); on British Columbia, see Dianne Newell, *Tangled Webs of History: Indians and the Law in Canada's Pacific Coast Fisheries* (Toronto: University of Toronto Press, 1999 [1993]), chaps. 6 and 7. On the history of whale watching in the region, see F. Scott A. Murray, "'Cashing In on Whales': Cetaceans as Symbol and Commodity along the Northern Pacific Coast, 1959–2008" (master's thesis, Simon Fraser University, 2009).
7. On the role of elephants in modern zoos, see Elizabeth Hanson, *Animal Attractions: Nature on Display in American Zoos* (Princeton, NJ: Princeton University Press, 2002), 44–45.

8. Capture statistics drawn, with corrections, from Michael A. Bigg and Allen A. Wolman, "Live-Capture Killer Whale (*Orcinus orca*) Fishery, British Columbia and Washington, 1962–1973," *Journal of the Fisheries Research Board of Canada* 32, no. 7 (1975): 1215–1216; Erich Hoyt, *Orca: The Whale Called Killer* (Camden East, ON: Camden House, 1990 [1981]), 226–243. Not to be relied on, because it contains numerous errors, is Edward A. Asper and Lanny H. Cornell, "Live Capture Statistics for the Killer Whale (*Orcinus Orca*) 1961–1976 in California, Washington, and British Columbia," *Aquatic Mammals* 5, no. 1 (1977): 21–26.
9. Whaling estimate drawn from Robert C. Rocha Jr., Phillip J. Clapham, and Yulia Ivashenko, "Emptying the Oceans: A Summary of Industrial Whaling Catches in the 20th Century," *Marine Fisheries Review* 76, no. 4 (2014): 37–48; on the dolphin by-catch, see Michael Gosliner, "The Tuna-Dolphin Controversy," in *Conservation and Management of Marine Mammals*, ed. John R. Twiss Jr. and Randall R. Reeves (Washington, DC: Smithsonian Institution Press, 1999), 120–155; and M. Blake Butler, "Fishing on Porpoise: The Origins, Struggles, and Successes of the Tuna-Dolphin Controversy" (master's thesis, University of Victoria, 2017); on the commercial harvest of killer whales, see Hoyt, *Orca*, 226–232.
10. Diana Starr Cooper, *Night after Night* (Washington, DC: Island Press, 1994), 125–127. On the history of elephants in the circus, see Susan Nance, *Entertaining Elephants: Animal Agency and the Business of the American Circus* (Baltimore: Johns Hopkins University Press, 2013).
11. Hoyt, *Orca*; Ric O'Barry, *Behind the Dolphin Smile: One Man's Campaign to Protect the World's Dolphins* (San Rafael, CA: Earth Aware, 2012 [1988]); Susan G. Davis, *Spectacular Nature: Corporate Culture and the Sea World Experience* (Berkeley: University of California Press, 1997); "A Whale of a Business," *Frontline* (PBS, November 11, 1997); David Kirby, *Death at SeaWorld: Shamu and the Dark Side of Killer Whales in Captivity* (New York: St. Martin's Press, 2012); Sandra Pollard, *Puget Sound Whales for Sale: The Fight to End Orca Hunting* (Charleston, SC: History Press, 2014); John Hargrove, *Beneath the Surface: Killer Whales, SeaWorld, and the Truth beyond* Blackfish (New York: St. Martin's Press, 2015); David Neiwert, *Of Orcas and Men: What Killer Whales Can Teach Us* (New York: Overlook Press, 2015). One account that does examine the connection between orca capture and culture change, if only through a single animal, is Mark Leiren-Young, *The Killer Whale Who Changed the World* (Vancouver: Greystone, 2016).
12. See, for example, Kurkpatrick Dorsey, *Whales and Nations: Environmental Diplomacy on the High Seas* (Seattle: University of Washington Press, 2013).
13. D. Graham Burnett, *The Sounding of the Whale: Science and Cetaceans in the Twentieth Century* (Chicago: University of Chicago Press, 2012), 329, 530–533; Frank Zelko, "From Blubber and Baleen to Buddha of the Deep: The Rise of the Metaphysical Whale," *Society and Animals* 20, no. 1 (2012): 95–96.

14. According to one estimate, by spring 1971, twenty million people had seen whales caught in Puget Sound alone. See Ted Griffin, "About Whales," *Puget Soundings: Environment and Legislation Special Issue* (1971), 20. On Sea World's attendance, see Jeff Mart, "Cosmic Plot: The Last Catch of Killer Whales in Puget Sound," *Oceans*, May 1976, 56–59.
15. Activists influenced by captive whales include Greenpeace leaders Paul Spong and Bob Hunter. See Frank Zelko, *Make It a Green Peace! The Rise of Countercultural Environmentalism* (New York: Oxford University Press, 2013), 175–177.
16. Dale W. Rice, "Stomach Contents and Feeding Behavior of Killer Whales in the Eastern North Pacific," *Norsk Hvalfangst-Tidende* 2 (1968): 35–38.
17. Attenborough quoted in Matt Adam Williams, "Securing Nature's Future," *The Ecologist*, April 4, 2013, http://www.theecologist.org/blogs_and_comments/commentators/other_comments/1874015/securing_natures_future.html (accessed August 9, 2017).

CHAPTER 1

1. *The Natural History of Pliny*, trans. John Bostock and H. T. Riley (London: Henry G. Bohn, 1855), 2:365–366.
2. Ibid., 366; see also Giuseppe Notarbartolo-Di-Sciara, "Killer Whale, *Orcinus Orca*, in the Mediterranean Sea," *Marine Mammal Science* 3, no. 4 (1987): 356.
3. Although most observers today consider "orca" a term of endearment, it was traditionally considered more fearsome than "killer whale." See, for example, Charles M. Scammon, "The Orca," *Overland Monthly* 9, no. 1 (1872): 52–57.
4. Robin W. Baird, "Predators, Prey, and Play: Killer Whales and Other Marine Mammals," *Journal of the American Cetacean Society* 40, no. 1 (2011): 54–57; John K. B. Ford and Graeme Ellis, *Transients: Mammal-Hunting Killer Whales of British Columbia, Washington, and Southeast Alaska* (Vancouver: University of British Columbia Press, 1999), 27; John K. B. Ford, *Marine Mammals of British Columbia* (Victoria: Royal BC Museum, 2014), 30.
5. Carl Linnaeus, *Systemae Naturae*, 10th ed. (1758), 1:77.
6. John K. B. Ford, Graeme M. Ellis, and Kenneth C. Balcomb, *Killer Whales: The Natural History and Genealogy of Orcinus Orca in British Columbia and Washington State* (Seattle: University of Washington Press, 1994), 69.
7. Robin W. Baird, *Killer Whales of the World: Natural History and Conservation* (Stillwater, MN: Voyageur Press, 2002), 8–9.
8. Killer whale specialist John Ford estimates Granny's birthdate at around 1936. Ford, personal communication, October 4, 2016.
9. Hal Whitehead and Luke Rendell, *The Cultural Lives of Whales and Dolphins* (Chicago: University of Chicago Press, 2015), 193–194.
10. Lance G. Barrett-Lennard, Craig O. Matkin, John W. Durban, Eva L. Saulitis, and David Ellifrit, "Predation of Gray Whales and Prolonged Feeding on Submerged

Carcasses by Transient Killer Whales at Unimak Island, Alaska," *Marine Ecology Progress Series* 421 (2011):229–241; Eva Saulitis, *Into Great Silence: A Memoir of Discovery and Loss among Vanishing Orcas* (Boston: Beacon Press, 2013), 84.

11. Lance Barrett-Lennard, "Future Challenges" (presentation at the Moby Doll Symposium, Saturna Island, BC, May 25, 2013).
12. "Biologist: Orca Attacks on Gray Whales Up in California Bay," *Washington Post*, April 29, 2017. For an example of the sea-wolf theme, see Peter Knudtson, *Orca: Visions of the Killer Whale* (San Francisco: Sierra Club Books, 1996).
13. Magnus quoted in Roberts, *Unnatural History of the Sea*, 176; Charles M. Scammon, *The Marine Mammals of the Northwestern Coast of North America* (New York: Dover, 1968 [1874]), 89.
14. Frank M. Kelley, "The Giant Orca," *Victoria Colonist*, April 1948, GR 111, box 30, folder 3, G. Clifford Carl Papers, British Columbia Provincial Archives, Victoria, BC.
15. William T. Hornaday, *The American Natural History: A Foundation of Useful Knowledge of the Higher Animals of North America* (New York: Charles Scribner's Sons, 1910), 148–149.
16. Robert F. Scott, *Scott's Last Expedition* (London: John Murray, 1923 [1913]), 73–75. For his part, Ponting had little doubt that the whales "turned about with the deliberate intention of attacking me." Herbert G. Ponting, *The Great White South* (London: Duckworth, 1921), 84.
17. "'Killer' Whale No Mean Foe," *Clovis News*, February 5, 1914; Garrett P. Serviss, "The Dreadful Killer Whale," *The Bee* (Omaha), July 10, 1914.
18. Ryan Tucker Jones, *Empire of Extinction: Russians and the North Pacific's Strange Beasts of the Sea, 1741–1867* (New York: Oxford University Press, 2014), 77.
19. Tom Mead, *The Killers of Eden: The Killer Whales of Twofold Bay* (London: Angus and Robertson, 1961); Danielle Clode, *Killers in Eden: The Story of a Rare Partnership between Men and Killer Whales* (Crows Nest, NSW: Allen and Unwin, 2002). See also Whitehead and Rendell, *Cultural Lives of Whales and Dolphins*, 141–145.
20. John K. B. Ford et al., "Shark Predation and Tooth Wear in a Population of Northeastern Pacific Killer Whales," *Aquatic Biology* 11 (2011): 213–224.
21. Ford and Ellis, *Transients*, 18–19.
22. This does not include steelhead—an anadromous salmonid that can spawn multiple times.
23. Ford, Ellis, and Balcomb, *Killer Whales*, 19.
24. Ibid., 19–26.
25. Bruce Obee and Graeme Ellis, *Guardians of the Whales: The Quest to Study Whales in the Wild* (Vancouver: Whitecap Books, 1992), 33.
26. Author interview with John Ford, March 18, 2016.
27. The most outspoken proponent of viewing orca society as a model for people was Paul Spong. See Rex Weyler, *Song of the Whale* (Garden City, NY: Doubleday,

1986), 114–119. See also Joan McIntyre, ed., *Mind in the Waters: A Book to Celebrate the Consciousness of Whales and Dolphins* (New York: Charles Scribner's Sons, 1974).

28. Ford and Ellis, *Transients*, 83.
29. Lance Barrett-Lennard, "Killer Whale Evolution: Populations, Ecotypes, Species, Oh My!," *Journal of the American Cetacean Society* 40, no. 1 (2011): 48–53; Rudiger Riesch et al., "Cultural Traditions and the Evolution of Reproductive Isolation: Ecological Speciation in Killer Whales?," *Biological Journal of the Linnean Society* 106, no. 1 (2012): 1–17. For examples of authors who compare killer whale cultures to racial differences, see Knudtson, *Orca*, 6, 28–30; Ford, Ellis, and Balcomb, *Killer Whales*, 17–18; Obee and Ellis, *Guardians of the Whales*, xvi.
30. Peter B. Moyle, "Historical Abundance and Decline of Chinook Salmon in the Central Valley Region of California," *North American Journal of Fisheries Management* 18, no. 3 (1998): 487–521; Jim Myers, "Frame of Reference: Understanding the Distribution of Historical Chinook Salmon Populations," (unpublished presentation, Northwest Fisheries Center); Baird, *Killer Whales of the World*, 130–131.
31. Ford, Ellis, and Balcomb, *Killer Whales*, 21, 30.
32. On Coast Salish social organization, see Wayne Suttles, *Coast Salish Essays* (Seattle: University of Washington Press, 1987), esp. 15–25. On the patterns of indigenous fishing, see Newell, *Tangled Webs*, chap. 2.
33. Film at Tulalip Tribes Hibulb Cultural Center, Tulalip, WA (viewed August 17, 2015).
34. Lissa K. Wadewitz, *The Nature of Borders: Salmon, Boundaries, and Bandits on the Salish Sea* (Seattle: University of Washington Press, 2012), chap. 1; Wayne Suttles, *The Economic Life of the Coast Salish of Haro and Rosario Straits* (New York: Garland, 1974); Lutz, *Makúk*, 64–66.
35. Robert Webb, *On the Northwest: Commercial Whaling in the Pacific Northwest, 1790–1967* (Vancouver: University of British Columbia Press, 1988), xvi, 16–23; Joshua L. Reid, *The Sea Is My Country: The Maritime World of the Makahs, an Indigenous Borderlands People* (New Haven, CT: Yale University Press, 2015).
36. Eugene Arima and Alan Hoover, *The Whaling People of the West Coast of Vancouver Island and Cape Flattery* (Victoria: Royal BC Museum, 2011), 59; Matt Miletich, "Here Comes the Nootka Sailor," *Seattle Times*, February 11, 1962; Scammon, *Marine Mammals*, 92.
37. Author interview with Richard Hunt, June 7, 2017.
38. Terry Glavin, *This Ragged Place: Travels across the Landscape* (Vancouver: New Star Books, 1996), 172.
39. Knudtson, *Orca*, 10; Neiwert, *Of Orcas and Men*, 15; Arima and Hoover, *Whaling People*, 28.
40. Barry Gough, *Juan de Fuca's Strait: Voyages in the Waterway of Forgotten Dreams* (Madeira Park, BC: Harbour Publishing, 2012).

41. Vancouver quoted in ibid., 196.
42. William G. Robbins, "Nature's Northwest" in *The Great Northwest: The Search for Regional Identity*, ed. William G. Robbins (Corvallis: Oregon State University Press, 2001), 162.
43. Webb, *On the Northwest*, 75–76.
44. Log of the *William Hamilton* quoted in ibid., 122.
45. Scammon, *Marine Mammals*, 90.

CHAPTER 2

1. "Small Whale Sighted Near Resort Beach," *Oregon Daily Journal*, October 12, 1931; James H. McCool, "Sea Monster Believed to Be Whale Ascends River to Columbia Slough," *Morning Oregonian*, October 13, 1931; "Sea Monster Is Native of Japan, Expert Says," *Oregon Daily Journal*, October 15, 1931.
2. "Ethelbert, Portland's Famous Whale, Gives 8000 Spectators Real Thrills," *Morning Oregonian*, October 18, 1931; "The Christmas Tale of What Happened to Ethelbert, Oscar and Paddlewing," *Ogden Standard Examiner*, December 20, 1931.
3. "Whale Slows Up, Indicating It May Be Ill," *Oregon Daily Journal*, October 20, 1931; "Death for Whale Asked by Society," *Oregon Daily Journal*, October 22, 1931.
4. "Harpoon Ends Life of Gay Whale," *Oregon Daily Journal*, October 24, 1931.
5. William Moyes, "Playful Whale Done to Death," *Morning Oregonian*, October 25, 1931.
6. George Shepherd, "Killer Whale in Slough at Portland, Oregon," *Journal of Mammalogy* 13, no. 2 (1932): 171–172.
7. "Slough Whale Slayers Held," *Oregon Daily Journal*, October 25, 1931; "End of Ethelbert," *Time*, January 4, 1932.
8. On knowing nature through work, see Richard White, "'Are You an Environmentalist or Do You Work for a Living?': Work and Nature," in *Uncommon Ground: Rethinking the Human Place in Nature*, ed. William Cronon (New York: Norton, 1995), 171–185.
9. Carlos Schwantes, *The Pacific Northwest: An Interpretive History* (Lincoln: University of Nebraska Press, 2000), 144–153.
10. Webb, *On the Northwest*, 124–128.
11. Lutz, *Makúk*, 198–201; Kurkpatrick Dorsey, *The Dawn of Conservation Diplomacy: US-Canadian Wildlife Protection Treaties in the Progressive Era* (Seattle: University of Washington Press, 1998), chaps. 4 and 5.
12. Edward A. Preble, *A Biological Survey of the Pribilof Islands, Alaska* (Washington, DC: Government Printing Office, 1923), 117–118.
13. G. Dallas Hanna, "What Becomes of the Fur Seals?," *Science* 55, no. 1428 (1922): 505–507.
14. "Porpoise Hunting—Thrilling Sport and Paying Business," *Seattle Daily Times*, September 19, 1926.

15. Robert J. Browning, *Fisheries of the North Pacific: History, Species, Gear, and Processes* (Anchorage: Alaska Northwest Publishing, 1974); Diane Newell, ed., *The Development of the Pacific Salmon Canning Industry: A Grown Man's Game* (Kingston, ON: McGill-Queen's University Press, 1989); Newell, *Tangled Webs*, 62–65.
16. William G. Robbins and Katrine Barber, *Nature's Northwest: The North Pacific Slope in the Twentieth Century* (Tucson: University of Arizona Press, 2011), 22–23.
17. Newell, *Tangled Webs*, 16.
18. Wadewitz, *Nature of Borders*, 83–84, 118–119. Tensions grew after the US failure to ratify the 1908 Inland Fisheries Treaty. See Dorsey, *Dawn of Conservation Diplomacy*, 19–21.
19. Ford, *Marine Mammals of British Columbia*, 55.
20. Newell, *Tangled Webs*, 95–100.
21. "O, See the Whale!," *Seattle Daily Times*, July 31, 1908; Victor B. Scheffer and John W. Slipp, "The Whales and Dolphins of Washington State with a Key to the Cetaceans of the West Coast of North America," *American Midland Naturalist* 39, no. 2 (1948): 275.
22. "They Steer by Ear," *Popular Mechanics*, December 1941, 34–38.
23. Author interview with Ford, March 18, 2016.
24. Webb, *On the Northwest*, 150–151, 177–182.
25. Ibid., 169–170.
26. Ibid., 222–236.
27. Ibid., 242–247.
28. Richard White, *The Organic Machine: The Remaking of the Columbia River* (New York: Hill and Wang, 1995), chap. 3; Schwantes, *Pacific Northwest*, 385–388.
29. Robbins and Barber, *Nature's Northwest*, 115.
30. David Zimmerman, *Maritime Command Pacific: The Royal Canadian Navy's West Coast Fleet in the Early Cold War* (Vancouver: University of British Columbia Press, 2015), chap. 1.
31. Newell, *Tangled Webs*, 194–195.
32. Ford and Ellis, *Transients*, 81.
33. GR 111, box 30, folder 3, Carl Papers.
34. G. Clifford Carl, "A School of Killer Whales Stranded at Estevan Point, Vancouver Island," *Provincial Museum of Natural History and Anthropology: Report for the Year 1945* (Victoria, BC: Department of Education, 1946): 21–28. DNA analysis later determined that the twenty animals were offshore killer whales. Ford, et al., "Shark Predation," 213–224.
35. "A Request to Lightkeepers," GR 111, box 30, folder 3, Carl Papers; Carl quoted in "Killer Whale Seen Off V.I.," *Vancouver News-Herald*, August 2, 1947.
36. Burnett, *Sounding of the Whale*, chap. 3; council quoted in Dorsey, *Whales and Nations*, 41; A. Remington Kellogg, "Whales: Giants of the Sea," *National Geographic* 77, no. 1 (1940): 35–90.

37. "Record of Movements of Killer Whales," Portlock Point, May 1946 GR 111, box 30, folder 3, Carl Papers.
38. "Record of Movements of Killer Whales," Scarlett Point, 1947, GR 111, box 30, folder 3, Carl Papers.
39. Pike to Carl, May 31, 1950, GR 111, box 30, folder 3, Carl Papers.
40. M. Wylie Blanchet, *The Curve of Time* (Vancouver: Whitecap Books, 1996 [1961]), 95–97.
41. "G.I.'s Help Fishermen Kill Whales," *Seattle Times*, September 22, 1954.
42. "Iceland: Killing the Killers," *Time*, October 4, 1954.
43. "Air Force Guns to Shoot Whales," *Seattle Times*, October 16, 1955; "Killer Whales Destroyed," *Naval Aviation News*, December 1956, 19.
44. Masaharu Nishiwaki and Chikao Handa, "Killer Whales Caught in the Coastal Waters Off Japan for Recent 10 Years," *Scientific Report of the Whales Research Institute* 13 (1958): 85–96, quotation on 90; Robert L. Pitman, "Killer Whale," in *Encyclopedia of the Antarctic*, ed. Beau Riffenburgh (New York: Routledge, 2007), 1:572.
45. Webb, *On the Northwest*, 259–278.
46. Author interview with Joel Eilertsen, July 25, 2017.
47. Author interview with Harry Hole, July 25, 2017.
48. Gordon C. Pike and Ian B. MacAskie, *Marine Mammals of British Columbia* (Ottawa: Fisheries Research Board of Canada, 1969), 20–21.
49. Mark V. Barrow Jr., *Nature's Ghosts: Confronting Extinction from the Age of Jefferson to the Age of Ecology* (Chicago: University of Chicago Press, 2009), 318.
50. Marty Loken, "'Enemy No. 1' Killing Off Harbor Seals," *Seattle Times*, August 23, 1970.
51. Hazlewood quoted in Scott Wallace and Brian Gisborne, *Basking Sharks: The Slaughter of BC's Gentle Giants* (Vancouver: New Star Books, 2006), 51.
52. Hubbard to Whitmore, October 11, 1957, document in author's possession, courtesy of John Ford, Department of Fisheries and Oceans (hereafter DFO).
53. G. E. Moore, "Notes of Meeting Held at Campbell River, July 16, 1960," document in author's possession, courtesy of John Ford, DFO.
54. Gordon C. Pike, "Meeting to Discuss the Importance of Killer Whales in the Campbell River Economy," July 20, 1960, document in author's possession, courtesy of John Ford, DFO.
55. "Report of Committee of Special Investigation—Killer Whales," July 27, 1960, and Levelton to Moore, "RE: Killer Whale Control Program," June 20, 1961, both documents in author's possession, courtesy of John Ford, DFO; author interview with Stan Palmer, July 19, 2017 (emphasis in original).
56. Daniel Francis and Gil Hewlett, "They Shoot Orcas Don't They," *Tyee*, May 13, 2008.

CHAPTER 3

1. Paula Becker and Alan J. Stein, *The Future Remembered: The 1962 Seattle World's Fair and Its Legacy* (Seattle: Seattle Center Foundation, 1962); Joel Connelly, "Century 21 Introduced Seattle to Its Future," *Seattle P-I*, April 15, 2002.

2. As historian Jeffrey C. Sanders notes, by the early 1960s, Seattle had grown from a city dependent on resource extraction to a leader in the "postwar, high-tech consumer and military-industrial economy." See Sanders, *Seattle and the Roots of Urban Sustainability: Inventing Ecotopia* (Pittsburgh: Pittsburgh University Press, 2010), 36; on Seattle's shifting animal history, see Frederick L. Brown, *The City Is More Than Human: An Animal History of Seattle* (Seattle: University of Washington Press, 2016).
3. "Downtown for People," *Seattle Times*, October 29, 1961 (Sunday supplement).
4. Mark C. Keyes, foreword in Edward I. Griffin, *Namu: Quest for the Killer Whale* (Seattle: Gryphon West, 1982), xi–xii.
5. James S. Griffin, *More Than Luck: A Memoir* (Bloomington, IN: iUniverse Press, 2011), chap. 1.
6. Griffin, *Namu*, 5.
7. Scheffer and Slipp, "Whales and Dolphins," 277–289; Enos Bradner, "Blackfish Still a Mystery," *Seattle Times*, March 19, 1944.
8. Author interview with Gary Spilman, July 17, 2016.
9. Griffin communication to author, February 1, 2016.
10. Griffin, *Namu*, 9.
11. Author interview with Ted Griffin, July 2, 2013.
12. *US Navy Diving Manual* (Washington, DC: Navy Department, 1959), 166; "Ask Andy: Is There Really a Killer Whale?," *Seattle Times*, December 13, 1961.
13. Rachel Carson, *The Sea around Us* (New York: Oxford University Press, 1989 [1951]). Carson followed up four years later with *The Edge of the Sea* (New York: Houghton Mifflin, 1998 [1955]).
14. Charles K. Moore, "Puget Sound's Skin-Diving 'Boom,'" *Seattle Times*, September 14, 1958.
15. Kennedy quoted in Jacob Darwin Hamblin, *Oceanographers and the Cold War: Disciples of Marine Science* (Seattle: University of Washington Press, 2005), 155.
16. Gregg Mitman, *Reel Nature: America's Romance with Wildlife on Film* (Seattle: University of Washington Press, 1999), 159–166.
17. Craig Phillips, *The Captive Sea: Life behind the Scenes of the Great Modern Oceanariums* (Philadelphia: Chilton Books, 1964), 106–108.
18. Griffin, *Namu*, 1–2. This fantasy compares to the experiments detailed in Charles Foster, *Being a Beast: Adventures across the Species Divide* (New York: Metropolitan Books, 2016).
19. Griffin communication to author, May 7, 2013.
20. Author interview with Gary Keffler, April 27, 2016.
21. Ibid.
22. Griffin, *Namu*, 13–14.
23. Dave Stephens, *Ivar: The Life and Times of Ivar Haglund* (Seattle: Dunhill, 1986).
24. "Gerti: 300 Spotters Report Seeing Sea Lion," *Seattle Times*, August 8, 1963; Griffin communication to author, July 2, 2013.

25. Griffin, "Killer Whales," manuscript notes, August 4, 1966, Ted Griffin Papers, University of Victoria, Special Collections, Victoria, BC.
26. "Another View of Puget Sound's Salmon Decline," *Seattle Times*, October 7, 1960.
27. Griffin, "Namu" (unpublished manuscript), 57–59, Griffin Papers.
28. Phillips, *Captive Sea*, 259, 106–107; Norris to Carl, September 11, 1957; Carl to Norris, January 30, 1958, box 14, folder 15, Carl Papers.
29. Joseph J. Cook and William L. Wisner, *Killer Whale! A Factual Account of the Fiercest Creature of the Ocean—the Orca* (New York: Dodd, Mead, 1963), 42–52; "Surf Bathers Set Record Escaping Whale," *Desert Sun* (Palm Springs), September 2, 1958.
30. Vinson Brown, "Lord of the Seas—the Killer Whale," *Healdsburg Tribune*, August 17, 1961.
31. "5,000 See Killer Whale Captured at Newport," *Los Angeles Times*, November 19, 1961; "Killer Whale in New Home at Marineland," *Los Angeles Times*, November 20, 1961; "Killer Whale Succumbs," *Palos Verdes News*, November 22, 1961.
32. "Captured Whale Dies—of Old Age," *Los Angeles Times*, November 21, 1961; Frank Brocato, interview, "Historical Chronology," *Frontline*, http://www.pbs.org/wgbh/pages/frontline/shows/whales/etc/cron.html (accessed August 9, 2017).
33. DNA analysis later revealed that Wanda was a member of the offshore killer whale population, whose diet of sleeper sharks causes extensive tooth wear.
34. Brocato interview.
35. Hoyt, *Orca*, 18–19.
36. Author interview with Griffin, October 10, 2013.
37. Cook and Wisner, *Killer Whale!*, 20.
38. Griffin, *Namu*, 23–26.
39. "Shark Taken for 'Walk' to Regain Its Strength," *Seattle Times*, June 7, 1964; "Water Baby: Snorki, Jr., Orphan Seal, Turns Trouper," *Seattle Times*, July 28, 1964.
40. *Report of the Task Force Review Committee on the Marine Mammal Biological Laboratory* (December 1965), box 3, RG 150, US National Archives, College Park, MD (hereafter USNA).
41. Victor B. Scheffer, *Adventures of a Zoologist* (New York: Charles Scribner's Sons, 1980), 111–112; George Y. Harry Jr., "Marine Mammal Research," *A History of the Northwest and Alaska Fisheries Center, 1931-1981: Fifty Years of Cooperation and Commitment*, eds. Rae R. Mitsuoka et al. (Seattle: NOAA Northwest and Alaska Fisheries Center, 1982), 63–80; Victor B. Scheffer, Clifford H. Fiscus, and Ethel I. Todd, *History of Scientific Study and Management of the Alaskan Fur Seal,* Callorhinus ursinus, *1786–1964* (Washington, DC: NOAA, 1984), 47.
42. Scheffer is paraphrased in Harry W. Higman, "Killer Whales Gave Boatmen Scary Moment," *Seattle Times*, July 26, 1959.
43. "Ship Leaves for Study of Killer Whales," *Seattle Times*, February 2, 1960; Rice, "Stomach Contents."
44. "Seattlites Land Big One—a Killer Whale," *Seattle Times*, April 4, 1961; Rice, "Stomach Contents."

45. Author interview with Griffin, July 2, 2013.
46. Griffin, *Namu*, 42–43.
47. Griffin, "Namu" (unpublished manuscript), chap. 1, Griffin Papers.
48. Griffin, *Namu*, 49.

CHAPTER 4

1. Author interview with Josef Bauer, May 22, 2016; "Wounded Whale Swims In on Leash," *Vancouver Sun*, July 17, 1964.
2. "Wounded Whale Swims In on Leash," *Vancouver Sun*, July 17, 1964.
3. *Vancouver Aquarium Newsletter* 8, no. 6 (October 1964).
4. Newell, *Tangled Webs*, 126; David J. Mitchell, *W. A. C. Bennett and the Rise of British Columbia* (Vancouver: Douglas and McIntyre, 1983), chaps. 8 and 9.
5. Murray Newman, *Life in a Fishbowl: Confessions of an Aquarium Director* (Vancouver: Douglas and McIntyre, 1994), chaps. 1 and 2. On the history of Stanley Park, see Sean Kheraj, *Inventing Stanley Park: An Environmental History* (Vancouver: University of British Columbia Press, 2013).
6. Newman, *Life in a Fishbowl*, 16–19.
7. Author interview with Terry McLeod, April 25, 2015.
8. Newman, *Life in a Fishbowl*, 71–74.
9. Ibid., 83.
10. Dorsey, *Whales and Nations*, 292 (appendix); Hoyt, *Orca*, 231.
11. Leiren-Young, *Killer Whale*, 29.
12. *Vancouver Aquarium Newsletter* 8, no. 6 (October 1964); author interview with Bauer, May 22, 2016; Leiren-Young, *Killer Whale*, ix.
13. Author interview with Bauer, May 22, 2016.
14. "Life or Death Fight in Strait: Small Boat Battles 5-Ton Whale," *Vancouver Sun*, July 16, 1964.
15. *Vancouver Aquarium Newsletter* 8, no. 6 (October 1964).
16. "Wounded Whale Swims In on Leash," *Vancouver Sun*, July 17, 1964.
17. "Killer Whale to Quit Dock for Jericho Beach Pen," *Vancouver Sun*, July 18, 1964.
18. "Docked Hound Dog Faces Wacky Trip across Harbor," *Vancouver Sun*, July 20, 1964; *Vancouver Aquarium Newsletter* 8, no. 6 (October 1964).
19. Newman's comments made at "Moby Doll Symposium: Reflections on Change," May 25, 2013, Saturna Island.
20. Ian Smith, "Killer Whale Sought by Undersea Garden," *Victoria Colonist*, July 19, 1964; Ian Smith, "Killer Whale Could Be Trained," *Victoria Colonist*, July 21, 1964.
21. "Docked Hound Dog Faces Wacky Trip Across Harbor," *Vancouver Sun*, July 20, 1964; "Vancouver Won't Part with Its 'Moby Doll,'" *Victoria Times*, July 22, 1964.
22. See, for example, Leiren-Young, *Killer Whale*.
23. "SPCA Indignant: Whale Hunters 'Too Cruel,'" *Victoria Colonist*, July 18, 1964; "Free the Killer," *Victoria Times*, July 21, 1964.

24. "Vancouver Won't Part with Its 'Moby Doll,'" *Victoria Times*, July 22, 1964; "Whale Will Bring Disaster to City," *Vancouver Sun*, July 22, 1964.
25. "Killer Whale to Quit Dock for Jericho Beach Pen," *Vancouver Sun*, July 18, 1964; "Killer Whale Called for Help," *Victoria Colonist*, July 19, 1964.
26. "Is Our Whale Calling for Poppa?," *Vancouver Sun*, July 20, 1964.
27. "Curator Tags Lady Killer: From Now On, It's Moby Doll," *Vancouver Sun*, July 22, 1964.
28. "Whale Takes a Pint-Sized Meal," *Vancouver Sun*, July 21, 1964; "Killer Whale Has a Lot to Sell, Whale of a Story to Tell," *Vancouver Sun*, July 22, 1964.
29. "Forgot She's Lady: Moby Doll Whistles Back," *Victoria Colonist*, July 30, 1964.
30. Alexandra Morton, *Listening to Whales: What the Orcas Have Taught Us* (New York: Ballantine, 2002), 85.
31. "Curator Tags Lady Killer: From Now On, It's Moby Doll," *Vancouver Sun*, July 22, 1964; "Moby Doll May Just Be Moby Dick," *Nanaimo Free Press*, July 23, 1964.
32. "Moby Doll the Killer Whale Is Not For Sale," *Nanaimo Free Press*, July 25, 1964.
33. Fred Allgood, "One Big Struggle, and Moby's Home," *Vancouver Sun*, July 25, 1964; "Moby Doll Proves Right Name," *Victoria Colonist*, July 28, 1964.
34. Newman, *Life in a Fishbowl*, 19.
35. Fred Allgood, "Moby Doll Sulk Delays Viewing," *Vancouver Sun*, July 28, 1964.
36. Keith Bradbury, "Hopeful Whale Watchers Walked, Swam, Flew," *Vancouver Sun*, July 27, 1964.
37. Mark Perry communication to author, May 1, 2013; author interview with Mark Perry, May 15, 2013.
38. Author interview with Bauer, May 22, 2016.
39. Fred Allgood, "Moby Doll Sulk Delays Viewing," *Vancouver Sun*, July 28, 1964; Allgood, "Moby Doll Blows Happy Tune with Her Whistling Jailer, *Vancouver Sun*, July 29, 1964; "Forgot She's Lady: Moby Doll Whistles Back," *Victoria Colonist*, July 30, 1964; "Curator Considers Stuffing Moby Doll," *Vancouver Sun*, July 31, 1964.
40. Griffin, *Namu*, 65–66; Griffin communication to author, June 14, 2017.
41. "Moby Doll: Four Fish Tempt Whale to Break Long Fast," *Seattle Times*, September 10, 1964; *Vancouver Aquarium Newsletter* 6, no. 8 (October 1964).
42. Griffin, *Namu*, 66–67.
43. Leiren-Young, *Killer Whale*, 112.
44. Author interview with Bauer, May 22, 2016.
45. Terry Hammond and Keith Bradbury, "Death of Moby Needless," *Vancouver Sun*, October 10, 1964; Newman quoted in Hoyt, *Orca*, 16; "Moby Doll's Death: Animal Lover 'Callous,' Says Expert," *Victoria Colonist*, October 11, 1964; "Moby Doll Dies—She's a He," *Victoria Colonist*, October 10, 1964.
46. Robert H. Forbes, "A Report from Vancouver," *Seattle Times* November 22, 1964; author interview with McLeod, April 25, 2015.
47. Author interview with Bauer, May 22, 2016; John Ford communication to author, April 4, 2016.

48. "Moby Distributed for Scientific Study," *Vancouver Sun*, October 13, 1964.
49. Newman quoted in Hoyt, *Orca*, 16.

CHAPTER 5

1. Griffin, "Namu" (unpublished manuscript), 61, Griffin Papers; Griffin, *Namu*, 52–53.
2. Griffin, "Namu" (unpublished manuscript), 48–51, Griffin Papers; author interview with Joan Grant, December 1, 2013.
3. Griffin, "The Killer Whale," manuscript notes, August 4, 1966, Griffin Papers; Griffin, *Namu*, 56–57.
4. Griffin, *Namu*, 54–56; Griffin communication to author, July 15, 2017.
5. Nat Cole, "Whale of a Sale Going," *Vancouver Sun*, June 25, 1965; Edward I. Griffin, "Making Friends with a Killer Whale," *National Geographic* 129, no. 3 (1966): 433–437.
6. Griffin, "Making Friends with a Killer Whale." Acoustic expert John Ford later used recordings to identify the whales' pod.
7. Newman, *Life in a Fishbowl*, 104–106.
8. Nat Cole, "Whale of a Sale Going," *Vancouver Sun*, June 25, 1965.
9. Griffin, "Namu" (unpublished manuscript), 71–73, Griffin Papers; author interview with Randall Babich, March 25, 2016.
10. Griffin, *Namu*, 70–72.
11. "Whale Hunt Off Vancouver Island Planned," *Seattle Times*, April 14, 1965.
12. "City Can't Afford $50,000 for Whales, Says Alderman," *Vancouver Sun*, June 25, 1965.
13. Griffin, *Namu*, 72–73.
14. Robert McGarvey and William Lechkobit to Ted Griffin, Bill of Sale, June 28, 1965, Griffin Papers.
15. "Seattle Buyer Ecstatic over Gorgeous Whale," *Vancouver Sun*, June 29, 1965.
16. "Joint Bidders Refused but Not Beaten—Yet," *Victoria Colonist*, June 27, 1965; "Seattle Buyer Ecstatic over Gorgeous Whale," *Vancouver Sun*, June 29, 1965.
17. Griffin, *Namu*, 86.
18. "Building Pen for Floating Killer Whale," *Nanaimo Free Press*, July 2, 1965.
19. Griffin, *Namu*, 82.
20. Ibid., 85.
21. "Gobbling Killer Ready for Trip," *Vancouver Sun*, July 6, 1965; "BC Killer Whale Namu Moved to Floating Pen," *Vancouver Sun*, July 9, 1965.
22. Griffin, *Namu*, 89.
23. Newman, *Life in a Fishbowl*, 105.
24. Griffin, *Namu*, 93.
25. "Move Seen to Free B.C. Killer Whale," *Nanaimo Free Press*, July 10, 1965.
26. "Law Eliminates Hazzard: Jack Backs Off Whale," *Vancouver Sun*, July 10, 1965; Emmett Watson, "Griffin's Whale Navy Still Nervous," *Seattle P-I*, July 13, 1965.

27. Griffin, *Namu*, 99–100.
28. Stanton H. Patty, "Whaling Station Lets Namu Have Safe-Conduct Pass," *Seattle Times*, July 13, 1965.
29. Emmett Watson, "Namu's a Killer—but Does He Lay Eggs or Give Milk?," *Seattle P-I*, July 1965.
30. Griffin, *Namu*, 103.
31. Emmett Watson, "Griffin's Whale Navy Still Nervous," *Seattle P-I*, July 13, 1965.
32. Hardwick quoted in Emmett Watson, "Namu, Wife, Kids Bound for Seattle," *Seattle P-I*, July 14, 1965.
33. Quoted in Daniel Francis and Gil Hewlett, *Operation Orca: Springer, Luna, and the Struggle to Save West Coast Killer Whales* (Madeira Park, BC: Harbour Press, 2007), 63.
34. Stan Patty, "Family Loyal: Mrs. Namu Spurns 'Home Wrecker,'" *Seattle Times*, July 16, 1965.
35. Author interview with Ford, March 18, 2016.
36. "Impound Whale, Union Urges," *Victoria Colonist*, July 15, 1965; "Pursue Namu, Fish Union Says," *Vancouver Sun*, July 15, 1965.
37. Emmett Watson, "Whose Whale? It's Still Ours," *Seattle P-I*, July 16, 1965.
38. Charles Russell, "Canada Cries . . . We Want a Whale," *Seattle P-I*, July 16, 1965.
39. "Legendary Trouble Dogs Namu Voyage," *Vancouver Sun*, July 19, 1965.
40. Author interview with Keffler, April 27, 2016.
41. Francis and Hewlett, *Operation Orca*, 64.
42. Jim Brahan, "Namu Will Die in Lonely Exile," *Victoria Colonist*, July 19, 1965.
43. "Whale's Duty Debated," *Vancouver Sun*, July 14, 1965; Griffin communication to author, July 1, 2013.
44. Fisheries Department director Thor Tollefson granted Griffin a thirty-day exemption to wean the whale from salmon. See "Namu Nearing US Border," *Victoria Colonist*, July 24, 1965.
45. "Yoo-Hoo to Namu the Whale," *Sports Illustrated*, July 26, 1965; Emmett Watson, "Namu in Pen, and There He'll Stay," *Seattle P-I*, July 17, 1965; Fergus Hoffman, "Namu Glides Up San Juan Channel," *Seattle P-I*, July 25, 1965.
46. Author interview with Mark Funk, May 25, 2017.
47. Author interview with Ralph Munro, June 11, 2013 (emphasis in original).
48. Emmett Watson, "Everett's Bid for First Look at Namu Fails," *Seattle P-I*, July 26, 1965; "Seattle Set for Whale of a Time," *Victoria Colonist*, 27 July 1965; Don Page, "Namu Is Here: Whale Navy in Seattle Waters," *Seattle P-I*, July 27, 1965.
49. "Ted Griffin, Namu Day Proclaimed," *Seattle Times* July 15, 1965; Griffin, "Namu" (unpublished manuscript), 114–115, Griffin Papers.

CHAPTER 6

1. "Whaler Hunts Seals," *San Luis Obispo County Telegram-Tribune*, January 28, 1966; "Killer Whale Harpooned Outside Morro's Harbor," *San Luis Obispo County Telegram-Tribune*, February 14, 1966.

2. Rice, "Stomach Contents," 37.
3. "Killer Whale Harpooned," *San Luis Obispo County Telegram-Tribune*, February 14, 1966.
4. Leslie Millin, "Seattle Goes Nutty over Captured Whale," *Victoria Times*, July 31, 1965.
5. "Heat Troubles Whale Namu, but He Eats," *Montreal Gazette*, August 3, 1965; "Admissions Repay Outlaw for Namu," *Seattle Times*, September 2, 1965.
6. Emmett Watson, "Scientists Join the Namu Rooters," *Seattle P-I*, July 25, 1965.
7. Griffin, "Namu" (unpublished manuscript), 126, Griffin Papers; "Killer? Namu Itches to Be Friendly," *Seattle P-I*, August 13, 1965.
8. Poulter to James Scripps, September 21, 1970, Griffin Papers.
9. "Namu Said in Good Spirits," *Spokane Daily Chronicle*, August 13, 1965; Griffin, *Namu*, 120–121.
10. John C. Lilly, *Man and Dolphin* (New York: Pyramid Books, 1961).
11. Sam Ridgway, *The Dolphin Doctor* (Dublin, NH: Yankee Books, 1987).
12. Griffin, *Namu*, 123–124.
13. Gil Paust, "Blowout for a Killer," *Argosy*, December 1965.
14. Stanton H. Patty, "Owner Swims 20 Minutes with Namu," *Seattle Times*, September 2, 1965.
15. Keyes, foreword in Griffin, *Namu*, xi–xii.
16. Griffin, *Namu*, 128.
17. Dorothy Sivo quoted in Tristan Baurick, "Namu: The Whale That Brought Hollywood to Kitsap," *Kitsap Sun*, December 27, 2016.
18. Griffin, *Namu*, 113; Griffin communication to author, May 19, 2013.
19. Author interview with Munro, June 11, 2013.
20. Griffin, *Namu*, 135.
21. Griffin communication to author, March 18, 2016.
22. Griffin, *Namu*, 140–142.
23. Ibid., 142–148.
24. "Dart from 'Copter Tranquillizes Whale," *Seattle Times*, July 15, 1965; "Killer? Namu Itches to Be Friendly," *Seattle P-I*, August 13, 1965.
25. Griffin communication to author, April 10, 2016; Griffin, "Namu" (unpublished manuscript), 178–179, Griffin Papers.
26. Griffin communication to author, April 9, 2016; John Ford communication to author, June 12, 2017.
27. Griffin, *Namu*, 152–153; "Bride for Namu Enroute to Cove," *Bremerton Sun*, November 2, 1965.
28. Griffin, *Namu*, 154.
29. Don Page, "Namu Left Waiting at Altar," *Seattle P-I*, November 3, 1965.
30. Don Page, "Namu Bridge: 1 'No-Mo,' 2nd Flutters," *Seattle P-I*, November 2, 1965.
31. Don Page, "Truth: Namu's 2 Mate Died," *Seattle P-I*, November 4, 1965; Griffin communication to author, May 23, 2013.
32. Griffin, *Namu*, chap. 16.

33. Griffin communication to author, May 13, 2013; "Killer Whale Bay Flown to San Diego," *Los Angeles Times*, December 21, 1965; Davis, *Spectacular Nature*, esp. chap. 6.
34. Donn R. Besselievre to Griffin, December 13, 1965, Griffin Papers.
35. Griffin to City Council, January 10, 1966, Cf 254775, Seattle Municipal Archives, Seattle, WA.
36. Stanton H. Patty, "Plans for Namu Home at Center Disclosed," *Seattle Times*, January 18, 1966; "Center Commission Acts: O.K. of Marine Park Urged," *Seattle Times*, January 14, 1966; Braman, statement to city council, January 31, 1966, Cf 254775, Seattle Municipal Archives.
37. Griffin, "Namu" (unpublished manuscript), 189–190, Griffin Papers; Louis R. Guzzo, "Give Us a Marineland!," *Seattle P-I*, August 1, 1965, "Interhigh Council Outlines Youth Center at Namu Site," *Seattle Times*, February 11, 1966.
38. Braman to Garth Marston, February 16, 1966, Arts Commission, 654-602, box 2, folder 32, Seattle Municipal Archives.
39. KING Broadcasting Company to Marston, February 17, 1966, Arts Commission, 654-602, box 2, folder 32, Seattle Municipal Archives.
40. John F. Lawrence, "A Tale of a Whale: Seattle's Namu Finds New Friends and Foes," *Wall Street Journal*, April 25, 1966.
41. Griffin communication to author, May 11, 2013. According to some accounts, navy researchers had used orca calls to frighten dolphins into swimming faster during speed trials. See Burnett, *Sounding of the Whale*, 601n134.
42. Griffin, "Making Friends with a Killer Whale," 418–446; Kellogg quoted on 446.
43. "Would You Pull for Whale or Actor?," *Desert Sun* (Palm Springs), March 11, 1966.
44. Byron Johnsrud, "Temperamental Namu Proving Stereotype of 'Movie Star,'" *Seattle Times*, April 1, 1966; Robert Lansing, "Namu: Nice Guy Killer Whale," *Los Angeles Times*, July 24, 1966; Lansing quoted in Tristan Baurick, "Namu: The Whale That Brought Hollywood to Kitsap," *Kitsap Sun*, December 27, 2016.
45. Griffin, "Namu" (unpublished manuscript), 201–206, Griffin Papers; Griffin, *Namu*, 192–196.
46. Jim Halpin, "The Namu Nobody Knows," *Seattle Magazine*, April 1966, 26–32 (emphasis in original).
47. Author interview with Funk, May 25, 2017 (emphasis in original).
48. "Namu Cuts Up for Children," *Seattle Times*, May 3, 1966.
49. Stanton H. Patty, "Leap Pays Dividends for Namu," *Seattle Times*, May 22, 1966. For a lyrical depiction of an elephant act, see Cooper, *Night after Night*, 123–136. Performances with land predators had deep roots in American culture, perhaps most famously with Grizzly Adams. See Jon T. Coleman, "The Shoemaker's Circus: Grizzly Adams and Nineteenth-Century Animal Entertainment," *Environmental History* 20, no. 4 (2015): 593–618.
50. Griffin, *Namu*, 207–208.
51. "Secret Removal: Namu Takes Final Journey," *Seattle Times*, July 11, 1965.

52. Wayne Johnson, "Namu Is Great; Movie Isn't," *Seattle Times*, August 2, 1966; Margaret Harford, "Saga of Killer Whale Offers Fine Movie Fare for Family," *Los Angeles Times*, August 5, 1966.
53. Griffin, *Namu*, 190, 208–209.

CHAPTER 7

1. Stanton H. Patty, "Letters about Namu Still Arriving," *Seattle Times*, August 9, 1966.
2. Griffin, *Namu*, 211–212; author interview with Joan Grant December 1, 2013.
3. Author interview with Griffin, July 2, 2013.
4. Bigg and Wolman, "Live-Capture Killer Whale (*Orcinus orca*) Fishery."
5. "Griffin Will Hunt New Killer Whale," *Seattle Times*, January 5, 1967; "Namu's Trainer Pursues New Whale Family," *Seattle Times*, February 5, 1967.
6. Griffin communication to author, May 7, 2013.
7. Author interview with Babich, March 25, 2016.
8. Bigg and Wolman, "Live-Capture Killer Whale (*Orcinus orca*) Fishery."
9. Griffin communication to author, May 13, 2013.
10. Griffin, "Namu" (unpublished manuscript), 223–224, 230–231, Griffin Papers; "Griffin Loses Whale; Hunt Is Recessed," *Seattle Times*, January 10, 1966; Stanton H. Patty, "Leap Pays Dividends for Namu," May 22, 1966.
11. Aldo Leopold, *A Sand County Almanac and Sketches Here and There* (New York: Oxford University Press, 1968 [1949]); Rachel Carson, *Silent Spring* (New York: Houghton Mifflin, 2002 [1962]); Mark H. Lytle, *The Gentle Subversive: Rachel Carson,* Silent Spring, *and the Rise of the Environmental Movement* (New York: Oxford University Press, 2007); Christopher C. Sellers, *Crabgrass Crucible: Suburban Nature and the Rise of Environmentalism in Twentieth-Century America* (Chapel Hill: University of North Carolina Press, 2012), 255–257. See also David Kinkela, *DDT and the American Century: Global Health, Environmental Politics, and the Pesticide That Changed the World* (Chapel Hill: University of North Carolina Press, 2011), esp. chap. 5.
12. Farley Mowat, *Never Cry Wolf* (Toronto: Emblem, 2009 [1963]), vi; Tina Loo, *States of Nature: Conserving Canada's Wildlife in the Twentieth Century* (Vancouver: University of British Columbia Press, 2006), 10.
13. Barrow, *Nature's Ghosts*.
14. Burnett, *Sounding of the Whale*, 464–485.
15. Michael L. Weber, *From Abundance to Scarcity: A History of U.S. Marine Fisheries Policy* (Washington, DC: Island Press, 2002), 104–105; "Whalers Have Poor Season," *Vancouver Sun*, September 6, 1967; Webb, *On the Northwest*, 284.
16. Dorsey, *Whales and Nations*, 216–218.
17. Farley Mowat, *A Whale for the Killing* (Vancouver: Douglas and McIntyre, 2012 [1972]), 133.

18. "Captive Whale Disappears in Pond," *Daily Illini*, February 8, 1967.
19. "Namu's Trainer Pursues New Whale Family," *Seattle Times*, February 5, 1967.
20. Griffin, "Namu" (unpublished manuscript), 219–220, Griffin Papers.
21. Griffin, "Namu" (unpublished manuscript), 221–222, Griffin Papers; Griffin communication to author; Griffin, personal communication, May 10, 2013.
22. Griffin, "Namu" (unpublished manuscript), 225–226, Griffin Papers.
23. Griffin, "Namu" (unpublished manuscript), 230–231, Griffin Papers; author interview with Keffler, April 27, 2016.
24. Griffin, "Namu" (unpublished manuscript), 233–234, Griffin Papers.
25. Lance Barrett-Lennard, "Future Challenges," Moby Doll Symposium, May 25, 2013, Saturna Island.
26. "2 Small Boats Capture 10 Big Sea Creatures," *Seattle Times*, February 16, 1967; "Griffin Out to Keep 3 of 10 Trapped Killer Whales," *Seattle Times*, February 16, 1967.
27. "Whale Tryouts?," *Bremerton Sun*, February 21, 1967.
28. "Griffin Set Back in Corralling Whales," *Seattle Times*, February 17, 1967; Don Hannula, "Female Whale Dies in Griffin's Pen," *Seattle Times*, February 2, 1967; Griffin, "Namu" (unpublished manuscript), 240–242, Griffin Papers.
29. Poulter to Scripps, September 21, 1970, Griffin Papers; Griffin, "Namu" (unpublished manuscript), 245–247, Griffin Papers.
30. "2 More Killer Whales Die in Griffin's Holding Pen," *Seattle Times*, February 28, 1967; Griffin, "Namu" (unpublished manuscript), 251–252, Griffin Papers; Griffin communication to author, April 9, 2016; "Two More Whales May Come Here," *Seattle Times*, March 1, 1967.
31. "Boys Picket Griffin's Whale Display," *Seattle Times*, March 5, 1967; Griffin communication to author, May 11, 23, 2013.
32. Griffin communication to author, May 19, 2013.
33. "San Diego Buys 2 Whales for $10,000," *Seattle Times*, March 9, 1967.
34. "B.C. to Get Whale of a Loan," *Seattle Times*, March 7, 1967.
35. "Whale of a Time Awaits Walter," *Saskatoon Star-Phoenix*, March 11, 1967; Bob Purcell, "4,000-Pound Killer Whale Takes Reporter for a Ride," *Vancouver Sun*, March 10, 1967; Griffin communication to author, May 23, 2013.
36. Don Hannula, "All Goes Whale During Historical Phone Call," *Seattle Times*, March 16, 1967; Kathy Tait, "Whale Calls Long Distance," *Province*, March 17, 1967; "Whales Run Up $150 Phone Bill," *Victoria Times*, March 21, 1967; Griffin communication to author, May 3, 2013.
37. Author interview with Mark Perry, May 15, 2013.
38. Griffin, *Namu*, 219–220.
39. Newman, *Life in a Fishbowl*, 106–108.
40. Griffin, "Namu" (unpublished manuscript), 260–261, Griffin Papers; Griffin communication to author, May 3, 2013.
41. Mowat, *Whale for the Killing*, 221–223.

42. "Walter Comes to Aquarium, and It's a Whale of a Sale," *Vancouver Sun*, March 20, 1967.
43. "Whale Calf Moved to Griffin Aquarium," *Seattle Times*, February 25, 1967; Griffin, *Namu*, 215–217; Mary G. McCormick-Ray communication to author, September 16, 2017.
44. Ray and Scheffer quoted in "For Love or Money," *Sports Illustrated*, March 6, 1967.
45. Scheffer, *Adventures of a Zoologist*, 39–40.
46. Harry, "Marine Mammal Research," 63–70; Scheffer, *Adventures of a Zoologist*, 53.
47. Rice, "Stomach Contents."
48. *Report of the Task Force Review Committee on the Marine Mammal Biological Laboratory* (December 1965), 29, box 3, RG 150, USNA; Poulter to Scheffer, April 27, 1967, Griffin Papers.
49. KIRO Radio Television, "Griffin's Whales," editorial, March 13, 1967.
50. Poulter to KIRO Radio Television, April 27, 1967, Griffin Papers.
51. Joan Griffin, "Griffins and Whales," *Puget Soundings*, May 1967.

CHAPTER 8

1. Himie Koshevoy, column, *Province*, September 25, 1967.
2. Ibid.
3. Author interview with McLeod, April 25, 2015.
4. Terry McLeod, "The Aquarium's Cetaceans," *Vancouver Aquarium Newsletter* 11, no. 8 (November 1967).
5. Author interview with McLeod, April 25, 2015.
6. Perry communication to author, April 7, 2016.
7. Author interview with Perry, May 15, 2013.
8. "Half Million See Aquarium," *Vancouver Sun*, September 7, 1967. In total, 696,071 people visited in 1967—an increase of 57 percent from 1966. See Newman, *Life in a Fishbowl*, 107–108.
9. "Guides Sought by Aquarium," *Vancouver Sun*, September 7, 1967.
10. Author interview with Perry, May 15, 2013.
11. Newman, *Life in a Fishbowl*, 191.
12. "Housing People, Whales Problem at Aquarium," *Vancouver Sun*, n.d.
13. "Skana Cut by Window," *Vancouver Sun*, January 22, 1968.
14. Author interview with McLeod, April 25, 2015.
15. George Dobie, "Skana Recovers from Cuts, but Will She Jump Again?," *Vancouver Sun*, January 23, 1967.
16. "Skana Frustrated Female, Needs Male, Hints Newman," *Vancouver Sun*, January 24, 1968.
17. Ryan Tucker Jones, "The Ecology of Revenge Socialism" (unpublished paper), 2016. Courtesy of Ryan Tucker Jones.

18. Weyler, *Song of the Whale*, 6–7.
19. Ibid., 6. See also, Lilly, *Man and Dolphin*; John C. Lilly, *The Mind of the Dolphin* (New York: Doubleday, 1967).
20. Weyler, *Song of the Whale*, 8–11.
21. James Dickerson, *North to Canada: Men and Women against the Vietnam War* (Westport, CT: Praeger, 1999); John Hagan, *Northern Passage: American Vietnam War Resisters in Canada* (Cambridge, MA: Harvard University Press, 2001); Zelko, *Make It a Green Peace!*, chap. 3.
22. Author interview with Perry, May 15, 2013.
23. Ibid.; author interview with McLeod, April 25, 2015.
24. Author interview with Don White, September 28, 2013.
25. Ibid.; "Dolphin Digs Reading Test," *Province*, December 6, 1967.
26. "Whale of a Brain: Skana's IQ Rises," *Vancouver Sun*, July 27, 1968.
27. Author interview with White, September 28, 2013.
28. Ibid.
29. Spong quoted in Weyler, *Song of the Whale*, 19.
30. Ibid., 23–24.
31. Paul Spong, "Adventures with Orcas," Saturna Island, May 3, 2014.
32. Author interview with White, September 28, 2013.
33. Alf Strand and Robin Taylor, "Students Invade Faculty Club at UBC," *Vancouver Sun*, October 25, 1968; Rex Weyler, *Greenpeace: How a Group of Ecologists, Journalists and Visionaries Changed the World* (Vancouver: Raincoast Books, 2004), 47–48.
34. Lorne Smith, "Radio Blackout Foils Bid to Gain City a Narwhal," *Vancouver Sun*, November 5, 1968.
35. "Aquarium Whale Dies in Pen," *Vancouver Sun*, November 13, 1968; "Whale Freed by Vandals," *Vancouver Sun*, February 14, 1969.
36. Weyler, *Song of the Whale*, 32–45.
37. McLeod to Ron Church, n.d. (August 1969?), in author's possession; "Skana's Aid Failed to Save Dolphin," *Vancouver Sun*, March 20, 1969.
38. Weyler, *Song of the Whale*, 46.
39. Author interview with McLeod, April 25, 2015.
40. "Whale Dies at Pier 56; Wood Blamed," *Seattle Times*, May 16, 1967.
41. Ron Percical, "Friend Wants Whale Freed," *Vancouver Sun*, June 4, 1969; Weyler, *Song of the Whale*, 53.
42. Zelko, *Make It a Green Peace!*, 168.
43. Moira Farrow, "He'd Skinnydip with Skana—but He Was Fired First," *Vancouver Sun*, June 16, 1969; afternoon edition, June 16, 1969.
44. Author interview with Perry, May 15, 2013.
45. "Dr. Spong Wails," *Georgia Straight*, June 26–July 2, 1969.
46. Weyler, *Song of the Whale*, 57–58.

CHAPTER 9

1. Author interview with Sonny Reid and Marie Reid, July 1, 2016.
2. Reid quoted in Brian Lee, "Pender Harbour's Pioneering Role in the Live Capture of Orcas, Part I," *Harbour Spiel*, December 2007, 11.
3. Max Wyman, "Killer Whale Cornered, Now to Land and Sell It," *Vancouver Sun*, February 23, 1967.
4. "Killer Whale Cornered," *Quebec Chronicle-Telegraph*, February 24, 1968; "Aquarium Gets Killer for $5,000," *Vancouver Sun*, February 24, 1968.
5. Author interview with Reid and Reid, July 1, 2016; Sonny Reid quoted in "Skana's Partner Suspected Female," *Vancouver Sun,* February 26, 1968.
6. Author interview with Reid and Reid, July 1, 2016.
7. "Howard Deplores New Fishing Laws," *Vancouver Sun*, September 9, 1969.
8. Author interview with Reid and Reid, July 1, 2016.
9. "One Whale of a Set," *North Island Gazette*, August 2, 1967. Now known as Orky I, the whale died at Marineland two years later.
10. "Aquarium Gets Killer for $5,000," *Vancouver Sun*, February 24, 1968.
11. McLeod communication to author, April 20, 2016.
12. "City Won't See Whale for While," *Vancouver Sun*, February 27, 1968.
13. "Sorry about That, Skana, New Whale Is a Lady, Too," *Vancouver Sun*, March 5(?), 1968.
14. "New Octopus and Other Specimens," *Vancouver Aquarium Newsletter* 12, no. 4 (April 1968).
15. John Ford communication to author, June 14, 2017.
16. Author interview with Reid and Reid, July 1, 2016.
17. Lee, "Pender Harbour's Pioneering Role in the Live Capture of Orcas, Part I," 12–13.
18. Author interview with Anne Clemence, July 7, 2017; "Seven Killer Whales Penned at Once," *Vancouver Sun*, April 27, 1968; "Killers Draw 'Fishy' Set," *Province*, April 29, 1968.
19. Reid quoted in "Killers Draw 'Fishy' Set," *Province*, April 29, 1968.
20. Author interview with McLeod, April 25, 2015; "Two Whales Bought by Marineland," *Vancouver Sun*, May 2, 1968; "US Aquariums Clean Out B.C.'s Captive Whale Stock," *Vancouver Sun*, May 9, 1968.
21. Author interview with Clemence, July 7, 2017.
22. Author interview with Reid and Reid, July 1, 2016.
23. Alf Strand, "Whales Act Like Fish Out of Water," *Vancouver Sun*, May 10, 1968.
24. Author interview with Clemence, July 7, 2017.
25. The second female performed for a decade in Redwood City before dying at an oceanarium in Japan. See Hoyt, *Orca*, 147.
26. "Aquarium's Whale Gets Pen Pals," *Vancouver Sun*, May 1, 1968. Handlers also referred to Irving as "Skookum Cecil."
27. "Latest Whale Sale Sets Record," *Province*, June 11, 1968.

28. Author interview with Graeme Ellis, March 17, 2016.
29. Author interview with Clemence, July 7, 2017.
30. "Whale Loss Dashes Tourism Hopes," *Vancouver Sun*, February 15, 1969.
31. Author interview with Ellis, March 17, 2016; Richard Blair, "Upstaged by Aquarium Boss, Killer Whale Flips Tail," *Vancouver Sun*, August 2, 1968.
32. Author interview with Perry, May 15, 2013.
33. Ellis quoted in Erich Hoyt, "The Whales Called 'Killer,'" *National Geographic* 166, no. 2 (1984): 222–223.
34. Ellis quoted in Obee and Ellis, *Guardians of the Whales*, 11–12; Hoyt, *Orca*, 99.
35. "Aquarium Loser: Whale Bolts from Open Pen," *Vancouver Sun*, August 26, 1968; Hoyt, *Orca*, 147.
36. Author interview with White, September 28, 2013.
37. Author interview with Reid and Reid, July 1, 2016.
38. "Property of the Aquarium: Whale Freed by Vandals," *Vancouver Sun*, February 14, 1969.
39. "Whale Loss Dashes Tourism Hopes," *Vancouver Sun*, February 15, 1969.
40. Author interview with Reid and Reid, July 1, 2016.
41. Moira Farrow, "B.C. Fishermen Capture 9 Whales," *Vancouver Sun*, December 12, 1969; Farrow, "Big Catch 'Complete Fluke,'" *Vancouver Sun*, December 13, 1969.
42. Isabel Gooldrup quoted in Brian Lee, "Pender Harbour's Pioneering Role in the Live Capture of Orcas, Part II," *Harbour Spiel*, January 2008, 12.
43. Author interview with Reid and Reid, July 1, 2016.
44. Ibid.
45. Reid communication to author, April 22, 2016.
46. Author interview with Reid and Reid, July 1, 2016.
47. Reid communication to author, April 22, 2016; author interview with Reid and Reid, July 1, 2016.
48. "Four Killer Whales Allowed to Go Free," *Vancouver Sun*, December 20, 1969; Hoyt, *Orca*, 70–71.
49. "A Whale of a Plane Ride for Calypso," *Sydney Morning Herald*, January 1, 1970.

CHAPTER 10

1. Author interview with Ric O'Barry, January 11, 2016.
2. Ibid.
3. Ibid.
4. "WOMETCO Woos Whale—Wow!," *Enterpriser* 11, no. 11 (1968): 3.
5. Author interview with O'Barry, January 11, 2016; Barry, *Behind the Dolphin Smile*.
6. Nigel Rothfels, *Savages and Beasts: The Birth of the Modern Zoo* (Baltimore: Johns Hopkins University Press, 2002), 45–50, 161–177.
7. On the early trade and transportation of marine mammals, see Phillips, *Captive Sea*, chap. 16. In April 1967, for example, the Seattle Marine Aquarium acquired

a bottlenose dolphin from Sea World—the main purchaser of orcas captured by Namu Inc. See "Killer Whale's New Companion to Frolic at Mall," *Seattle Times*, April 7, 1968.

8. On elephants and the establishment of American zoos, see Hanson, *Animal Attractions*, 61–69.
9. "Exciting Sea World Is Set for Geauga Lake Park Area," *Cleveland Press*, June 2, 1968.
10. Griffin communication to author, October 26, 2015.
11. Randall Babich communication to author, May 4, 2016; "Big Ones to Get Away: Killer Whales Are Rounded Up; Now Await Their Pen," *Seattle Times*, February 21, 1968.
12. Author interview with Babich, March 25, 2016.
13. Griffin, "Namu" (unpublished manuscript), 264–266, Griffin Papers.
14. Charles Rice, "Whale Rounded Up at Vaughn, Pod-ner," *Tacoma News Tribune*, February 21, 1968; Charles Rice, "Whales Are a Slippery Lot Out Thar in the Vaughn Bay Corral," *Tacoma News Tribune*, February 23, 1968.
15. Kenneth Gormly, "Thar She Blows," *Gig Harbor Gateway*, n.d.
16. Author interview with Kenneth Gormly, May 17, 2016.
17. Griffin, "Namu" (unpublished manuscript) , 269–272, Griffin Papers.
18. Ibid., 271–272.
19. Ibid.
20. Walter A. Evans, "Whale Travels Air Mail to New Home in East," *Seattle P-I*, April 1, 1968. Lupa died the following September. "NY Whale Dies in Pool," *Eugene Register-Guard*, September 7, 1968.
21. "Killer Whale to Take to Air—to Miami Seaquarium," *Seattle Times*, May 12, 1968.
22. Griffin communication to author, May 16, 2013; "Whale Dies in Dutch Captivity," *Vancouver Sun*, October 25, 1968.
23. Hamblin, *Oceanographers and the Cold War*; Joshua Horwitz, *War of the Whales: A True Story* (New York: Simon and Schuster, 2014), chap. 5.
24. Forrest G. Wood, *Marine Mammals and Man: The Navy's Porpoises and Sea Lions* (Washington, DC: Robert B. Luce, 1973).
25. Griffin communication to author, May 11, 2013.
26. Author interview with Griffin, July 2, 2013.
27. Griffin, "Namu" (unpublished manuscript), 274–275, Griffin Papers.
28. Griffin communication to author, May 13, 2013.
29. Griffin communication to author, April 28, 2016.
30. Author interview with Keffler, April 27, 2016.
31. Griffin, "Namu" (unpublished manuscript), 275–277, Griffin Papers.
32. Sam Ridgway communication to author, March 25, 2015.
33. Mitchell quoted in Griffin, "Namu" (unpublished manuscript), 278–279, Griffin Papers.
34. Griffin communication to author, May 16, 2013.

35. Mitman, *Reel Nature*, 57–58.
36. Griffin, "Namu" (unpublished manuscript), 284–286, Griffin Papers.
37. Jacob Darwin Hamblin, *Arming Mother Nature: The Birth of Catastrophic Environmentalism* (New York: Oxford University Press, 2013), 190.
38. Nixon quoted in J. Brooks Flippen, *Nixon and the Environment* (Albuquerque: University of New Mexico Press, 2000), 51; Barrow, *Nature's Ghosts*, 326–336.
39. House Bill No. 178, State of Washington, Griffin Papers.
40. Griffin, "Namu" (unpublished manuscript), 287–288, Griffin Papers. Although signed earlier, the contract with the lab was revealed in "Another Whale Hunt Planned," *Seattle Times*, December 4, 1969.
41. Griffin, "Namu" (unpublished manuscript), 287–288, Griffin Papers.
42. Griffin, "Killer Whale Captures in Washington State," notes, Griffin Papers; author interview with Keffler, April 27, 2016.
43. Griffin, "Namu" (unpublished manuscript), 289–290, Griffin Papers.
44. "Another Whale Hunt Planned," *Seattle Times*, December 4, 1969.

CHAPTER 11

1. Author interview with White, September 28, 2013.
2. White quoted in Hoyt, *Orca*, 116.
3. Wright quoted in ibid.
4. Author interview with Ellis, March 17, 2016.
5. Janis Ringuette, "Beacon Hill Park's Famous White Bear," Beacon Hill Park History, http://www.beaconhillparkhistory.org/articles/118_kermode.htm (accessed August 9, 2017).
6. G. Clifford Carl, "Albinistic Killer Whales in British Columbia," *Provincial Museum of Natural History and Anthropology: Report for the Year 1959* (Victoria, BC: Department of Education, 1960): 29–36.
7. "Record of Movements of Killer Whales" (Scarlett Point), 1947, GR 111, box 30, folder 3, Carl Papers.
8. "Albino Killer Whale Is Seen by Picknickers at Witty's," *Victoria Times*, August 4, 1947; Ray Wormald, "Alice, Albino Whale, Eludes Camera after Being Seen Off Trial Island," *Victoria Colonist*, October 14, 1950; Brian Loughnan, "Alice's White Infant Has Naturalists Agog," *Victoria Colonist,* March 29, 1952.
9. Humphry Davy, "Whale of a Sight in Store If Albino Alice Stays on Course," *Victoria Times*, April 23, 1955; Davy, "Alice, the Gay White Whale, Frolics with Two Playboy Companions," *Victoria Times*, January 20, 1958.
10. "Battle to the Death Witnessed by Tug Crew," *Nanaimo Free Press*, February 29, 1960.
11. "They Sometimes Live Up to Their 'Suspicious' Reputation," *Victoria Colonist*, July 21, 1964.
12. Carl to Kurt Cehak, October 16, 1964, box 30, folder 2, Carl Papers.

13. "King Fisherman: First Angler Chalks Up Crest Sweep," *Victoria Colonist*, October 30, 1960.
14. Hudson Blake, "Old Boathouse Like a Magnet," *Oak Bay News*, October 1, 1991.
15. "Marina Opens," *Oak Bay Leader*, April 8, 1964.
16. Ian Smith, "Killer Whale Sought by Undersea Garden," *Victoria Colonist*, July 19, 1964.
17. Ian Smith, "Killer Whale Could Be Trained," *Victoria Colonist*, July 21, 1964.
18. Author interview with John Colby, August 5, 2014.
19. Griffin communication to author, May 16, 2013; Griffin, "Namu" (unpublished manuscript), 282–283, Griffin Papers.
20. Bob Wright, notes from meeting with Griffin and Goldsberry, John F. Colby Papers, University of Victoria, Special Collections, Victoria, BC.
21. Author interview with Ellis, March 17, 2016.
22. Ibid.
23. Author interview with Reid and Reid, July 1, 2016.
24. Pat Dufour, "Rare Albino Killer Whale Netted," *Victoria Times*, March 3, 1970.
25. Ellis quoted in Hoyt, *Orca*, 115.
26. Interview of Tor Miller, 1989, in author's possession, courtesy of John Ford, DFO; author interview with Ellis, March 17, 2016.
27. "Friendly Fellow," newspaper clip, September 1969, file 619-A-4, folder 2, MSS. 1287, Murray A. Newman Fond, Vancouver Municipal Archives, Vancouver, BC.
28. Author interview with White, September 28, 2013.
29. Author interviews with White, September 28, 2013; February 1, 2015.
30. Author interview with White, September 28, 2013.
31. Pat Dufour, "Rare Albino Killer Whale Netted," *Victoria Times*, March 3, 1970.
32. Pat Dufour, "Cries of Protest Triggered by Capture of White Whale," *Victoria Times*, March 5, 1970.
33. "Complaints Squelched on Whales," *Tacoma News-Tribune*, March 6, 1970.
34. During the 1960 hearing, Haig-Brown had complained that during a 1932 trip through Johnstone Strait, he had seen five hundred to one thousand killer whales. G. E. Moore, "Notes of Meeting Held at Campbell River, July 16, 1960," courtesy of John Ford, DFO.
35. "Killer Whales: Few Save Many," *Victoria Times*, March 7, 1970.
36. Humphry Davy, "Ahab, Are You Listening?," *Victoria Times*, March 6, 1970.
37. Ford and Ellis, *Transients*, 20–21; Hoey, quoted in Hoyt, *Orca*, 119.
38. Bruce Jordan Bott, *Swim!* (Vancouver: Life Systems, 2008), 89–91.
39. Cameron quoted in Pat Dufour, "It's a Nose-to-Nose Affair for Haida, White Playmate," *Victoria Times*, March 25, 1970.
40. Alan Hoey, "The Albino Female Killer Whale," August 1972(?), Colby Papers.
41. "One Deformed: A Lot of Whale to Go Oak Bay," *Victoria Times*, March 16, 1970.
42. Author interview with White, September 28, 2013.
43. Hoyt, *Orca*, 118–120; author interview with White, February 1, 2015.

44. White quoted in Hoyt, *Orca*, 120.
45. Author interview with Ellis, March 17, 2016.
46. Author interview with White, September 28, 2013.
47. Author interview with Ford, March 18, 2016; author interview with White, September 28, 2013.
48. White quoted in Hoyt, *Orca*, 121.
49. "Famed Whales Freed," *Vancouver Sun*, October 28, 1970.
50. Author interview with White, September 28, 2013.

CHAPTER 12

1. Griffin, "Namu" (unpublished manuscript), 304–306, Griffin Papers.
2. Griffin communication to author, March 12, 2016.
3. "Whaling for Fun and Profit," *Townsend Leader*, January 29, 1970.
4. Don Page, "Wanted: Pod of Killers," *Seattle P-I*, February 19, 1970; "How to Handle a Killer Whale—Gently," *Seattle Times*, February 21, 1970.
5. Griffin, "Namu" (unpublished manuscript), 290–293, Griffin Papers.
6. Sharon E. Kingsland, *The Evolution of American Ecology, 1890–2000* (Baltimore: Johns Hopkins University Press, 2005); Stephen Bocking, *Ecologists and Environmental Politics: A History of Contemporary Ecology* (New Haven, CT: Yale University Press, 1997).
7. Brent Walth, *Fire at Eden's Gate: Tom McCall and the Oregon Story* (Portland: Oregon Historical Society Press, 1994); Don Brazier, *History of the Washington Legislature, 1965–1982* (Olympia: Washington State Senate, 2007), 14; Adam Rome, *The Genius of Earth Day: How a 1970 Teach-In Unexpectedly Made the First Green Generation* (New York: Hill and Wang, 2014).
8. Author interview with O'Barry, January 11, 2016.
9. O'Barry, *Behind the Dolphin Smile*, chap. 1.
10. Bonnie Jean Schein communication to author, October 20, 2016; Bonnie Jean Schein, "Damage to Environmental Resource," *West Seattle Herald*, April 23, 1970.
11. Biggs to Griffin, April 8, 1970, Game Department records, box 48, Washington State Archives, Olympia, WA (hereafter WSA); "Marine Mammals to Be Aired," Special News Release, May 27, 1970, Interim Committee on Game and Game Fish, Griffin Papers.
12. "Seattle Center Marine Life Park," First Study Report, April 1, 1970, Griffin Papers.
13. "Game Department," notes and correspondence, 1970, Griffin Papers.
14. "State Game Chief Proposes Regulations on Sea Mammas," *Seattle Times*, June 7, 1970; "Agenda," public hearing, June 6, 1970, Griffin Papers; notes on public hearing, Griffin Papers. See Victor B. Scheffer, *The Year of the Whale* (New York: Charles Scribner's Sons, 1969).
15. Susanne Schwartz, "Status of Killer Whales Stirs Controversy," *Seattle Times*, June 14, 1970.
16. Dorsey, *Whales and Nations*, 219–220.

17. Andreas to Magnuson, July 21, 1970; Magnuson to Rogers, July 24, 1970, box 1337, folder "INCO—Whale, 1/1/70," RG 59, USNA; Shelby Scates, *Warren G. Magnuson and the Shaping of Twentieth-Century America* (Seattle: University of Washington Press, 1997), 294–295.
18. Ford, Ellis, and Balcomb, *Killer Whales*, 21, 30.
19. Author interview with Perry, May 15, 2013; Griffin, "Namu" (unpublished manuscript), 294, Griffin Papers.
20. Griffin, "Namu" (unpublished manuscript), 296–301, Griffin Papers; author interview with Griffin, January 10, 2016.
21. Handwritten notes on 1970 capture, Griffin Papers; Griffin, "Namu" (unpublished manuscript), 296–301, Griffin Papers
22. Author interview with Griffin, July 2, 2013.
23. Ibid.
24. Griffin communication to author, June 2–3, 2016.
25. "Killer Whales Captured Off Whidbey Island," *Seattle Times*, August 8, 1970; "Killer Whales Stage 'Salt Water Ballet,'" *Seattle Times*, August 9, 1970; author interview with Griffin, July 2, 2013.
26. Harriet Ritvo, *The Animal Estate: The English and Other Creatures in the Victorian Age* (Cambridge, MA: Harvard University Press, 1987), 243–246.
27. Griffin communication to author, June 8, 2016.
28. Author interview with Terrell Newby, October 24, 2015.
29. "Penn Cove Roundup Recalled," *South Whidbey Record*, August 5, 2000.
30. Measurements were given in "Rocks Anchor Dead Whales Off Whidbey," *Seattle Times*, November 19, 1970; author interview with Griffin, July 2, 2013.
31. Author interview with Stone, June 1, 2017; author interview with Newby, October 24, 2015.
32. Author interview with Griffin, April 26, 2016 (emphasis in original).
33. Author interview with Newby, October 24, 2015.
34. Griffin communication to author, June 2, 2016.
35. Author interview with John Stone, June 1, 2017.
36. Griffin, "Namu" (unpublished manuscript), 307–308, Griffin Papers; "50 Whales Still in Net Enclosures," *Seattle Times*, August 14, 1970.
37. Author interview with Stone, June 1, 2017.
38. "Whale Drowns in Penn Cove Mishaps," *Seattle Times*, August 17, 1970; handwritten notes on 1970 capture, Griffin Papers.
39. "The Amiable Killers," *Bainbridge Review*, August 19, 1970.
40. "Scripps, Dr. Ray Oppose Whale Netting on Sound," *Seattle P-I*, August 19, 1970; "Whale Hunter Shocked at Criticism," *Seattle Times*, August 19, 1970.
41. "50 Pickets Protest Treatment of Whales," *Seattle Times*, August 23, 1970; Griffin, "Namu" (unpublished manuscript), 316, Griffin Papers.
42. Author interview with Griffin, July 2, 2013.
43. Griffin to Vodanovich, August 24, 1970, Griffin Papers.
44. Griffin to Kalich, September 3, 1970, Griffin Papers.

45. Hoyt, *Orca*, 241–242.
46. Goldsberry quoted in "Killer Whale Controversy," *Pacific Northwest Sea*, Fall 1970, 3–4.
47. Scheffer, letter to editor, *New York Times*, June 21, 1970.
48. Scheffer quoted in Delphine Haley, "The Great Killer Whale Dispute," *Northwest Today*, November 1, 1970.
49. Rice quoted in ibid.
50. Keyes quoted in ibid.
51. "Rocks Anchor Dead Whales Off Whidbey," *Seattle Times*, November 19, 1970; Griffin, "Namu" (unpublished manuscript), 321–323, Griffin Papers.
52. Emmitt Watson, "Maritime Mystery," *Seattle P-I*, November 24, 1970.
53. Author interview with Funk, May 25, 2017.
54. "Hunter 'Checking Into' Whale Deaths," *Seattle P-I*, November 23, 1970.
55. Ray and Wolman quoted in "Loss of Killer-Whale Carcasses Deplored," *Seattle Times*, November 25, 1970.
56. "Hunter 'Checking Into' Whale Deaths," *Seattle P-I*, November 23, 1970.
57. Terrell Newby, "Killer Whale Deaths Reported," *Pacific Search* 5, no. 1 (1971).
58. "Director Deplores Seattle Hunt Methods," *Victoria Colonist*, November 26, 1970.
59. Ibid.
60. Ibid.

CHAPTER 13

1. Minutes, Game Commission Hearing, April 10–11, 1972, Game Department records, box 17, WSA.
2. "Elizabeth Stanton Lay," obituary, *Seattle Times*, May 6, 2007; "In Memoriam: Elizabeth Stanton Lay, '32," *Reed Magazine*, November 2007 http://www.reed.edu/reed-magazine/in-memoriam/obituaries/november2007/elizabeth-stanton-lay-1932.html (accessed October 12, 2017).
3. Minutes, Game Commission Hearing, April 10–11, 1972, Game Department records, box 17, WSA.
4. Zelko, "From Blubber and Baleen."
5. Ron Fulkerson, "Gigi Going Home: Whale of a Return," *San Diego Union*, March 13, 1972.
6. Roger S. Payne and Scott McVay, "Songs of the Humpback Whales," *Science* 173, no. 3997 (1971): 585–597; Burnett, *Sounding of the Whale*, 628–630.
7. "Hickel Bans Imports of Whale Products," *New York Times*, November 24, 1970; Burnett, *Sounding of the Whale*, 524–525.
8. Brazier, *History of the Washington Legislature*.
9. Griffin communication to author, May 11, 2013.
10. "Ted Griffin Backs Whale Program," *Seattle Times*, January 29, 1971; Griffin to Cherberg, April 20, 1971, Griffin Papers.

11. "Whale Bill Is Finally Approved," *Seattle Times*, May 10, 1971; Wayland to Patterson, "Killer Whales," policy summary, January 16, 1980, Game Department records, box 48, WSA.
12. Weber, *From Abundance to Scarcity*, 124–125.
13. G. Carleton Ray and Frank M. Potter, "The Making of the Marine Mammal Protection Act of 1972," *Aquatic Mammals* 37, no. 4 (2011): 528–530; for the proceedings of the Skyland conference, see William E. Shevill, ed., *The Whale Problem* (Cambridge, MA: Harvard University Press, 1974); Dorsey, *Whales and Nations*, 220–222.
14. "U.S. Senate Acts on Whaling Moratorium Proposal," *Animal Welfare Institute Information Report* 20, no. 2 (1971): 381–383.
15. Norris quoted in Weber, *From Abundance to Scarcity*, 107.
16. George L. Small, subcommittee testimony, July 26, 1971, box 109, folder 37, Magnuson Papers, University of Washington, Special Collections, Seattle, WA.
17. Lyle Burt, "Killer Whales: How Many Are There?," *Seattle Times*, June 6, 1971.
18. "Summary: July 26, 1971 killer whale census," excerpted in Garry Garrison, "Killer Whale Management," January 1972, Game Department records, box 48, WSA; Griffin to Lauckhart, June 25, 1971, Griffin Papers; "Stand Up and Be Counted," *Seattle Times*, August 3, 1971.
19. Gary Garrison, "Killer Whale Management" (1971), Game Department records, box 48, WSA.
20. George J. Palo, "Notes on the Natural History of the Killer Whale *Orcinus orca* in Washington State," *Murrelet* 53, no. 2: (1971): 23; Keyes to Game Department, September 3, 1971, Game Department records, box 28, WSA.
21. Crouse to Griffin, August 20, 1971, Game Department records, box 48, WSA; Griffin to van Heel, August 4, 1971, Griffin Papers.
22. Jack Adkins and E. Reade Brown, "Killer Whale Trapping Operations" (1971), Game Department records, box 28, WSA.
23. The animals helped stock the company's new Ohio franchise, as well as replace the first Shamu, who died in August 1971. "Shamu Dies in California," *Seattle Times*, August 30, 1971.
24. Farrell to Brown, September 7, 1971, Game Department records, box 28, WSA; Garry Garrison, "Killer Whale Management" (n.d.), Game Department records, box 48, WSA.
25. Farrell to Stevens, September 7, 1971, Game Department records, box 28, WSA.
26. Don McGaffin, obituary, *Seattle P-I*, May 30, 2005; Excell quoted in Frank Chesley, "McGaffin, Donald Edward 'Don' (1926–2005)," HistoryLink.org, October 25, 2005, http://www.historylink.org/File/7489 (accessed July 30, 2017).
27. Author interview with Funk, May 25, 2017.
28. Don McGaffin, "Run Whales, Run," *Active Magazine*, September 21, 1971, Game Department records, box 28, WSA. Author Sandra Pollard has written that it was Griffin whom McGaffin struck. See Pollard, *Puget Sound Whales for Sale*, 112–114. When interviewed, however, Griffin had no recollection of ever encountering

McGaffin, and John Stone, who was working at the inn during the altercation, confirmed that Griffin was not present. Griffin communication to author, March 25, 2015; author interview with Stone, June 1, 2017.

29. "Wanted" poster, Griffin Papers.
30. Ted Griffin, "About Whales," 20.
31. Author interview with Grant, December 1, 2013; Griffin, "Namu" (unpublished manuscript), 327, Griffin Papers.
32. Griffin, "Namu" (unpublished manuscript), 328, 316–317, Griffin Papers; author interview with Griffin, July 2, 2013.
33. Griffin, *More Than Luck*, 20–23.
34. Author interview with Grant, December 1, 2013.
35. Zelko, *Make It a Green Peace!*, chap. 4.
36. John Colby, Henderson Bay/No. Rosedale notes, March 1972, Colby Papers.
37. John Colby interview with Terry Newby, March 11, 1972, Colby Papers.
38. "Six at Purdy: Trapped Killer Whales Freed," *Seattle Times*, March 14, 1972; "Seattle Aquarium Whales," *Seattle Times*, April 15, 1972. The total number of whales in this catch is unclear. Newspapers reported that seven orcas were caught in this capture and six released, but Colby's notes indicated nine animals, as did Goldsberry's later report. See Cornell and Asper, "Live Capture Statistics," 23.
39. "State Hearing Set Tuesday on Killer Whales," *Seattle Times*, April 9, 1972.
40. Minutes, Game Commission Hearing, April 10–11, 1972, Game Department records, box 17, WSA.
41. Ibid.
42. Ibid.
43. Ibid.
44. Ibid.
45. Ibid.
46. Ibid.
47. Ibid.
48. Lyle Burt, "Whale Trapping Permits Delayed," *Seattle Times*, April 12, 1972.
49. C. J. Skreen, "Let's Hear It for Our Friend—the Whale," *Seattle Times*, May 21, 1972. The documentary ran several more times in the following weeks.
50. Minutes, Game Commission Hearing, May 22–23, 1972, Game Department records, box 17, WSA; "Namu, Inc., Gets Permits to Catch 4 Killer Whales," *Seattle Times*, July 18, 1972; Griffin, "Namu" (unpublished manuscript), 337, Griffin Papers.
51. Griffin communication to author, October 5, 2015.
52. Griffin, "Namu" (unpublished manuscript), 340, Griffin Papers.
53. Author interview with Griffin, July 2, 2013.
54. Griffin communication to author, March 9, 2016.
55. Author interview with Griffin, July 2, 2013.
56. Keffler communication to author, March 5, 2016.

CHAPTER 14

1. Author interview with Jeff Foster, April 25, 2016.
2. Ibid.
3. Griffin to Goldsberry, deed of sale, Griffin Papers.
4. Crouse to Goldsberry, June 14, 1972, Game Files AR, WSA.
5. "Foundation to Aid Environmental Groups," *Seattle Times*, September 29, 1972.
6. Author interview with Foster, April 25, 2016.
7. Ibid.
8. John Colby communication to author, July 9, 2016.
9. Dorsey, *Whales and Nations*, 209.
10. Walter Sullivan, "Cry of the Vanishing Whale Heeded in Stockholm," *New York Times*, June 9, 1972; "Conference Approves 10 Year Whaling Ban," *Washington Post*, June 10, 1972.
11. "US Call for Whaling Halt Rejected by World Body," *Washington Post*, June 30, 1972.
12. Hickel quoted in "US Loses Its Bid for a 10-Year Ban on Whale Hunting," *New York Times*, June 30, 1972.
13. Garmatz quoted in Weber, *From Abundance to Scarcity*, 125.
14. " 'Malign Neglect' of Ocean Mammals," *Seattle Times*, June 7, 1972; Flippen, *Nixon and the Environment*, 178–179; Ray and Potter, "Making the Marine Mammal Protection Act," 534–543.
15. Barrow, *Nature's Ghosts*, 339–341.
16. Author interview with Bob Hofman and Mike Gosliner, July 1, 2015; Magnuson speech, box 109, folder 37, Magnuson Papers.
17. Evans to Nixon, November 17, 1972, box 2S-2-753, folder "Killer Whales 1973," Evans Papers, WSA.
18. Robert Hofman, "The Continuing Legacies of the Marine Mammal Commission and Its Committee of Scientific Advisors on Marine Mammals," *Aquatic Mammals* 25, no. 1 (2009): 97–98.
19. Author interview with Hofman and Gosliner, July 1, 2015.
20. Weyler, *Greenpeace*, 126.
21. "Sea World Draws 460,000," *San Diego Evening Tribune*, January 8, 1965.
22. Davis, *Spectacular Nature*, esp. chap. 6.
23. Author interview with Jim Antrim, June 25, 2016.
24. Ibid.
25. Ibid.
26. "New Owners: 2 Whales Sought for Seattle Aquarium," *Seattle Times*, February 17, 1973.
27. John Colby, "Meeting with Don Goldsberry notes," January 23, 1973, Colby Papers.
28. "New Owners: 2 Killer Whales Sought for Marine Aquarium," *Seattle Times*, February 27, 1973.

29. Rusty Yerxa, “Hardship Hearing for Hunter—and Whales,” *Seattle Times*, February 28, 1973; Colby, personal communication to author, July 7, 2016.
30. Evans to Dent, March 9, 1973, box 2S-2-753, folder “Killer Whales 1973,” Evans Papers, WSA.
31. Class letters from “Miss Bow,” in box 2S-2-753, folder “Killer Whales 1973,” Evans Papers, WSA.
32. “Save the Killer Whale!,” *Seattle P-I*, March 18, 1973, box 109, folder 38, Magnuson Papers.
33. Powell to Walker, March 20, 1973, box 2S-2-753, folder “Killer Whales 1973,” Evans Papers, WSA.
34. Dent to Evans, April 3, 1973, box 2S-2-753, folder “Killer Whales 1973,” Evans Papers, WSA.
35. “Fact Sheet on Killer Whales and Sea World Permit Application,” March(?) 1973, box 109, folder 38, Magnuson Papers.
36. Ibid.
37. Author interview with Foster, April 25, 2016
38. Babich to Griffin, October 28, 1973; Griffin to Babich, January 14, 1974, Griffin Papers.

CHAPTER 15

1. John Colby, “Taku ” (unpublished manuscript), 1974, Colby Papers.
2. Ibid.
3. The US Navy had previously placed radio harnesses on two orcas in Hawaii but never intended to release them. C. A. Bowers and R. S. Henderson, *Project Deep Ops: Deep Object Recovery with Pilot and Killer Whales* (San Diego: Naval Undersea Center, 1972), courtesy of Sam Ridgway.
4. John Colby communication to author, August 14, 2017; author interview with Colby, May 8, 2012.
5. Michael A. Bigg, “An Assessment of Killer Whale (*Orcinus orca*) Stocks Off Vancouver Island, British Columbia,” *Report of the International Whaling Commission* 32 (1982): 656–658.
6. Dan Joling, “Mammal-Eating ‘Transient’ Orcas May Be Named after Researcher,” *Seattle Times*, November 26, 2012. For examples of this celebration of Bigg’s career, see Morton, *Listening to Whales*, 65–70; and Obee and Ellis, *Guardians of the Whales*, 19–24.
7. Etienne Benson, *Wired Wilderness: Technologies of Tracking and the Making of Modern Wildlife* (Baltimore: Johns Hopkins University Press, 2010).
8. Sam Ridgway, ed., *Mammals of the Sea: Biology and Medicine* (Springfield, IL: Charles C. Thomas, 1972), 131.
9. Webb, *On the Northwest*, 266–267, 275, 286; Dale W. Rice, “Gordon C. Pike, 1922–1968,” obituary, *Journal of Mammalogy* 51, no. 2 (1970): 434.

10. *Vancouver Aquarium Newsletter* 8, no. 6 (October 1964); Leiren-Young, *Killer Whale*, 120–121.
11. Newman, *Life in a Fishbowl*, 157–159.
12. Michael A. Bigg, Ian B. MacAskie, and Graeme Ellis, *Preliminary Report: Abundance and Movements of Killer Whales Off Eastern and Southern Vancouver Island with Comments on Management* (St. Anne de Bellevue: Arctic Biological Station, 1976), in author's possession, courtesy of John Ford, DFO.
13. Weyler, *Song of the Whale*, 72–73. On this connection between the counterculture and indigenous politics, see Sherry L. Smith, *Hippies, Indians, and the Fight for Red Power* (New York: Oxford University Press, 2012).
14. Hoyt, *Orca*, 44–46.
15. Spong quoted in Leiren-Young, *Killer Whale*, 139. See also Weyler, *Song of the Whale*, chaps. 4 and 5. For a photograph of Spong playing the flute for Haida at Sealand of the Pacific in 1971 or 1972, see Zelko, *Make It a Green Peace!* (mislabeled as Spong's 1974 visit to the Vancouver Aquarium).
16. Weyler, *Song of the Whale*, 114–119; Mowat quoted on 119.
17. Hoyt, *Orca*, 107.
18. Weyler, *Song of the Whale*, 142–144; Robert Hunter, *Warriors of the Rainbow: A Chronicle of the Greenpeace Movement from 1971 to 1979* (Amsterdam: Greenpeace International, 2011 [1979]), 139–141.
19. Weyler, *Song of the Whale*, 145–147; Canadian Embassy to Ottawa, April 22, 1974, box 376, file 20-18-5, Library and Archives Canada, Ottawa, ON (courtesy of Ryan Tucker Jones).
20. Wright to Colby et al., memorandum, December 18, 1972, Colby Papers.
21. Hourston to Wright, Whale Capture Permit, March 2, 1973, Colby Papers; author interview with Angus Matthews, May 24, 2017.
22. "Whaling-Hunting Permit Renewed for Sealand," *Victoria Times*, June 28, 1973.
23. Colby communication to author, March 29, 2016.
24. Colby, "August 6, 1973 Pedder Bay Whale Catch," Sealand of the Pacific log, Colby Papers.
25. Colby communication to author, March 29–30, 2016.
26. Ibid.
27. Ibid.
28. Author interview with Colby, May 8, 2012.
29. Colby, "Pedder Bay Whale Feeding Program," log, Colby Papers.
30. John Colby, photo log of 1973 Pedder Bay catch; Colby notes, "Outline A," Colby Papers.
31. John Colby, photo log of Pedder Bay Capture, 1973; Colby notes, "Outline A," Colby Papers.
32. Bigg quoted in Ford, Ellis, and Balcomb, *Killer Whales*, 14.
33. Pat Dufour, "Whales in Research Swim to Be Wired for Sound," *Victoria Times*, August 28, 1973.

34. Author interview with Matthews, May 24, 2017.
35. Jon McDermott, "Whales Undergoing Studies," *Beaver County Times*, October 17, 1973.
36. Author interview with Colby, May 8, 2012.
37. Colby communication to author, July 19, 2016; Colby notes, "Outline A," 1974, Colby Papers.
38. Colby to Hourston, June 19, 1974, Colby Papers.
39. Bigg quoted in Ford, Ellis, and Balcomb, *Killer Whales*, 14.
40. "Whale Census," *Victoria Colonist*, August 8, 1974.
41. Author interview with Ellis, March 17, 2016.
42. Ibid.
43. Don White, "Let's Not Lose Our Remaining Killer Whales," *Vancouver Sun*, April 12, 1975.
44. Erickson to Hofman, October 3, 1975, "Killer Whale Workshop Report," box 11, folder "Killer Whale Group," Douglas Chapman Papers, University of Washington, Special Collections, Seattle, WA.
45. Morton, *Listening to Whales*, 69–70.
46. Bigg and Wolman, "Live-Capture Killer Whale (*Orcinus orca*) Fishery." In the January 1977 volume of *Aquatic Mammals*, Sea World officials Lanny Cornell and Edward Asper published an article contesting Bigg's statistics, but their data should not be relied upon, because they contain numerous errors. See Asper and Cornell, "Live Capture Statistics."
47. Weyler, *Greenpeace*, chap. 7.
48. Author interview with Matthews, May 24, 2017 (emphasis in original).
49. "Greenpeace Joins Protest," *Vancouver Sun*, August 21, 1975.
50. "B.C. Declares Moratorium on Capture of Killer Whales," *Vancouver Sun*, September 13, 1975; Weyler, *Song of the Whale*, 173–174.
51. Author interview with White, February 1, 2015.

CHAPTER 16

1. Author interview with Karen Ellick (formerly Karen Munro), October 10, 2014.
2. Ibid.; Affidavit of William H. Oliver, March 10, 1976, Case C76-57T, box 35, RG 21, USNA-Pacific Alaska Region.
3. Affidavit of Pennie Oliver, March 10, 1976, Case C76-57T, box 35, RG 21, USNA-Pacific Alaska Region.
4. Author interview with Munro, June 11, 2013.
5. Ibid.
6. Mike Layton, "Five Whales Captured: Hundreds Watch at Olympia Harbor," *Seattle P-I*, March 8, 1976.
7. "Application for Public Display Permit," October 15, 1973, Case C76-57T, box 35, RG 21, USNA-Pacific Alaska Region.

8. Eric Nalder, "State Partly Responsible for Whale Permit," *Seattle P-I*, March 11, 1976.
9. Kerry Webster, "Whale Hunters Ask Renewal of Permit," *Tacoma News Tribune*, February 20, 1974.
10. "Give the Whale a Break," *Tacoma News Tribune*, February 26, 1974.
11. Warren G. Magnuson, "Targets of Annihilation," *Spokesman Review*, March 24, 1974.
12. "Magnuson Asks NOAA Deny Whale Permits," *Bellingham Herald*, April 24, 1974.
13. Robert Schoning, US Department of Commerce, Permit No. 22, May 7, 1974, in Case C76-57T, box 35, RG 21, USNA-Pacific Alaska Region; "Killer Whale Capture Okayed in Puget Sound," *Olympian*, May 21, 1974.
14. "Toward Extinction," *Seattle P-I*, May 29, 1974; Shoning quoted in Stephen Green, "Killer Whales Not in Peril, US Aide Says," *Seattle P-I*, June 9, 1974.
15. Author interview with Foster, April 25, 2016.
16. Ibid.
17. Colby, personal communication to author, June 12, 2017.
18. Author interview with Foster, April 25, 2016.
19. Colby to Goldsberry, January 27, 1976, Colby Papers.
20. Author interview with Foster, April 25, 2016.
21. Colby communication to author, June 23, 2013.
22. Schoning to Erickson, June 18, 1975, Permit No. 98, in Case C76-57T, box 35, RG 21, USNA-Pacific Alaska Region.
23. Erickson to Killer Whale Project Participants, June 23, 1975, in Case C76-57T, box 35, RG 21, USNA-Pacific Alaska Region.
24. Julie Emery, "Whale Sanctuary 'Most Practical,' Says Writer," *Seattle Times*, March, 14, 1976; Jeff Mart, "Cosmic Plot: The Last Catch of Killer Whales in Puget Sound," *Oceans*, May 1976, 56–59.
25. It was Henderson Inlet, immediately to the east of Budd Inlet.
26. Affidavit of Samuel G. Arvan, March 10, 1976, Case C76-57T, box 35, RG 21, USNA-Pacific Alaska Region.
27. Author interview with Foster, April 25, 2016.
28. Munro quoted in "Captured Whales May Be Released," *Spokesman Review*, March 9, 1976; author interview with Munro, June 11, 2013.
29. Cohen, *Treaties on Trial*, esp. chap. 1.
30. "Captured Whales May Be Released," *Spokesman Review*, March 9, 1976; "Six Whales Wait," *Seattle P-I*, March 9, 1976.
31. Jeffrey Craig Sanders, "Animal Trouble and Urban Anxiety: Human-Animal Interaction in Post–Earth Day Seattle," *Environmental History* 16, no. 2 (2011): 247.
32. "Killer-Whale Capture Called 'Normal,'" *Seattle Times*, March 10, 1976.
33. "State's Sad Harvest Should Be Halted," *Seattle P-I*, March 10, 1976.
34. Eric Nalder, "State Partly Responsible for Whale Permit," *Seattle P-I*, March 11, 1976.

35. Author interview with Munro, June 11, 2013.
36. Ibid. (emphasis in original).
37. Although no explosives were dropped from the plane, this claim has continued to appear in accounts of the event. See, for example, Weyler, *Greenpeace*, 176.
38. Testimony of Dennis Ohlde, Case C76-57T, box 35, RG 21, USNA-Pacific Alaska Region.
39. "US Appeals Judge Blocks Whales' Freedom," *Bremerton Sun*, March 13, 1976.
40. Author interview with Colby, September 10, 2012.
41. "Killer Whale Escapes as Protests Mount," *Seattle P-I*, March 10, 1976.
42. Author interview with Colby, September 10, 2012.
43. Keyes quoted in Erik Lacitis, "2 of 5 Captured Whales Rip Netting, Swim to Freedom," *Seattle Times*, March 14, 1976.
44. "A Captive Whale Heads to Freedom," *Seattle P-I*, March 15, 1976.
45. Author interview with Foster, April 25, 2016.
46. Ibid.
47. Ibid.
48. Erik Lacitis, "2 of 5 Captured Whales Rip Netting, Swim to Freedom," *Seattle Times*, March 14, 1976.
49. Frank Hewlett, "Whale-Haven Bill Gains Speed," *Seattle Times*, March 16, 1976.
50. Mary Murdach, Fred Wiepke, and G. D. Graham, letters to editor, *Seattle P-I*, March 20, 1976.
51. Virginia Bencel, letter to editor, *Seattle P-I*, March 20, 1976; Maxine J. Banker, letter to editor, *Seattle Times*, March 16, 1976.
52. Ryoji Mihara, letter to editor, *Seattle Times*, March 16, 1976.
53. Aubrey T. Dunham, letter to editor, *Seattle Times*, March 16, 1976.
54. A. G. Schille, letter to editor, *Seattle P-I*, March 20, 1976.
55. Bruce Johansen, "Sea World to Hold Off on Whale Captures," *Seattle Times*, March 24, 1976.
56. Author interview with Griffin, July 2, 2013.
57. "Captured Whales May Be Released," *Spokesman Review*, March 9, 1976; author interviews with Griffin, July 2, 2013; December 4, 2016.
58. Bruce Johansen, "Sea World to Hold Off on Whale Captures," *Seattle Times*, March 24, 1976.
59. Scheffer quoted in Eric Nalder, "US Changes Stand on Whale Captures," *Seattle P-I*, March 29, 1976.
60. "Whale of a Party," invitation, Colby Papers; author interview with Ellick, October 10, 2014.
61. Author interview with Munro, June 11, 2013.
62. Author interview with Foster, April 25, 2016; Ford and Ellis, *Transients*.
63. Author interview with Colby, July 1, 2012.
64. Balcomb, notes from logbook, April 26, 1976, https://www.whaleresearch.com/single-post/2016/04/26/Encounter-4-The-Release-of-T13-T14-April-26-1976

(accessed August 10, 2017); author interview with Colby, July 1, 2012. An account of Erickson's research appears in Benson, *Wired Wilderness*, 157–159.

65. Author interview with Munro, June 11, 2013.
66. Ibid.

CHAPTER 17

1. Newman, *Life in a Fishbowl*, 191; "Famed Whale Succumbs," *Spokane Daily Chronicle*, October 6, 1980.
2. Newman, *Life in a Fishbowl*, 192.
3. Ibid., 192–196; Moira Farrow, "Hyak's New Friends Settling In," *Vancouver Sun*, December 22, 1980.
4. Author interview with Foster, April 25, 2016; Moira Farrow, "Finna Turns Out Be a Lot of Bull," *Vancouver Sun*, January 24, 1981.
5. "Sea World 'Interested' in Defiance Park," *Seattle P-I*, March 11, 1976; "New Show in Town," *Seattle Times*, May 18, 1977.
6. Colby communication to author, August 22, 2016.
7. "Fat Whale Gets Big Welcome," *San Diego Union*, October 18, 1976.
8. Richard Ellis, *Men and Whales* (New York: Knopf, 1991), 39–42, 472.
9. Finley, *All The Fish in the Sea*, 161–162.
10. "Killing the Killers," *Time*, October 4, 1954.
11. Jóhann Sigurjónsson and Stephen Leatherwood, "The Icelandic Live-Capture Fishery for Killer Whales, 1976–1988," *Journal of the Marine Research Institute* 11 (1988): 308–309.
12. Author interview with Foster, April 25, 2016.
13. Sigurjónsson and Leatherwood, "Icelandic Live-Capture Fishery for Killer Whales."
14. Denise A. Carabet, "$49 Million Offered for Sea World," *San Diego Union*, November 6, 1976; Antrim communication to author, October 22, 2016.
15. Author interview with Foster, April 25, 2016.
16. Ibid.
17. Ibid.; author interview with Antrim, June 25, 2016; Antrim communication to author, August 24, 2016.
18. Bigg, MacAskie, and Ellis, *Preliminary Report: Abundance and Movements of Killer Whales*; John Ford communication to author, October 21, 2016.
19. Horwitz, *War of the Whales*, chaps. 4 and 5; author interview with Kenneth Balcomb, July 12, 2017.
20. Author interview with Balcomb, July 12, 2017; Balcomb quoted in Rice, "Stomach Contents," 36.
21. Author interview with Balcomb, July 12, 2017.
22. Ibid.
23. Author interview with Rick Chandler, December 1, 2013; author interview with Balcomb, July 12, 2017.

24. Rick Chandler, Camille Goebel, and Ken Balcomb, "Who Is That Killer Whale? A New Key to Whale Watching," *Pacific Search* 11, no. 7 (1977): 25–35; author interview with Chandler, December 1, 2013.
25. "Orca Excursion," advertisement by Orca Survey Coordination Center, *Pacific Search* 11, no. 7 (1977): 52; John L. Pedrick Jr. (NOAA) to Kenneth C. Balcomb III., April 25, 1978, courtesy of Ken Balcomb.
26. Author interview with Balcomb, July 12, 2017.
27. Weyler, *Greenpeace*, 504–505.
28. Author interview with Antrim, June 25, 2016.
29. Soon after, on February 26, 1978, a Japanese team made its first live capture, netting five orcas. See Hoyt, *Orca*, 204.
30. Author interview with Foster, April 25, 2016.
31. Ibid.
32. Warren G. Magnuson, press release, July 21, 1980, box 11, folder "Killer Whale Group," Chapman Papers.
33. Author interview with Ellis, June 5, 2017; M. V. Ivashin, "USSR, Progress Report on Cetacean Research, June 1979–May 1980," *Report of the International Whaling Commission* 31 (1981): 221–226.
34. Ford communication to author, June 10, 2017; Michael A. Bigg, "An Assessment of Killer Whale (*Orcinus orca*) Stocks off Vancouver Island, British Columbia," *Report of the International Whaling Commission* 32 (1982): 655–666.
35. Newman, *Life in a Fishbowl*, 194–195.
36. Jóhann Sigurjónsson, "Killer Whale Census Off Iceland during October 1982," *Report of the International Whaling Commission* 34 (1984): 609–612.

CHAPTER 18

1. Stephen Hume, "Sealand Gets Permission for Two Whales," *Times-Colonist*, June 16, 1982.
2. Stephen Hume, "Haida Welfare Priority to Sealand Proposal," *Victoria Times-Colonist*, June 25, 1982; Hume, "Haida Still Speaks the Lingo of That Old Whale Gang of His," *Victoria Times-Colonist*, September 3, 1982.
3. M. Mader, "Speaking Up," letter to editor, *Victoria Times-Colonist*, September 13, 1982; Bob Nixon, "Who Speaks for Killer Whales?," *Victoria Times-Colonist*, September 25, 1982.
4. Author interview with Sam Ridgway, June 21, 2016.
5. Ibid.
6. Bowers and Henderson, *Project Deep Ops*, 31, courtesy of Sam Ridgway; Ridgway, ed., *Mammals of the Sea*, 726–729, figure 10-48.
7. Author interview with Ellis, March 17, 2016.
8. Author interview with Colby, May 8, 2012.
9. Perry communication to author, June 19, 2013.

10. Ibid. (emphasis in original).
11. Alan Hoey, "The Albino Female Killed Whale" (August 1972?), Colby Papers.
12. Author interview with Ellis, March 17, 2016.
13. Transcript of interview with Bob Wright, August 22, 1972(?), Colby Papers.
14. Colby communication to author, September 13, 2016; John Colby, "Sexual Activities in the Killer Whales of Sealand of the Pacific, Victoria, B.C." (unpublished manuscript), December 1973, Colby Papers.
15. Rob Watters et al., Sealand log for Chimo, October–November 1972, Colby Papers; Colby communication to author, September 14, 2016.
16. Rob Watters et al., Sealand log for Chimo, October–November 1972, Colby Papers.
17. Interview of Tor Miller, 1989, courtesy of John Ford, DFO.
18. "Killer Whale Blues Piped Away," *Ellensburg Daily Record*, November 21, 1972; author interview with Colby, July 1, 2012.
19. Paul Horn, *Inside II*, liner notes, 1972.
20. "Killer Whale Blues Piped Away," *Ellensburg Daily Record*, November 21, 1972.
21. Paul Jeune, *Miracle: The Story of a Baby Killer Whale* (Victoria, BC: Sealand of the Pacific, 1978).
22. Author interview with Matthews, May 24, 1977.
23. Jeune, *Miracle*.
24. Author interview with Matthews, May 24, 2017; Jim Gibson, "Next Three Days Tests 'Miracle,'" *Victoria Colonist*, August 12, 1977.
25. "Ailing Whale Victim of Movie Influence," *Victoria Colonist*, August 13, 1977.
26. Julie MacDonald, "Miracle, Lucky One," letter to editor, *Victoria Colonist*, August 20, 1977.
27. Jeune, *Miracle*; Tony Akehurst, "Save Whale," letter to editor, *Victoria Colonist*, August 31, 1977.
28. Bigg quoted in Hoyt, *Orca*, 200; "Teething Miracle Sets Tongues Wagging," *Victoria Colonist*, August 19, 1977.
29. Jeune, *Miracle*.
30. Morton, *Listening to Whales*, 144–145.
31. Al Forrest, "Adored Killer Whale Performed to the End," *Victoria Times-Colonist*, January 13, 1982.
32. Pat Dufour, "Was Miracle's Net Cut?," *Victoria Times-Colonist*, January 14, 1982.
33. Author interview with Matthews, May 24, 2017.
34. "A Special Friend," editorial, *Victoria Times-Colonist*, January 14, 1982.
35. Robert Hunter, column, *Vancouver Sun*, October 25, 1974; Hoyt, *Orca*, 201–202.
36. Author interview with Matthews, May 24, 2017.
37. Stephen Hume, "Whale Catch 'Like Lions in Cages,'" *Victoria Times-Colonist*, June 23, 1982; "State Taking Whale Gripe to Ottawa," *Victoria Times-Colonist*, July 30, 1982.
38. Craig Tomashoff, "Canadian Aquarium Defends Whale-Capture Plan," *Seattle Times*, August 11, 1982.

39. "Feds Clear Whale Trade-Off," *Victoria Times-Colonist*, August 28, 1982.
40. Stephen Hume, "Haida Still Speaks the Lingo of That Old Whale Gang of His," *Victoria Times-Colonist*, September 3, 1982; Hume, "Catch Canadian Whale, US Urges," *Victoria Times-Colonist*, August 7, 1982.
41. "Whales Diverted from Sealand Net," *Victoria Times-Colonist*, September 17, 1982.
42. Author interview with Matthews, May 24, 2017.
43. Paul Watson, "Who Speaks for Killer Whales?," *Victoria Times-Colonist*, September 25, 1982.
44. Pat Dufour, "Haida's Dead," *Victoria Times-Colonist*, October 4, 1982.
45. Perry communication to author, June 19, 2013.
46. "Victoria Speaks in a Loud Way over Haida and Sealand," *Victoria Times-Colonist*, October 16, 1982.
47. Mrs. B. Dunstan, letter to editor, *Victoria Times-Colonist*, October 19, 1982.

CHAPTER 19

1. Briggs quoted in Morton, *Listening to Whales*, 237. Both animals died soon after, and without its matriarch, A4 pod split into two groups. See Ford, Ellis, and Balcomb, *Killer Whales*, 52.
2. Author interview with Jim Borrowman, July 8, 2017.
3. Hoyt, *Orca*, 19.
4. Author interview with Borrowman, July 8, 2017; Hoyt, *Orca*, 197–198; Hoyt and Jim Borrowman recounted the expedition in "Diving with Orcas," *Diver* 5, no. 8 (1979): 20–23.
5. Hoyt and Borrowman, "Diving with Orcas," 20–23.
6. Author interview with Ford, September 26, 2016; author interview with Borrowman, July 8, 2017; Obee and Ellis, *Guardians of the Whales*, 58–61.
7. Roselia Broyles, "'Whale-Watchers' Enjoy Tour into Robson Bight," *Seattle Times*, June 28, 1983.
8. Morton, *Listening to Whales*, 203.
9. Author interview with Borrowman, July 8, 2017.
10. Author interview with Munro, December 7, 2016; Ralph Munro, "Overeager Boaters Endanger Future of Orcas in Sound," letter to editor, *Seattle Times*, August 28, 1985.
11. Cornell and Munro quoted in David Freed, "Hearing Set in Seattle Tuesday on Sea World Killer-Whale Plan," *Seattle Times*, August 10, 1983.
12. Weyler, *Song of the Whale*, 262–263; author interview with Munro, December 7, 2016.
13. Author interview with Antrim, May 26, 2017.
14. Author interview with Munro, June 11, 2013; "Sea World Gets OK to Capture Killer Whales for Display, Study," *Seattle Times*, November 1, 1983.

15. Obee and Ellis, *Guardians of the Whales*, 38–40; author interview with Ellis, June 5, 2017.
16. Weyler, *Song of the Whale*, 264–265; "100 Orcas to Be Taken," *Sound of Prince William*, May 23, 1984; author interview with Munro, June 11, 2013.
17. David Freed, "Federal Judge Voids Sea World's Permit to Capture Whales," *Los Angeles Times*, January 22, 1985. Sea World won a partial victory in appellate court the following year, but by that time the company had decided the political costs of the project were too high. See Jim Schachter, "Both Sides Say Ruling on Whales Was Victory," *Los Angeles Times*, June 21, 1986.
18. Obee and Ellis, *Guardians of the Whales*, 43–45; Saulitis, *Into Great Silence*, 91.
19. Kirby, *Death at SeaWorld*, 43; "Rare Newborn Killer Whale Thrives at Sea World Park," *Bellingham Herald*, January 7, 1986.
20. Nina Easton, "The Death of Marineland," *Los Angeles Times Magazine*, August 9, 1987, 6–10, 23–26; Gary Hanauer, "The Killing Tanks," *Penthouse*, October 1989, 40–41. Cornell also arranged to borrow a female named Gudrun from the Harderwijk Dolphinarium for breeding in Orland. Cornell to F. B. den Herder, March 31, 1987, "Inside Sea World," *Frontline*, http://www.pbs.org/wgbh/pages/frontline/shows/whales/SeaWorld (accessed August 9, 2017); "Female Killer Whale Flown In for Breeding," *Ocala Star-Banner*, November 19, 1987.
21. Foster communication to author, July 11, 2017; author interview with Antrim, June 25, 2016.
22. Author interview with Antrim, September 28, 2016.
23. Author interview with Foster, April 25, 2016.
24. Ibid.
25. Author interview with Antrim, September 28, 2016; Jovanovich quoted in Hanauer, "Killing Tanks," 60. On the injuries to trainers in this period, see Kirby, *Death at SeaWorld*, 175–177.
26. Author interview with Foster, April 25, 2016.
27. Ibid.
28. Kirby, *Death at SeaWorld*, chap. 12.
29. Eric Nalder, "Activists Want Whales to Be Wild Again," *Seattle Times*, January 12, 1992.
30. Kirby, *Death at SeaWorld*, 139–145.
31. Kenneth Brower, *Freeing Keiko: The Journey of a Killer Whale from* Free Willy *to the Wild* (New York: Gotham Books, 2005), 18–21.
32. See Schuler-Donner interview on "A Whale of a Business," *Frontline*, http:www.pbs.org/wghb/pages/frontline/shows/whales (accessed August 9, 2017).
33. Tony Perry, "Plan to 'Free Willy' Will Not Have a Harmonious Ending," *Los Angeles Times*, November 1, 1993; Jessica Seigel, "Whale Lovers Debating Wisdom of Freeing Keiko," *Chicago Tribune*, March 7, 1994.
34. Kirby, *Death at SeaWorld*, 241–244.
35. Ibid., 261–263.

36. "Free Willy: The Denouement," *New York Times Magazine*, September 6, 1998; Donald G. McNeil Jr., "Keiko Makes It Clear: His '*Free Willy*' Was Just a Role," *New York Times*, November 6, 2001.
37. Mark A. Simmons, *Killing Keiko: The True Story of Free Willy's Return to the Wild* (Orlando, FL: Callinectes Press, 2014).
38. Author interview with Jim Horton, August 15, 2016.
39. Kirby, *Death at SeaWorld*, 270–281.
40. Author interview with Ford, March 18, 2016; John Ford, "Evolution of an Orcateer," *Vancouver Sun*, April 21, 2001.
41. Author interview with Ford, March 18, 2016; Ford, Ellis, and Balcomb, *Killer Whales*, 21.
42. Author interview with Ford, March 18, 2016.
43. Ford quoted in Obee and Ellis, *Guardians of the Whales*, 68.
44. Newman, *Life in a Fishbowl*, 207.
45. Author interview with Ford, September 26, 2016; author interview with Annelise Sorg, March 22, 2014.
46. Author interview with Ford, September 26, 2016.
47. Ibid.
48. For a good account of these deliberations, see Francis and Hewlett, *Operation Orca*.
49. Griffin quoted in Eric Sorensen, "A New Course for Orcas," *Seattle Times*, July 14, 2002.
50. Author interview with Ford, September 26, 2016.
51. Author interview with Borrowman, August 1, 2017.
52. Borrowman conversation to author, July 20, 2017; Carla Wilson, "Springer the Orca, Rescued 15 Years Ago, Spotted with New Calf," *Victoria Times-Colonist*, July 14, 2017.
53. Spong's presentation at Springer's Fifteenth Anniversary Commemoration, Whale Interpretive Center, Telegraph Cove, BC, July 22, 2017.
54. Michael Parfit and Suzanne Chisolm, *The Lost Whale: The True Story of an Orca Named Luna* (New York: St. Martin's Press, 2013).
55. Author interview with Ford, September 26, 2016.
56. "Heartsick Tribe Says Goodbye to Luna," *Olympian*, March 15, 2006.
57. *Recovery Plan for Southern Resident Killer Whales* (Seattle: National Marine Fisheries Service, 2008), II-52–55.
58. Ford, Ellis, and Balcomb, *Killer Whales*, 45; author interview with Ford, March 18, 2016.
59. Craig Welch, "Feds Make Dramatic Move to Save Orcas: Puget Sound Pods Put on Endangered-Species List," *Seattle Times*, November 16, 2005.
60. "Recovery Plan for Southern Resident Killer Whales," I-1.

EPILOGUE

1. Christopher Frizzelle, "It's Time to Free Lolita, a Puget Sound Killer Whale That's Been Held Captive in Florida for 45 Years," *Stranger*, September 30, 2015.
2. Author interview with Funk, May 25, 2017 (emphasis in original).
3. Scheffer quoted in Paula Bock, "Whale Watcher: Biologist Ken Balcomb Hopes Liberating Lolita Will Unlock Mysteries of the Orcas," *Seattle Times*, October 23, 1994.
4. Author interview with Griffin, January 9, 2016.
5. Author interview with O'Barry, January 11, 2016.
6. Pacific Whale Watch Association, https://www.pacificwhalewatchassociation.com/ (accessed July 25, 2017).
7. Jason Gunter quoted in http://vault.sierraclub.org/sierra/200911/whales.aspx (accessed July 13, 2017). Special thanks to my sons, Ben and Nate, for finding the "Namu the Orca" stuffy at the Painter's Lodge gift shop.
8. Craig Welch, "Orca Killed by Satellite Tag Leads to Criticism of Scientific Practices," *National Geographic*, October 6, 2016, http://www.nationalgeographic.com.au/animals/orca-killed-by-satellite-tag-leads-to-criticism-of-science-practices.aspx (accessed October 12, 2017).
9. Rob Shaw, Gordon Hoekstra, and Stephanie Ip, "All Five Conditions Met for B.C.'s Approval of Kinder Morgan Pipeline: Christy Clark," *Vancouver Sun*, January 12, 2017.
10. Teresa M. Mongillo et al., *Exposure to a Mixture of Toxic Chemicals: Implications for the Health of Endangered Southern Resident Killer Whales* (Seattle: Northwest Fisheries Science Center, November 2016).
11. Jon T. Coleman, *Vicious: Wolves and Men in America* (New Haven, CT: Yale University Press, 2004), 202.
12. Frank Zelko, "The Whale That Inspired Greenpeace," https://blog.oup.com/2013/09/greenpeace-origin-killer-whale-skana/ (accessed August 10, 2017).

Bibliography

ARCHIVES AND OTHER PUBLIC REPOSITORIES

British Columbia Provincial Archives, Victoria, BC
Central Florida University, Special Collections, Orlando, FL
Library and Archives Canada, Ottawa, ON
Oak Bay Municipal Archives, Oak Bay, BC
Smithsonian Institution Archives, Washington, DC
Seattle Municipal Archives, Seattle, WA
University of Victoria, Special Collections, Victoria, BC
University of Washington, Special Collections, Seattle, WA
US National Archives, College Park, MD.
Vancouver Municipal Archives, Vancouver, BC
Washington State Archives, Olympia, WA

INTERVIEWS

Antrim, Jim (June 25, 2016; September 28, 2016; May 26, 2017)
Babich, Randall (March 25, 2016)
Balcomb, Kenneth (July 12, 2017)
Bauer, Josef (May 22, 2016)
Borrowman, Jim (July 8, 2017)
Chandler, Rick (December 1, 2013)
Clemence, Anne (July 7, 2017)
Colby, John (July 1, 2012; September 10, 2012; August 5, 2014)
Cook, Kathryn (May 9, 2013)
Eilertsen, Joel (July 25, 2017)
Ellick, Karen (October 10, 2014)

Ellis, Graeme (March 17, 2016; June 5, 2017)
Ford, John (March 18, 2016; September 26, 2016)
Foster, Jeff (April 25, 2016)
Funk, Mark (May 25, 2017)
Gormly, Kenneth (May 17, 2016)
Grant, Joan (December 1, 2013)
Griffin, Ted (July 1–3, 2013; October 10, 2013; January 9–10, 2016; April 26, 2016; December 4, 2016)
Hofman, Bob, and Jim Gosliner (July 1, 2015)
Hole, Harry (July 25, 2017)
Horton, Jim (August 15, 2016)
Hunt, Richard (June 7, 2017)
Keffler, Gary (April 27, 2016)
Kelsey, Elin (July 9, 2013)
Matthews, Angus (May 24, 2017)
McLeod, Terry (April 25, 2015)
Munro, Ralph (June 11, 2013; December 7, 2016)
Nachtigall, Paul (August 12, 2016)
Nessel, Susan (July 17, 2017)
Newby, Terrell (October 24, 2015)
O'Barry, Ric (January 11, 2016)
Palmer, Stan (July 19, 2017)
Perry, Mark (May 15, 2013)
Ray, G. Carleton (October 9, 2017)
Reid, Sonny, and Marie Reid (July 1, 2016)
Ridgway, Sam (June 21, 2016)
Shore, Valerie (November 17, 2016)
Sorg, Annelise (March 22, 2014)
Spilman, Gary (July 17, 2016)
Stone, John (June 1, 2017)
White, Don (September 28, 2013; February 1, 2015)

PRINTED PRIMARY SOURCES

Andrews, Roy C. *Whale Hunting with Gun and Camera*. New York: Appleton, 1916.
Asper, Edward A., and Lanny H. Cornell. "Live Capture Statistics for the Killer Whale (*Orcinus Orca*) 1961–1976 in California, Washington, and British Columbia." *Aquatic Mammals* 5, no. 1 (1977): 21–26.
Baird, Robin W. 2001. "Status of Killer Whales, *Orcinus orca*, in Canada." *Canadian Field Naturalist* 115, no. 4 (2001): 676–701.
Bigg, Michael A. "An Assessment of Killer Whale (*Orcinus orca*) Stocks Off Vancouver Island, British Columbia." *Report of the International Whaling Commission* 32 (1982): 655–666.

Bigg, Michael A., and Allen A. Wolman. "Live-Capture Killer Whale (*Orcinus orca*) Fishery, British Columbia and Washington, 1962–1973." *Journal of the Fisheries Research Board of Canada* 32, no. 7 (1975): 1213–1221.

Blanchet, M. Wylie. *The Curve of Time*. Vancouver: Whitecap Books, 1996 [1961].

Bott, Bruce J. *Swim!* Vancouver: Life Systems, 2008.

Bowers, C. A., and R. S. Henderson, *Project Deep Ops: Deep Object Recovery with Pilot and Killer Whales*. San Diego: Naval Undersea Center, 1972.

Callenbach, Ernest. *Ecotopia: The Notebooks and Reports of William Weston*. New York: Heyday Books, 2005 [1975].

Cameron, William M. "Killer Whales Stranded Near Masset." *Progress Reports of the Pacific Coast Stations* 49 (1941): 17.

Carl, G. Clifford. "Albinistic Killer Whales in British Columbia." *Provincial Museum of Natural History and Anthropology: Report for the Year 1959* (Victoria, BC: Department of Education, 1960): 29–36.

———. "A School of Killer Whales Stranded at Estevan Point, Vancouver Island." *Provincial Museum of Natural History and Anthropology: Report for the Year 1945* (Victoria, BC: Department of Education, 1946): 21–28.

Carson, Rachel. *The Edge of the Sea*. New York: Houghton Mifflin, 1998 [1955].

———. *The Sea around Us*. New York: Oxford University Press, 1989 [1951].

———. *Silent Spring*. New York: Houghton Mifflin, 2002 [1962].

Chandler, Rick, Camille Goebel, and Ken Balcomb. "Who Is That Killer Whale? A New Key to Whale Watching." *Pacific Search* 11, no. 7 (1977): 25–35.

Christensen, Ivar. "Preliminary Report on the Norwegian Fishery for Small Whales: Expansion of Norwegian Whaling to Arctic and Northwest Atlantic Waters and Norwegian Investigations of the Biology of Small Whales." *Journal of the Fisheries Research Board of Canada* 32, no. 7 (1975): 1083–1093.

Clarke, James. *Man Is the Prey*. New York: Stein and Day, 1969.

Cook, Joseph J., and William L. Wisner. *Killer Whale! A Factual Account of the Fiercest Creature of the Ocean—the Orca*. New York: Dodd, Mead, 1963.

Cornell, Lanny H. "Puget Sound Already Is a Killer Whale Sanctuary." *Pacific Search* 9, no. 1 (1974): 16–18.

Crowley, Walt. *Rites of Passage: A Memoir of the Sixties in Seattle*. Seattle: University of Washington Press, 2010.

Cummings, William C., and Paul O. Thompson. "Gray Whales, *Eschrichtius robustus,* Avoid the Underwater Sounds of Killer Whales, *Orcinus orca*." *Fishery Bulletin* 69, no. 3 (1971): 525–530.

Dahlheim, Marilyn E. "A Review of the Biology and Exploitation of the Killer Whale, *Orcinus orca*, with Comments on Recent Sighting from Antarctica." *Report of the International Whaling Commission* 31 (1981): 541–546.

Easton, Nina. "The Death of Marineland." *Los Angeles Times Magazine*, August 9, 1987, 6–10, 23–26.

Fish, James F., and John S. Vania. "Killer Whale, *Orcinus orca,* Sounds Repel White Whales, *Delphinapterus leucas*." *Fishery Bulletin* 69, no. 3 (1971): 531–535.

Griffin, Edward I. "Making Friends with a Killer Whale." *National Geographic* 129, no. 3 (1966): 418–446.

———. *Namu: Quest for the Killer Whale*. Seattle: Gryphon West, 1982.

Griffin, Edward I., and Donald G. Goldsberry. "Notes on the Capture, Care, and Feeding of the Killer Whale, *Orcinus orca*, at the Seattle Aquarium." *International Zoo Yearbook* 8 (1968): 206–208.

Griffin, James S. *More Than Luck: A Memoir*. Bloomington, IN: iUniverse, 2011.

Hanauer, Gary. "The Killing Tanks." *Penthouse*, October 1989, 39–62.

Hornaday, William T. *The American Natural History: A Foundation of Useful Knowledge of the Higher Animals of North America*. New York: Charles Scribner's Sons, 1910.

Hoyt, Erich. *Orca: The Whale Called Killer*. Camden East, ON: Camden House, 1990 [1981].

———. "The Whales Called 'Killer.'" *National Geographic* 166, no. 2 (1984): 220–237.

Hoyt, Erich, and Jim Borrowman. "Diving with Orcas." *Diver* 5, no. 8 (1979): 20–23.

Hunter, Robert. *Warriors of the Rainbow: A Chronicle of the Greenpeace Movement from 1971 to 1979*. Amsterdam: Greenpeace International, 2011 [1979].

Ivashin, M. V. "USSR, Progress Report on Cetacean Research, June 1979–May 1980." *Report of the International Whaling Commission* 31 (1981): 221–226.

Jeune, Paul. *Miracle: The Story of a Baby Killer Whale*. Victoria, BC: Sealand of the Pacific, 1978.

Kellogg, A. Remington. "Whales: Giants of the Sea." *National Geographic* 77, no. 1 (1940): 35–90.

Leopold, Aldo. *A Sand County Almanac and Sketches Here and There*. New York: Oxford University Press, 1968 [1949].

Lilly, John C. *Man and Dolphin*. New York: Pyramid Books, 1961.

———. *The Mind of the Dolphin*. New York: Doubleday, 1967.

McIntyre, Joan, ed. *Mind in the Waters: A Book to Celebrate the Consciousness of Whales and Dolphins*. New York: Charles Scribner's Sons, 1974.

McVay, Scott. "Last of the Great Whales." *Scientific American* 215, no. 2 (1966): 13–21.

Mead, Tom. *The Killers of Eden: The Killer Whales of Twofold Bay*. London: Angus and Robertson, 1961.

Merle, Robert. *The Day of the Dolphin*. New York: Simon and Schuster, 1969.

Mikhalev, Y. A., M. V. Ivashin, V. P. Savusin, and F. E. Zelenaya. "The Distribution and Biology of the Killer Whales in the Southern Hemisphere." *Report of the International Whaling Commission* 31 (1981): 551–566.

Mitsuoka, Rae R., Roger E. Pearson, Laura J. Rutledge, and Samuel Waterman, eds. *A History of the Northwest and Alaska Fisheries Center, 1931-1981: Fifty Years of Cooperation and Commitment*. Seattle: National Marine Fisheries Service, 1982.

Moore, Patrick. *Confessions of a Greenpeace Dropout: The Making of a Sensible Environmentalist*. Vancouver: Beatty Street, 2010.

Morton, Alexandra. *Listening to Whales: What the Orcas Have Taught Us*. New York: Ballantine, 2002.

Mowat, Farley. *Never Cry Wolf.* Toronto: Emblem, 2009 [1963].

———. *A Whale for the Killing.* Vancouver: Douglas and McIntyre, 2012 [1972].

The Natural History of Pliny. Translated by John Bostock and H. T. Riley. 6 vols. London: Henry G. Bohn, 1855.

"Naval War Declared against Killer Whales." *Science News Letter*, June 16, 1956, 374.

Newby, Terrell. "Killer Whale Deaths Reported." *Pacific Search* 5, no. 1 (1971).

Newman, Murray. *Life in a Fishbowl: Confessions of an Aquarium Director.* Vancouver: Douglas and McIntyre, 1994.

———. *People, Fish, and Whales: The Vancouver Aquarium Story.* Madeira Park, BC: Harbour Publishing, 2006.

Newman, Murray, and Patrick McGeer. "The Capture and Care of a Killer Whale, Orcinus Orca, in British Columbia." *Zoologica, Scientific Contributions of the New York Zoological Society* 51, no. 2 (1965): 59–70.

Nishiwaki, Masaharu, and Chikao Handa. "Killer Whales Caught in the Coastal Waters Off Japan for Recent 10 Years." *Scientific Report of the Whales Research Institute* 13 (1958): 85–96.

Norris, Kenneth S. *The Porpoise Watcher.* New York. W. W. Norton, 1974.

———, ed. *Whales, Dolphins, and Porpoises.* Berkeley: University of California Press, 1966.

O'Barry, Ric. *Behind the Dolphin Smile: One Man's Campaign to Protect the World's Dolphins.* San Rafael, CA: Earth Aware, 2012 [1988].

Payne, Roger S., and Scott McVay. "Songs of the Humpback Whales." *Science* 173, no. 3997 (1971): 585–597.

Phillips, Craig. *The Captive Sea: Life behind the Scenes of the Great Modern Oceanariums.* Philadelphia: Chilton Books, 1964.

Pike, Gordon C., and Ian B. MacAskie. *Marine Mammals of British Columbia.* Ottawa: Fisheries Research Board of Canada, 1969.

Preble, Edward A. *A Biological Survey of the Pribilof Islands, Alaska.* Washington, DC: Government Printing Office, 1923.

Ponting, Herbert G. *The Great White South.* London: Duckworth, 1921.

Recovery Plan for Southern Resident Killer Whales. Seattle: National Marine Fisheries Service, 2008.

Rice, Dale W. "Stomach Contents and Feeding Behavior of Killer Whales in the Eastern North Pacific." *Norsk Hvalfangst-Tidende* 2 (1968): 35–38.

Ridgway, Sam. *The Dolphin Doctor.* Dublin, NH: Yankee Books, 1987.

———, ed. *Mammals of the Sea: Biology and Medicine.* Springfield, IL: Charles C. Thomas, 1972.

Saulitis, Eva. *Into Great Silence: A Memoir of Discovery and Loss among Vanishing Orcas.* Boston: Beacon Press, 2013.

Scammon, Charles M. *The Marine Mammals of the Northwestern Coast of North America.* New York: Dover, 1968 [1874].

———. "The Orca." *Overland Monthly* 9, no. 1 (1872): 52–57.

Scheffer, Victor B. *Adventures of a Zoologist*. New York: Charles Scribner's Sons, 1980.
———. *The Year of the Whale*. New York: Charles Scribner's Sons, 1969.
Scheffer, Victor B., and John W. Slipp. "The Whales and Dolphins of Washington State with a Key to the Cetaceans of the West Coast of North America." *American Midland Naturalist* 39, no. 2 (1948): 257–337.
Scott, Walter F. *Scott's Last Expedition*. London: John Murray, 1923 [1913].
Shepherd, George. "Killer Whale in Slough at Portland, Oregon." *Journal of Mammalogy* 13, no. 2 (1932): 171–172.
Shevill, William E. *The Whale Problem*. Cambridge, MA: Harvard University Press, 1974.
Sigurjónsson, Jóhann. "Killer Whale Census Off Iceland during October 1982." *Report of the International Whaling Commission* 34 (1984): 609–612.
Sigurjónsson, Jóhann, and Stephen Leatherwood. "The Icelandic Live-Capture Fishery for Killer Whales, 1976–1988." *Journal of the Marine Research Institute* 11 (1988): 308–309.
Simmons, Mark A. *Killing Keiko: The True Story of Free Willy's Return to the Wild*. Orlando, FL: Callinectes Press, 2014.
Singer, Peter. *Animal Liberation*. New York: HarperCollins, 2009 [1975].
Spradley, James, ed. *Guests Never Leave Hungry: The Autobiography of James Sewid, a Kwakiutl Indian*. Kingston, ON: McGill-Queen's University Press, 1989.
True, Frederick W. "Notes on a Killer Whale (Genus Orcinus) from the Coast of Maine." *Proceedings of the National Museum* 27, no. 1357 (1904): 227–234.
US Navy Diving Manual. Washington, DC: Navy Department, 1959.
Watson, Paul. *Ocean Warrior: My Battle to End the Illegal Slaughter on the High Seas*. Toronto: Key Porter Books, 1994.
Wood, Forrest G. *Marine Mammals and Man: The Navy's Porpoises and Sea Lions*. Washington, DC: Robert B. Luce, 1973.

SECONDARY SOURCES

Alagona, Peter S. *After the Grizzly: Endangered Species and the Politics of Place in California*. Berkeley: University of California Press, 2013.
Arnold, David F. *The Fisherman's Frontier: People and Salmon in Southeast Alaska*. Seattle: University of Washington Press, 2008.
Baird, Robin W. *Killer Whales of the World: Natural History and Conservation*. Stillwater, MN: Voyageur Press, 2002.
Barrett-Lennard, Lance G., Craig O. Matkin, John W. Durban, Eva L. Saulitis, and David Ellifrit. "Predation of Gray Whales and Prolonged Feeding on Submerged Carcasses by Transient Killer Whales at Unimak Island, Alaska." *Marine Ecology Progress Series* 421 (2011): 229–241.
Baron, David. *The Beast in the Garden: The True Story of a Predator's Deadly Return to Suburban America*. New York: W. W. Norton, 2005.

Barrow, Mark V., Jr. *Nature's Ghosts: Confronting Extinction from the Age of Jefferson to the Age of Ecology*. Chicago: University of Chicago Press, 2009.

Becker, Paula, and Alan J. Stein. *The Future Remembered: The 1962 Seattle World's Fair and Its Legacy*. Seattle: Seattle Center Foundation, 1962.

Bender, Daniel E. *The Animal Game: Searching for Wilderness at the American Zoo*. Cambridge, MA: Harvard University Press, 2016.

Benson, Etienne. *Wired Wilderness: Technologies of Tracking and the Making of Modern Wildlife*. Baltimore: Johns Hopkins University Press, 2010.

Bocking, Stephen. *Ecologists and Environmental Politics: A History of Contemporary Ecology*. New Haven, CT: Yale University Press, 1997.

Bolster, W. Jeffrey. "Opportunities in Marine Environmental History." *Environmental History* 11, no. 3 (2006): 567–597.

Borstelmann, Thomas. *The 1970s: A New Global History from Civil Rights to Economic Inequality*. Princeton, NJ: Princeton University Press, 2011.

Brazier, Don. *History of the Washington Legislature, 1965–1982*. Olympia: Washington State Senate, 2007.

Brower, Kenneth. *Freeing Keiko: The Journey of a Killer Whale from* Free Willy *to the Wild*. New York: Gotham Books, 2005.

Brown, Frederick L. *The City Is More Than Human: An Animal History of Seattle*. Seattle: University of Washington Press, 2016.

Brown, Kate. *Plutopia: Nuclear Families, Atomic Cities, and the Great Soviet and American Plutonium Disasters*. New York: Oxford University Press, 2013.

Browning, Robert J. *Fisheries of the North Pacific: History, Species, Gear, and Processes*. Anchorage: Alaska Northwest Publishing, 1974.

Bruner, Bernd. *Bears: A Brief History*. New Haven, CT: Yale University Press, 2008.

Bsumek, Erika Marie, David Kinkela, and Mark Atwood Lawrence, eds. *Nation-States and the Global Environment: New Approaches to International Environmental History*. New York: Oxford University Press, 2013.

Burnett, D. Graham. *The Sounding of the Whale: Science and Cetaceans in the Twentieth Century*. Chicago: University of Chicago Press, 2012.

———. *Trying Leviathan: The Nineteenth-Century New York Court Case That Put the Whale on Trial and Challenged the Order of Nature*. Princeton, NJ: Princeton University Press, 2010.

Busch, Briton Cooper. *The War against the Seals: A History of the North American Seal Fishery*. Kingston, ON: McGill-Queen's University Press, 1985.

Clode, Danielle. *Killers in Eden: The Story of a Rare Partnership between Men and Killer Whales*. Crows Nest, NSW: Allen and Unwin, 2002.

Cohen, Fay G. *Treaties on Trial: The Continuing Controversy over Northwest Indian Fishing Rights*. Seattle: University of Washington Press, 1986.

Colby, Jason M. "The Whale and the Region: Killer Whale Capture and the Remaking of the Pacific Northwest." *Journal of the Canadian Historical Association* 24, no. 2 (2013): 425–454.

———. "Cetaceans in the City: Orca Captivity and Environmental Politics in Vancouver." In *Animal Metropolis: Histories of Human-Animal Relations in Urban Canada*, edited by Joanna Dean, Darcy Ingram, and Christabelle Sethna, 285–308. Calgary, AB: University of Calgary Press, 2017.

———. "Change in Black and White: Orca Bodies and the New Pacific Northwest." In *Animals and History*, edited by Susan Nance, 19–37. Syracuse, NY: Syracuse University Press, 2015.

Coleman, Jon T. *Here Lies Hugh Glass: A Mountain Man, a Bear, and the Rise of the American Nation*. New York: Hill and Wang, 2012.

———. "The Shoemaker's Circus: Grizzly Adams and Nineteenth-Century Animal Entertainment." *Environmental History* 20, no. 4 (2015): 593–618.

———. *Vicious: Wolves and Men in America*. New Haven, CT: Yale University Press, 2004.

Cooper, Diana Starr. *Night after Night*. Washington, DC: Island Press, 1994.

Coté, Charlotte. *Spirits of Our Whaling Ancestors: Revitalizing Makah and Nuu-chah-nulth Traditions*. Seattle: University of Washington Press, 2010.

Cowie, Jefferson. *Stayin' Alive: The 1970s and the Last Days of the Working Class*. New York: New Press, 2010.

Cronon, William, ed. *Uncommon Ground: Rethinking the Human Place in Nature*. New York: W. W. Norton, 1995.

Davis, Janet M. *The Circus Age: Culture and Society under the American Big Top*. Chapel Hill: University of North Carolina Press, 2002.

———. *The Gospel of Kindness: Animal Welfare and the Making of Modern America*. New York: Oxford University Press, 2016.

Davis, Lance E., Robert E. Gallman, and Karen Gleiter. *In Pursuit of Leviathan: Institutions, Productivity, and Profits in American Whaling, 1816–1906*. Chicago: University of Chicago Press, 1998.

Davis, Susan G. *Spectacular Nature: Corporate Culture and the Sea World Experience*. Berkeley: University of California Press, 1997.

Dawson, Michael. *Selling British Columbia: Tourism and Consumer Culture, 1890–1970*. Vancouver: University of British Columbia Press, 2005.

Dean, Joanna, Darcy Ingram, and Christabelle Sethna, eds. *Animal Metropolis: Histories of Human-Animal Relations in Urban Canada*. Calgary, AB: University of Calgary Press, 2017.

De Wall, Frans. *Are We Smart Enough to Know How Smart Animals Are?* New York: W. W. Norton, 2016.

Dickerson, James. *North to Canada: Men and Women against the Vietnam War*. Westport, CT: Praeger, 1999.

Dolin, Eric Jay. *Leviathan: The History of Whaling in America*. New York: W. W. Norton, 2008.

Donaldson, Sue, and Will Kymlicka. *Zoopolis: A Political Theory of Animal Rights*. New York: Oxford University Press, 2011.

Dorsey, Kurkpatrick. *The Dawn of Conservation Diplomacy: U.S.-Canadian Wildlife Protection Treaties in the Progressive Era.* Seattle: University of Washington Press, 1998.

———. *Whales and Nations: Environmental Diplomacy on the High Seas.* Seattle: University of Washington Press, 2013.

Drake, Brian. *Loving Nature, Fearing the State: Environmentalism and Anti-government Politics before Reagan.* Seattle: University of Washington Press, 2013.

Dummitt, Chris. *The Manly Modern: Masculinity in Postwar Canada.* Vancouver: University of British Columbia Press, 2008.

Dunlap, Thomas. *DDT: Scientists, Citizens, and Public Policy.* Princeton, NJ: Princeton University Press, 1981.

———. *Faith in Nature: Environmentalism as Religious Quest.* Seattle: University of Washington Press, 2004.

———. *Nature and the English Diaspora: Environment and History in the United States, Canada, Australia, and New Zealand.* Cambridge: Cambridge University Press, 1999.

Ellis, Richard. *The Empty Ocean: Plundering the World's Marine Life.* Washington, DC: Shearwater Books, 2003.

———. *Men and Whales.* New York: Knopf, 1991.

Epstein, Charlotte. *The Power of Words in International Relations: Birth of an Anti-whaling Discourse.* Cambridge, MA: MIT Press, 2008.

Estes, James A., Douglas P. Demaster, Daniel F. Doak, Terrie M. Williams, and Robert L. Brownell Jr., eds. *Whales, Whaling, and Ocean Ecosystems.* Berkeley: University of California Press, 2006.

Findlay, John M., and Kenneth Coates, eds. *Parallel Destinies: Canadian-American Relations West of the Rockies.* Seattle: University of Washington Press, 2002.

Finley, Carmel. *All the Boats on the Ocean: How Government Subsidies Led to Global Overfishing.* Chicago: University of Chicago Press, 2017.

———. *All the Fish in the Sea: Maximum Sustainable Yield and the Failure of Fisheries Management.* Chicago: University of Chicago Press, 2011.

Flippen, J. Brooks. *Conservative Conservationist: Russell E. Train and the Emergence of American Environmentalism.* Baton Rouge: Louisiana State University Press, 2006.

———. *Nixon and the Environment.* Albuquerque: University of New Mexico Press, 2000.

Ford, John K. B. *Marine Mammals of British Columbia.* Victoria: Royal BC Museum, 2014.

Ford, John K. B., and Graeme Ellis. *Transients: Mammal-Hunting Killer Whales of British Columbia, Washington, and Southeast Alaska.* Vancouver: University of British Columbia Press, 1999.

Ford, John K. B., Graeme M. Ellis, and Kenneth C. Balcomb. *Killer Whales: The Natural History and Genealogy of Orcinus Orca in British Columbia and Washington State.* Seattle: University of Washington Press, 1994.

Ford, John K. B., Graeme M. Ellis, Lance G. Barrett-Lennard, Alexandra B. Morton, R. S. Palm, and Kenneth C. Balcomb. "Dietary Specialization in Two Sympatric Populations of Killer Whales (*Orcinus orca*) in Coastal British Columbia and Adjacent Waters." *Canadian Journal of Zoology* 76, no. 8 (1998): 1456–1471.

Ford, John K. B., Graeme M. Ellis, Craig O. Matkin, Michael H. Wetklo, Lance G. Barrett-Lennard, and Ruth E. Withler. "Shark Predation and Tooth Wear in a Population of Northeastern Pacific Killer Whales." *Aquatic Biology* 11, no. 1 (2011): 213–224.

Foster, Charles. *Being a Beast: Adventures across the Species Divide*. New York: Metropolitan Books, 2016.

Francis, Daniel, and Gil Hewlett. *Operation Orca: Spring, Luna, and the Struggle to Save West Coast Killer Whales*. Madeira Park, BC: Harbour Publishing, 2007.

Fraser, David. *Understanding Animal Welfare: The Science in Its Cultural Context*. Boston: Wiley-Blackwell, 2008.

Gastil, Raymond, and Barnett Singer. *The Pacific Northwest: Growth of a Regional Identity*. Jefferson, NC: McFarland, 2010.

Glavin, Terry. *The Last Great Sea: A Voyage through the Human and Natural History of the North Pacific Ocean*. Vancouver: Greystone, 2000.

———. *This Ragged Place: Travels across the Landscape*. Vancouver: New Star Books, 1996.

Granatstein, J. L. *Yankee Go Home? Canadians and Anti-Americanism*. Toronto: HarperCollins Canada, 1996.

Grier, Katherine C. *Pets in America: A History*. Chapel Hill: University of North Carolina Press, 2006.

Hagan, John. *Northern Passage: American Vietnam War Resisters in Canada*. Cambridge, MA: Harvard University Press, 2001.

Hamblin, Jacob Darwin. *Arming Mother Nature: The Birth of Catastrophic Environmentalism*. New York: Oxford University Press, 2013.

———. *Oceanographers and the Cold War: Disciplines of Marine Science*. Seattle: University of Washington Press, 2005.

———. *Poison in the Well: Radioactive Waste in the Oceans at the Dawn of the Nuclear Age*. Newark, NJ: Rutgers University Press, 2008.

Hanson, Elizabeth. *Animal Attractions: Nature on Display in American Zoos*. Princeton, NJ: Princeton University Press, 2002.

Hargrove, John. *Beneath the Surface: Killer Whales, SeaWorld, and the Truth beyond* Blackfish. New York: St. Martin's Press, 2015.

Harmon, Alexandra. *Indians in the Making: Ethnic Relations and Indian Identities around Puget Sound*. Berkeley: University of California Press, 1998.

Harris, Cole. *The Resettlement of British Columbia: Essays on Colonialism and Geographical Change*. Vancouver: University of British Columbia Press, 1997.

Harter, John-Henry. *New Social Movements, Class, and the Environment: A Case Study of Greenpeace Canada*. Newcastle upon Tyne, UK: Cambridge Scholars, 2011.

Hays, Samuel P. *Beauty, Health, and Permanence: Environmental Politics in the United States, 1955–1985*. Cambridge: Cambridge University Press, 1987.

Hearne, Vicki. *Adam's Task: Calling Animals by Name*. New York: Skyhorse, 2007 [1986].

———. *Animal Happiness: A Moving Exploration of Animals and Their Emotions*. New York: Skyhorse, 1994.

Herzog, Hal. *Some We Love, Some We Hate, Some We Eat: Why It's So Hard to Think Straight about Animals*. New York: Harper, 2010.

Hofman, Robert. "The Continuing Legacies of the Marine Mammal Commission and Its Committee of Scientific Advisors on Marine Mammals." *Aquatic Mammals* 25, no. 1 (2009): 94–129.

Horwitz, Joshua. *War of the Whales: A True Story*. New York: Simon and Schuster, 2014.

Howkins, Adrian. *Frozen Empires: An Environment History of the Antarctic Peninsula*. New York: Oxford University Press, 2016.

Isitt, Benjamin. *Militant Minorities: British Columbia Workers and the Rise of a New Left, 1948–1972*. Toronto: University of Toronto Press, 2011.

Jones, Ryan Tucker. *Empire of Extinction: Russians and the North Pacific's Strange Beasts of the Sea, 1741–1867*. New York: Oxford University Press, 2014.

———. "Running into Whales: The History of the North Pacific from below the Waves." *American Historical Review* 118, no. 2 (2013): 349–377.

Kheraj, Sean. *Inventing Stanley Park: An Environmental History*. Vancouver: University of British Columbia Press, 2013.

King, Barbara J. *How Animals Grieve*. Chicago: University of Chicago Press, 2013.

———. *Personalities on the Plate: The Lives and Minds of the Animals We Eat*. Chicago: University of Chicago Press, 2017.

Kingsland, Sharon E. *The Evolution of American Ecology, 1890–2000*. Baltimore: Johns Hopkins University Press, 2005.

Kinkela, David. *DDT and the American Century: Global Health, Environmental Politics, and the Pesticide That Changed the World*. Chapel Hill: University of North Carolina Press, 2011.

Kirby, David. *Death at SeaWorld: Shamu and the Dark Side of Killer Whales in Captivity*. New York: St. Martin's Press, 2012.

Kirk, Andrew G. *Counterculture Green: The Whole Earth Catalog and American Environmentalism*. Lawrence: University Press of Kansas, 2007.

Klingle, Matthew. *Emerald City: An Environmental History of Seattle*. New Haven, CT: Yale University Press, 2009.

Knudtson, Peter. *Orca: Visions of the Killer Whale*. San Francisco: Sierra Club Books, 1996.

Kostash, Myrna. *Long Way from Home: The Story of the Sixties Generation in Canada*. Toronto: James Lorimer, 1980.

Leiren-Young, Mark. *The Killer Whale Who Changed the World.* Vancouver: Greystone, 2016.

Lichatowich, James A. *Salmon without Rivers: A History of the Pacific Salmon Crisis.* Washington, DC: Island Press, 2001.

Loo, Tina. *States of Nature: Conserving Canada's Wildlife in the Twentieth Century.* Vancouver: University of British Columbia Press, 2006.

Lutz, John. *Makúk: A New History of Aboriginal-White Relations.* Vancouver: University of British Columbia Press, 2009.

Lytle, Mark H. *The Gentle Subversive: Rachel Carson,* Silent Spring, *and the Rise of the Environmental Movement.* New York: Oxford University Press, 2007.

Mason, Jennifer. *Civilized Creatures: Urban Animals, Sentimental Culture, and American Literature, 1850–1900.* Baltimore: Johns Hopkins University Press, 2005.

Matsui, Kenichi. *Native Peoples and Water Rights: Irrigation, Dams, and the Law in Western Canada.* Kingston, ON: McGill-Queen's University Press, 2009.

McEvoy, Arthur. *The Fisherman's Problem: Ecology and Law in the California Fisheries, 1850–1980.* New York: Cambridge University Press, 1986.

Mitchell, David J. *W. A. C. Bennett and the Rise of British Columbia.* Vancouver: Douglas and McIntyre, 1983.

Mitman, Gregg. *Reel Nature: America's Romance with Wildlife on Film.* Seattle: University of Washington Press, 1999.

———. *The State of Nature: Ecology, Community, and American Social Thought, 1900–1950.* Chicago: University of Chicago Press, 1992.

Mongillo, Teresa, Gina M. Ylitalo, Linda D. Rhodes, Sandie M. O'Neill, Dawn P. Noren, and M. Bradley Hanson. *Exposure to a Mixture of Toxic Chemicals: Implications for the Health of Endangered Southern Resident Killer Whales.* Seattle: National Marine Fisheries Service, 2016.

Moyle, Peter B. "Historical Abundance and Decline of Chinook Salmon in the Central Valley Region of California." *North American Journal of Fisheries Management* 18, no. 3 (1998): 487–521.

Nance, Susan. *Entertaining Elephants: Animal Agency and the Business of the American Circus.* Baltimore: Johns Hopkins University Press, 2013.

———, ed. *The Historical Animal.* Syracuse, NY: Syracuse University Press, 2015.

Nash, Linda. *Inescapable Ecologies: A History of Environment, Disease, and Knowledge.* Berkeley: University of California Press, 2006.

Nash, Roderick F. *The Rights of Nature: A History of Environmental Ethics.* Madison: University of Wisconsin Press, 1989.

Neiwert, David. *Of Orcas and Men: What Killer Whales Can Teach Us.* New York: Overlook Press, 2015.

Newell, Diane, ed. *The Development of the Pacific Salmon Canning Industry: A Grown Man's Game.* Kingston, ON: McGill-Queen's University Press, 1989.

———. *Tangled Webs of History: Indians and the Law in Canada's Pacific Coast Fisheries.* Toronto: University of Toronto Press, 1993.

Newell, Diane, and R. E. Ommer, eds. *Fishing Places, Fishing Peoples: Traditions and Issues in Canadian Small-Scale Fisheries.* Toronto: University of Toronto Press, 1999 [1993].

Obee, Bruce, and Graeme Ellis. *Guardians of the Whales: The Quest to Study Whales in the Wild.* Vancouver: Whitecap Books, 1992.

Palmer, Bryan D. *Canada's 1960s: The Ironies of Identity in a Rebellious Era.* Toronto: University of Toronto Press, 2009.

Parfit, Michael, and Suzanne Chisolm. *The Lost Whale: The True Story of an Orca Named Luna.* New York: St. Martin's Press, 2013.

Pollard, Sandra. *Puget Sound Whales for Sale: The Fight to End Orca Hunting.* Charleston, SC: History Press, 2014.

Price, Jennifer. *Flight Maps: Adventures with Nature in Modern America.* New York: Basic Books, 1999.

Raibmon, Paige. *Authentic Indians: Episodes of Encounter from the Late-Nineteenth-Century Northwest Coast.* Durham, NC: Duke University Press, 2005.

Ray, G. Carleton, and Frank M. Potter Jr. "The Making of the Marine Mammal Protection Act of 1972." *Aquatic Mammals* 37, no. 4 (2011): 520–552.

Reid, Joshua L. *The Sea Is My Country: The Maritime World of the Makahs, an Indigenous Borderlands People.* New Haven, CT: Yale University Press, 2015.

Reiss, Diana. *The Dolphin in the Mirror: Exploring Dolphin Minds and Saving Dolphin Lives.* New York: Houghton Mifflin Harcourt, 2011.

Riesch, Rudiger, Lance G. Barrett-Lennard, Graeme E. Ellis, John K. B. Ford, and Volker B. Deecke. "Cultural Traditions and the Evolution of Reproductive Isolation: Ecological Speciation in Killer Whales?" *Biological Journal of the Linnean Society* 106, no. 1 (2012): 1–17.

Ritvo, Harriet. *The Animal Estate: The English and Other Creatures in the Victorian Age.* Cambridge, MA: Harvard University Press, 1987.

Robbins, William G., ed. *The Great Northwest: The Search for Regional Identity.* Corvallis: Oregon State University Press, 2001.

Robbins, William G., and Katrine Barber. *Nature's Northwest: The North Pacific Slope in the Twentieth Century.* Tucson: University of Arizona Press, 2011.

Roberts, Callum. *The Unnatural History of the Sea.* Washington, DC: Island Press, 2007.

Rocha, Robert C., Jr., Phillip J. Clapham, and Yulia Ivashenko. "Emptying the Oceans: A Summary of Industrial Whaling Catches in the 20th Century." *Marine Fisheries Review* 76, no. 4 (2014): 37–48.

Rome, Adam. *The Genius of Earth Day: How a 1970 Teach-In Unexpectedly Made the First Green Generation.* New York: Hill and Wang, 2014.

Rothenberg, David. *Thousand Mile Song: Whale Music in a Sea of Sound.* New York: Basic Books, 2008.

Rothfels, Nigel. *Savages and Beasts: The Birth of the Modern Zoo*. Baltimore: Johns Hopkins University Press, 2002.

Rozwadowski, Helen M. "Arthur C. Clarke and the Limitations of the Ocean as a Frontier." *Environmental History* 17, no. 3 (2012): 578–602.

———. *Fathoming the Ocean: The Discovery and Exploration of the Deep Sea*. Cambridge, MA: Belknap Press, 2005.

Sanders, Jeffrey Craig. "Animal Trouble and Urban Anxiety: Human-Animal Interaction in Post–Earth Day Seattle." *Environmental History* 16, no. 2 (2011): 226–261.

———. "The Battle for Fort Lawton: Competing Environmental Claims in Postwar Seattle." *Pacific Historical Review* 77, no. 2 (2008): 203–235.

———. *Seattle and the Roots of Urban Sustainability: Inventing Ecotopia*. Pittsburgh: University of Pittsburgh Press, 2010.

Scates, Shelby. *Warren G. Magnuson and the Shaping of Twentieth-Century America*. Seattle: University of Washington Press, 1997.

Scheffer, Victor B., Clifford H. Fiscus, and Ethel I. Todd. *History of Scientific Study and Management of the Alaskan Fur Seal,* Callorhinus ursinus, *1786–1964*. Washington, DC: NOAA, 1984.

Schwantes, Carlos. *The Pacific Northwest: An Interpretive History*. Lincoln: University of Nebraska Press, 2000.

Schwantes, Carlos, and John M. Findlay, eds. *Power and Place in the North American West*. Seattle: University of Washington Press, 1999.

Sellers, Christopher C. *Crabgrass Crucible: Suburban Nature and the Rise of Environmentalism in Twentieth-Century America*. Chapel Hill: University of North Carolina Press, 2012.

Shaffer, Marguerite S., and Phoebe S. K. Young. *Rendering Nature: Animals, Bodies, Place, Politics*. Philadelphia: University of Pennsylvania Press, 2015.

Shepard, Paul. *The Others: How Animals Made Us Human*. Washington, DC: Island Press, 1996.

Shoemaker, Nancy. "Whale Meat in American History." *Environmental History* 10, no. 2 (2005): 269–294.

Smith, P. "Branding Cascadia: Considering Cascadia's Conflicting Conceptions: Who Gets to Decide?" *Canadian Political Science Review* 2, no. 2 (2008): 57–83.

Smith, Sherry L. *Hippies, Indians, and the Fight for Red Power*. New York: Oxford University Press, 2012.

Spiegel, Marjorie. *The Dreaded Comparison: Human and Animal Slavery*. New York: Mirror Books, 1996.

Steinberg, Ted. *Down to Earth: Nature's Role in American History*. New York: Oxford University Press, 2008.

Stephens, Dave. *Ivar: The Life and Times of Ivar Haglund*. Seattle: Dunhill, 1986.

Suttles, Wayne. *Coast Salish Essays*. Seattle: University of Washington Press, 1987.

———. *The Economic Life of the Coast Salish of Haro and Rosario Straits.* New York: Garland, 1974.

Taylor, Joseph E. *Making Salmon: An Environmental History of the Northwest Fisheries Crisis.* Seattle: University of Washington Press, 2001.

Thompson, John, and Stephen Randall. *Canada and the United States: Ambivalent Allies.* 2nd ed. Athens: University of Georgia Press, 1997.

Thomson, Jennifer. "Surviving the 1970s: The Case of Friends of the Earth." *Environmental History* 22, no. 2 (2017): 235–256.

Thrush, Coll. *Native Seattle: Histories from the Crossing-Over Place.* Seattle: University of Washington Press, 2008.

Turner, James Morton. *The Promise of Wilderness: American Environmental Politics since 1964.* Seattle: University of Washington Press, 2012.

Twiss, John R., Jr., and Randall R. Reeves, eds. *Conservation and Management of Marine Mammals.* Washington, DC: Smithsonian Institution Press, 1999.

Wade, Paul R., Vladimir N. Burkanov, Marilyn E. Dahlheim, Nancy A. Friday, Lowell W. Fritz, Thomas R. Loughlin, Sally A. Mizroch, Marcia M. Muto, and Dale W. Rice. "Killer Whales and Marine Mammal Trends in the North Pacific—a Re-examination of Evidence for Sequential Megafauna Collapse and the Prey-Switching Hypothesis." *Marine Mammal Science* 23, no. 4 (2007): 766–802.

Wadewitz, Lissa K. *The Nature of Borders: Salmon, Boundaries, and Bandits on the Salish Sea.* Seattle: University of Washington Press, 2012.

Wallace, David Rains. *Neptune's Ark: From Ichthyosaurs to Orcas.* Berkeley: University of California Press, 2007.

Wallace, Scott, and Brian Gisborne. *Basking Sharks: The Slaughter of BC's Gentle Giants.* Vancouver: New Star Books, 2006.

Walth, Brent. *Fire at Eden's Gate: Tom McCall and the Oregon Story.* Portland: Oregon Historical Society Press, 1994.

Webb, Robert. *On the Northwest: Commercial Whaling in the Pacific Northwest, 1790–1967.* Vancouver: University of British Columbia Press, 1988.

Weber, Michael L. *From Abundance to Scarcity: A History of U.S. Marine Fisheries Policy.* Washington, DC: Island Press, 2002.

Weyler, Rex. *Greenpeace: How a Group of Ecologists, Journalists and Visionaries Changed the World.* Vancouver: Raincoast Books, 2004.

———. *Song of the Whale.* Garden City, NY: Doubleday, 1986.

White, Richard. *The Organic Machine: The Remaking of the Columbia River.* New York: Hill and Wang, 1996.

Whitehead, Hal, and Luke Rendell. *The Cultural Lives of Whales and Dolphins.* Chicago: University of Chicago Press, 2015.

Whitehead, Hal, and R. Reeves, "Killer Whales and Whaling: The Scavenging Hypothesis." *Biology Letters* 1, no. 4 (2005): 415–418.

Wilkinson, Charles. *Messages from Frank's Landing: A Story of Salmon, Treaties, and the Indian Way*. Seattle: University of Washington Press, 2000.

Yaffee, Steven Lewis. *The Wisdom of the Spotted Owl: Policy Lessons for a New Century*. Washington, DC: Island Press, 1994.

Zelko, Frank. "From Blubber and Baleen to Buddha of the Deep: The Rise of the Metaphysical Whale." *Society and Animals* 20, no. 1 (2012): 91–108.

———. *Make It a Green Peace! The Rise of Countercultural Environmentalism*. New York: Oxford University Press, 2013.

Zimmerman, David. *Maritime Command Pacific: The Royal Canadian Navy's West Coast Fleet in the Early Cold War*. Vancouver: University of British Columbia Press, 2015.

UNPUBLISHED SECONDARY SOURCES

Barrett-Lennard, L. G. "Population Structure and Mating Patterns of Killer Whales (*Orcinus orca*) as Revealed by DNA Analysis." PhD diss., University of British Columbia, 2000.

Bigg, Michael A., Ian B. MacAskie, and Graeme Ellis. *Preliminary Report: Abundance and Movements of Killer Whales Off Eastern and Southern Vancouver Island with Comments on Management*. St. Anne de Bellevue: Arctic Biological Station, Quebec, 1976.

Block, Tina. "Everyday Infidels: A Social History of Secularism in the Postwar Pacific Northwest." PhD diss., University of Victoria, 2006.

Butler, M. Blake. "Fishing on Porpoise: The Origins, Struggles, and Successes of the Tuna-Dolphin Controversy." Master's thesis, University of Victoria, 2017.

Ford, John K. B. "Call Traditions and Dialects of Killer Whales (*Orcinus orca*) in British Columbia." PhD diss., University of Victoria, 1984.

Jones, Ryan Tucker. "The Ecology of Revenge Socialism." Unpublished paper, 2016.

Malcolm, Christopher Duncan. "The Current State and Future Prospects of Whale-Watching Management, with Special Emphasis on Whale-Watching in British Columbia, Canada." PhD diss., University of Victoria, 2003.

Murray, F. Scott A. "'Cashing In on Whales': Cetaceans as Symbol and Commodity along the Northern Pacific Coast, 1959–2008." Master's thesis, Simon Fraser University, 2009.

Riera, Amalis. "Patterns of Seasonal Occurrence of Sympatric Killer Whale Lineages in Waters Off Southern Vancouver Island and Washington State, as Determined by Passive Acoustic Monitoring." Master's thesis, University of Victoria, 2012.

Tucker, Brian. "Inventing the Salish Sea: Exploring the Performative Act of Place Naming of the Pacific Coast of North America." Master's thesis, University of Victoria, 2010.

Warkentin, Traci L. "Captive Imaginations: Affordances for Ethics, Agency and Knowledge-Making in Whale-Human Encounters." PhD diss., York University, 2007.

Werner, Mark T. "What the Whale Was: Orca Cultural Histories in British Columbia since 1964." Master's thesis, University of British Columbia, 2008.

Index

Figures are noted with an italic *f* after the page number.